Proceedings of the
THIRD INTERNATIONAL
WORKSHOP ON NUDE MICE

Volume 1

Invited Lectures
Infection
Immunology

INTERNATIONAL PLANNING COMMITTEE

Margaret Holmes, Ph.D. (Australia)
Cluff Hopla, Ph.D. (U.S.A. - ICLAS Representative)
Berenice Kindred, Ph.D. (West Germany)
Tatsuji Nomura, M.D. (Japan)
Nakaaki Ohsawa, M.D. (Japan)
Norman Reed, Ph.D. (U.S.A.), Chairman
Jørgen Rygaard, M.D. (Denmark)
Bernard Sordat, M.D. (Switzerland)

LOCAL ORGANIZING COMMITTEE

Patricia Crowle, Ph.D. (Bozeman, Montana)
Jim Cutler, Ph.D. (Bozeman, Montana)
Beppino Giovanella, Ph.D. (Houston, Texas)
Carl Hansen, Ph.D. (Bethesda, Maryland)
David Houchens, Ph.D. (Columbus, Ohio)
John Jutila, Ph.D. (Bozeman, Montana)
Henry Outzen, Ph.D. (Bar Harbor, Maine)
Norman Reed, Ph.D. (Bozeman, Montana), Chairman
William Weidanz, Ph.D. (Philadelphia, Pennsylvania)

Workshop Secretaries:
Jerrie Beyrodt, Patricia Healow, Lenora Thornley

G. J. EATON, P$_H$D.

Proceedings of the THIRD INTERNATIONAL WORKSHOP ON NUDE MICE

Volume 1

Invited Lectures

Infection

Immunology

Montana State University
Bozeman, Montana
September 6-9, 1979

Edited by
NORMAN D. REED

GUSTAV FISCHER NEW YORK • STUTTGART

Norman D. Reed, Ph.D.
Department of Microbiology
Montana State University
Bozeman, Montana 59717

Library of Congress Cataloging in Publication Data
International Workshop on Nude Mice (3rd : 1979 :
  Montana State University, Bozeman)
  Proceedings of the third International Workshop on
Nude Mice, Montana State University, Bozeman, Montana,
September 6-9, 1979.
  Includes bibliographies and index.
  Contents: v. 1. Invited lectures, infection, and
immunology—v. 2. Oncology.
  1. Tumors—Immunological aspects—Congresses.
2. Nude mouse—Congresses. 3. Immunological deficiency
syndromes—Congresses. 4. Communicable diseases—
Immunological aspects—Congresses. I. Reed, Norman D.,
1935-   II. Title. [DNLM: 1. Mice, Nude—Congresses.
2. Animals, Laboratory—Congresses. W3 IN9327 3rd
1979p / QY 60.R6 I61 1979p]
QR188.6.I57   1979   616.99'2'0072   82-2400
ISBN 0-89574-104-0 (v. 1)   AACR2
ISBN 0-89574-105-9 (v. 2)

© 1982 Gustav Fischer New York, Inc.

This work is subject to copyright.
All rights are reserved, whether the whole or part of the
material is concerned, specifically those of translation,
reprinting, re-use of illustrations, broadcasting, reproduction
by photocopying machine or similar means, and storage in data
banks.

Printed in the United States of America.

ISBN 0-89574-105-9 Gustav Fischer New York
ISBN 3-437-30345-7 Gustav Fischer Stuttgart

PREFACE

In October, 1973, the First International Workshop on Nude Mice was held in Åarhus, Denmark. At the time of the first workshop, the nude mouse was not yet available to many laboratories. This first workshop and the published proceedings made apparent the usefulness of nude mice in biomedical research—especially in the fields of immunology and oncology—and the special needs for raising athymic nude mice.

By the time the Second International Workshop on Nude Mice was held in Tokyo, Japan, in October, 1976, the nude mouse was a popular research tool widely used in laboratories around the world. The theme of the second workshop was "The Potentialities and Limitations of the Nude Mouse." From the work discussed at the second workshop it was clear that many advances in immunology and oncology had been made using nude mice and that nude mice were becoming popular and useful tools in infectious disease research.

The Third International Workshop on Nude Mice, also organized around the theme "The Potentialities and Limitations of Nude Mice," was held September 6–9, 1979, in Bozeman, Montana. During this 3-day workshop there were many excellent presentations and much useful discussion among participants. The workshop brought together scientists from many disciplines to discuss current work with nude mice. In addition, participants were made aware of the nu^{str} mutation in mice, two independent mutations—rnu and rnu^{nz}—which provide nude, athymic rats, and also a mutation causing hairlessness and immunodeficiency in guinea pigs.

All three international workshops on nude mice were arranged under the auspices of the International Council for Laboratory Animal Science (formerly the International Committee on Laboratory Animals).

Norman D. Reed

ACKNOWLEDGMENTS

The Third International Workshop on Nude Mice was arranged under the auspices of ICLAS—The International Council for Laboratory Animal Science.

The generous financial support provided by many organizations is gratefully acknowledged; the names of these organizations are in the list of benefactors.

All members of the International Planning Committee made important contributions to the workshop. Special thanks are due to Tatsuji Nomura and Nakaaki Ohsawa for helping with the planning of the workshop during visits to the United States. Similarly, each member of the Local Organizing Committee was generous with time and talent.

Special thanks go to Jerrie Beyrodt, Patricia Crowle, Jim Cutler, Patricia Healow, Sharon Reed, and Lenora Thornley for giving attention to details, being unselfish, and being patient with a disorganized organizer.

BENEFACTORS

Abbott Laboratories, North Chicago, Illinois
Allied Mills, Inc., Chicago, Illinois
American Cancer Society—Montana Division, Billings, Montana
Battelle Memorial Institute, Columbus, Ohio
Cardinal Fine Wine Company and United Vintners, Bozeman, Montana
GL. Bomholtgard Ltd., Ry, Denmark
Central Institute for Experimental Animals, Kawasaka, Japan
Food and Drug Administration, Bethesda, Maryland
    Bureau of Biologics
    Bureau of Drugs
The Germfree Laboratories, Inc., Miami, Florida
GIBCO Animal Resources Laboratories, Madison, Wisconsin
Harlan Industries, Inc., Indianapolis, Indiana
The International Council for Laboratory Animal Science (ICLAS)
Lab Products, Inc., Rochelle Park, New Jersey
Montana State University, Bozeman, Montana
National Institutes of Health, Bethesda, Maryland
    National Institute of Allergy and Infectious Diseases
    National Cancer Institute
    Fogarty International Center
OLAC Ltd., Bicester, Oxfordshire, England
Paxton Processing Co., Inc., Paxton, Illinois
Ralston Purina Company, St. Louis, Missouri

LIST OF PARTICIPANTS

Betty J. Abbott, Screening Section, DEB, DTP, DCT, National Cancer Institute, Blair Building, Room 524A, 8300 Colesville Road, Silver Spring, Maryland 20910.
Tadahsi Arai, Research Institute for Chemobiodynamics, Chiba University, 1-8-1, Inohana, Chiba 280, Japan.
Gunther Bastert, Universitäts-Frauenklinik, Theodor-Stern-Kai 7, D-6000 Frankfurt/Main 70, West Germany.
Gillian Beattie, Department of Chemistry, Q058, University of California at San Diego, La Jolla, California 92093.
Robert E. Bellet, The Fox Chase Cancer Center, 7701 Burholme Avenue, Philadelphia, Pennsylvania 19111.
Michael V. Berridge, Wellington Cancer and Medical Research Institute, Clinical School of Medicine, Wellington, New Zealand.
Bradford O. Brooks, Department of Preventive Medicine, New York State College of Veterinary Medicine, Cornell University, P.O. Box 786, Ithaca, New York 14850.
Karsten Buschard, Pathological-Anatomical Institute, Kommunehospitalet, DK-1399 Copenhagen K, Denmark.
Elizabeth Cant, Department of Surgery, Funders Medical Centre, Bedford Park, South Australia 5042.
Peter J. A. Capel, Department of Medicine, Division of Nephrology, University Hospital, University of Nijmegen, Gart Grooteplein 16, Nijmegen, The Netherlands.
Ralph Clayman, Box 394 Memorial Building, University of Minnesota, Minneapolis, Minnesota 55455.
James Crowell, Jr., National Institutes of Health, Comparative Pathology Section, Building 28A, Room 111, 9000 Rockville Pike, Bethesda, Maryland 20205.
Patricia K. Crowle, Department of Biology, Montana State University, Bozeman, Montana 59717.
Jim E. Cutler, Department of Microbiology, Montana State University, Bozeman, Montana 59717.
Kiron M. Das, Department of Medicine, Albert Einstein College of Medicine, 1300 Morris Park Avenue, Ullman Building, Room 605, Bronx, New York 10461.
Koweth A. Davidson, Biology Division, Oak Ridge National Laboratory, P.O. Box Y, Oak Ridge, Tennessee 37830.
David D. Drutz, University of Texas Health Science Center, Department of Medicine, 7703 Floyd Curl Drive, San Antonio, Texas 78284.
Gordon J. Eaton, The Institute for Cancer Research, The Fox Chase Cancer Center, 7701 Burholme Avenue, Philadelphia, Pennsylvania 19111.
Horst Eichholz, 6000 Frankfurt/Main 50, Weilbrunn Str. 26, West Germany.
Wendall Farrow, Life Sciences, Inc., 2900 72nd Street North, St. Petersburg, Florida 33710.
Michael F. W. Festing, Medical Research Council Laboratory Animals Centre, Woodmansterne Road, Carshalton, Surrey SM5 4EF, England.
A. Howard Fieldsteel, Life Sciences Division, SRI International, 333 Ravenswood Avenue, Menlo Park, California 94025. (deceased)
John F. Finerty, Building 5, Room 235, Laboratory of Microbial Immunity, National Institute of Allergy and Infectious Diseases, National Institutes of Health, Bethesda, Maryland 20205.
Jørgen Fogh, Human Tumor Cell Laboratory, Sloan-Kettering Institute for Cancer Research, 145 Boston Post Road, Rye, New York 10580.

List of Participants

Hans Fortmeyer, Tierversuchsanlage des Klinikum der J.W. Goethe-Universität, Theodor-Stern-Kai 7, D-6000 Frankfurt/Main 70, West Germany.
Masahide Fujita, Department of Oncologic Surgery, Research Institute for Microbial Diseases, Osaka University, Yamada-kami, Suita City 565, Osaka, Japan.
Brenda L. Gallie, The Ontario Cancer Institute, Wellesley Hospital, 160 Wellesley Street East, Toronto, Ontario, Canada M4Y IJ3.
James W. Gautsch, Department of Cellular and Developmental Immunology, The Research Institute of Scripps Clinic, 10666 N. Torrey Pines Road, La Jolla, California 92037.
M. Eric Gershwin, Section of Rheumatology—Clinical Immunology, Department of Internal Medicine, TB 192, University of California, Davis, California 95616.
Beppino C. Giovalenna, Stehlin Foundation for Cancer Research, 777 St. Joseph Professional Building, Houston, Texas 77002.
Fernando Guiliani, Department of Chemistry, Q-058, University of California at San Diego, La Jolla, California 92093.
Jun-ichi Hata, Department of Pathology, Tokai University School of Medicine, Bohseidai, Isehara, Kanagawa, 259-11, Japan.
Lawrence Helson, Memorial Sloan-Kettering Cancer Center, 1275 York Avenue, New York, New York 10021.
J. Michael Holland, Biology Division, Oak Ridge National Laboratory, Oak Ridge, Tennessee 37830.
Richard Hong, Department of Pediatrics, University of Wisconsin Center for Health Sciences, 600 Highland Avenue, K4/434, Madison, Wisconsin 53792.
Cluff Hopla, Department of Zoology, 730 VanVleet Oval, Room 222, The University of Oklahoma, Norman, Oklahoma 73069.
David P. Houchens, Battelle Memorial Institute, 505 King Avenue, Columbus, Ohio 43201.
Richard Ikeda, Clinical Laboratory, 3195 Folson Blvd., Sacramento, California 95816.
Shunji Ikeuchi, 2-31-3 Okusawa Setagayaku, Tokyo 158, Japan.
Hiroshi Iwai, Central Institute for Experimental Animals, 1430 Nogawa, Takatsu-ku, Kawasaki 213, Japan.
Grete Jacobsen, Pathological-Anatomical Institute, Herlev Hospital, DK-2730 Herlev, Denmark.
Jeanne Joyce, University of Southern California, Department of Pathology, 2025 Zonal Avenue, Los Angeles, California 90033.
John Jutila, Department of Microbiology, Montana State University, Bozeman, Montana 59717.
Masao Kamiya, Department of Parasitology, Faculty of Veterinary Medicine, Hokhaido University, Sapporo, Japan.
Eiji Kawamura, Department of Surgery, The Kitasato Institute Hospital, 5-9-1, Skirokane, Minato-ku, Tokyo 108, Japan.
Lieselotte Kemper, Department of Pathology, Box 3712, Duke University Medical Center, Durham, North Carolina 27705.
Tasneem Khwaja, USC Comprehensive Cancer Center, 1720 Zonal Avenue, Los Angeles, California 90333.
Untae Kim, Department of Pathology, Roswell Park Memorial Institute, New York State Department of Health, 666 Elm Street, Buffalo, New York 14263.
Berenice Kindred, Institute for Immunology and Genetics, German Cancer Research Center, Im Neuvenheimerfeld 280, D-6900 Heidelberg 1, West Germany.
Aileen F. Knowles, Department of Chemistry, Q-058, School of Medicine, University of California at San Diego, La Jolla, California 92093.
Kenji Kohsaka, Department of Leprology, Research Institute for Microbial Diseases, Osaka University, Yamada-kami, Suita-shi, Osaka 565, Japan.
Akinori Kojima, Laboratory of Experimental Pathology, Aichi Cancer Research Institute, Chikusa, Nagoya 464, Japan.

List of Participants

Aurelia Koros, Allegheny General Hospital, Cancer Research Unit, 320 East North Avenue, Pittsburgh, Pennsylvania 15212.

Gerald G. Krueger, Division of Dermatology, Department of Medicine, University of Utah College of Medicine, 50 North Medical Drive, Salt Lake City, Utah 84132.

Takehisa Kurihara, Keiyu General Hospital, Department of Obstetrics and Gynecology, 47 Yamashita-cho, Nakaku, Yokohama 231, Japan.

Shih-shun Lee, Stehlin Foundation for Cancer Research, 777 St. Joseph Professional Building, Houston, Texas 77002.

Donald L. Lodmell, Laboratory of Persistent Viral Diseases, Rocky Mountain Laboratories, Hamilton, Montana 59840.

Claire Lofgren, University of California at San Francisco, Radiation Oncology Department, 1666 HSE, San Francisco, California 94143.

Bismarck B. Lozzio, University of Tennessee Department of Medical Biology, Memorial Research Center, Center for the Health Sciences, 1924 Alcoa Highway, Knoxville, Tennessee 37920.

Carmen B. Lozzio, University of Tennessee, Department of Medical Biology, Memorial Research Center, Center for the Health Sciences, 1924 Alcoa Highway, Knoxville, Tennessee 37920.

Kenneth J. McCormick, Stehlin Foundation for Cancer Research, 777 St. Joseph Professional Building, Houston, Texas 77002.

Emilio A. Machado, University of Tennessee, Department of Medical Biology, Memorial Research Center, Center for the Health Sciences, 1924 Alcoa Highway, Knoxville, Tennessee 37920.

M. Jean McManus, Department of Anatomy, Michigan State University, East Lansing, Michigan 48824.

Judith K. Manning, Department Medical Microbiology, University of Wisconsin Center for Health Sciences, Madison, Wisconsin 53792.

Kohji Maruo

Joe Mayo, Mammalian Genetics and Animal Production Section, National Cancer Institute, Frederick Cancer Research Center, Room 132, Building 201, Frederick, Maryland 21701.

J. Gabriel Michael, Department of Microbiology, University of Cincinnati Medical Center, 231 Bethesda Avenue, Cincinnati, Ohio 45267.

Ralf-Thomas Michel, 6082 Walldorf, Rothwiesenring 2, West Germany.

Tom E. Miller, Department of Medicine, University of Auckland School of Medicine, Auckland Hospital, Park Road, Auckland 3, New Zealand.

Nagahiro Minato, Department of Microbiology and Immunology, Albert Einstein College of Medicine, 1300 Morris Park Avenue, Bronx, New York 10461.

Graham F. Mitchell, Laboratory of Immunoparasitology, Walter and Eliza Hall Institute, Victoria 3050, Australia.

Albert New, Laboratory Animal Science, National Cancer Institute, National Institutes of Health, Building 13, Room 2E55, Bethesda, Maryland 20205.

Hideo Nishimura, Central Institute for Experimental Animals, 1430 Nogawa, Takatsu, Kawasaki 213, Japan.

Tatsuji Nomura, Central Institute for Experimental Animals, 1430 Nogawa, Takatsu, Kawasaki 213, Japan.

Nakaaki Ohsawa, The Third Department of Internal Medicine, Faculty of Medicine, University of Tokyo, Hongo 7-3-1, Tokyo 113, Japan.

Rainhardt Osieka, West German Cancer Center, Innere Klinik (Tumorforschung), Hufelandstrasse 55, D-4300 Essen 1, West Germany.

Henry Outzen, The Jackson Laboratory, Bar Harbor, Maine 04609.

Artemio A. Ovejera, Battelle Columbus Laboratories, 505 King Avenue, Columbus, Ohio 43201.

List of Participants

Sang-Gi Paik, Albert Einstein College of Medicine, Department of Genetics, Chanin Building, Room 607, 1300 Morris Park Avenue, Bronx, New York 10461.
Amadeo J. Pesce, University of Cincinnati Medical Center, Department of Pathology, 231 Bethesda Avenue, Cincinnati, Ohio 45267.
Jacqueline Plowman, Blair Building, Room 515, Drug Evaluation Branch, DPT, National Cancer Institute, 8300 Colesville Rd., Silver Spring, Maryland 20910.
Stephen Potkay
Radmila B. Raikow, Cancer Research Unit, Allegheny General Hospital, 320 E. North Avenue, Pittsburgh, Pennsylvania 15212.
Carolyn Reed, Division of Laboratory Animal Medicine, Quad F, Building 960T, Stanford University, Stanford, California 94305.
Norman D. Reed, Department of Microbiology, Montana State University, Bozeman, Montana 59717.
Lola Reid, Department of Molecular Pharmacology, 601 Chanin Cancer Center, Albert Einstein College of Medicine, 1300 Morris Park Avenue, Bronx, New York 10461.
Dean Roberts, Immunotoxicology, National Center for Toxicological Research, HFT 163, Jefferson, Arkansas 72079.
Johan C. Romijn, Erasmus University Rotterdam, Department of Urology, Dr. Molewater Plein 40, Rotterdam, The Netherlands.
Jørgen Rygaard, Pathological-Anatomical Institute, Kommunehospitalet, DK-1399 Copenhagen K, Denmark.
Francis E. Sharkey, Department of Pathology, Hershey Medical Center, Pennsylvania State University, Hershey, Pennsylvania 17033.
Seung-il Shin, Albert Einstein College of Medicine, Department of Genetics, Chanin Building, Room 607, 1300 Morris Park Avenue, Bronx, New York 10461.
Leonard D. Shultz, The Jackson Laboratory, Bar Harbor, Maine 04609.
Benjamin Siegel, University of Oregon Health Sciences Center, Portland, Oregon 97201.
Rachel S. Simon, Department of Genetics, 607 Chanin Building, Albert Einstein College of Medicine, 1300 Morris Park Avenue, Bronx, New York 10461.
Clare Skov, Idaho State University, Pocatello, Idaho 83201.
Donald E. Slagel, Department of Surgery, Division of Neurosurgery, University of Kentucky, College of Medicine, Lexington, Kentucky 40536.
James Small, Comparative Medicine Branch, MD 19-01, National Institute of Environmental Health Science, P.O. Box 12233, Research Triangle Park, North Carolina 27709.
Bernard Sordat, Department of Immunology, Swiss Institute for Experimental Cancer Research, 1066 Epalinges-sur-Lausanne, Switzerland.
Mogens Spang-Thomsen, The University Institute of Pathological Anatomy, 11, Frederik V's Vej, DK-2100 Copenhagen Ø, Denmark.
Marlies Stark, Medizinisches Institut für Lufthygiene und Silikosenforschung an der Universität, Gurlittstrasse 53, D 4000 Dusseldorf 1, West Germany.
James Stragand, Department of Laboratory Medicine, M.D. Anderson Hospital and Tumor Institute, 6723 Bertner, Houston, Texas 77030.
Kenji Suzuki, Keiyu General Hospital, Department of Obstetrics and Gynecology, 47 Yamashito-cho, Nakaku, Yokohama 231, Japan.
Tatsuo Suzuki, Department of Immunology, The Kitasato Institute Hospital, 5-9-1, Shirkane, Minato-ku, Tokyo 108, Japan.
Norikazu Tamaoki, Department of Pathology, Tokai University School of Medicine, Bohseidai, Isehara, Kanagawa 259-11, Japan.
J. Terry Ulrich, Department 90D, Abbott Laboratories, North Chicago, Illinois 60031.
Marion G. Valerio, Litton Bionetics, Inc., 5516 Nicholson Lane, Kensington, Maryland 20795.
Jack E. Vanderlip, Campus Veterinarian/Coordinator, Office of Animal Resources, University of California at San Diego, M-014, La Jolla, California 92093.
Carol Vervaert, Department of Surgery, Box 3917, Duke University Medical Center, Durham, North Carolina 27710.

List of Participants

Jakob Visfeldt, The University Institute of Pathological Anatomy, 11, Frederik V's Vej, DK-2100 Copenhagen Ø, Denmark.
Seiji Waki, Department of Parasitology, School of Medicine, Gunma University, Maebashi 371, Gunma, Japan.
Peter D. Walzer, Department of Medicine, University of Cincinnati College of Medicine, 231 Bethesda Avenue, Cincinnati, Ohio 45267.
Shaw Watanabe, Pathology Division, National Cancer Center Research Institute, 5-1-1, Tsukiji, Chuo-ku, Tokyo 104, Japan.
William Weidanz, Department of Microbiology and Immunology, Hahnemann Medical College, 230 N. Broad St., Philadelphia, Pennsylvania 19102.
Robert S. Wells, Department of Pathology, University of Colorado Health Sciences Center, 4200 East Ninth Avenue, Denver, Colorado 80262.
Clifford W. Welsch, Department of Anatomy, Michigan State University, East Lansing, Michigan 48824.
Robert Whitney, Jr., Chief, Veterinary Resources Branch, Building 14-G, National Institutes of Health, Bethesda, Maryland 20014.
Barbara Witham, Gibco Animal Resources Laboratory, P.O. Box 4220, Madison, Wisconsin 53711.

OPENING REMARKS

It is important that you understand the reason for holding the Third ICLAS Workshop on Nude Mice in Bozeman, Montana. It will surprise some of you to learn that the first colony of athymic nude mice in the United States was established on the campus of Montana State University by our host, Norman Reed, and his colleagues. It is my understanding that the strain of nude mice for the NIH program in the Veterinary Resources Branch was obtained from the Montana State University colony. Furthermore, we realized that most of you knew the eastern part of the United States, but few of you had ever been in the hinterlands, especially into the intermountain area of the western United States. We thought the "Big Sky Country" would provide an esthetic backdrop for you to exchange viewpoints with each other and that you would not be diverted from the subject at hand.

The first workshop, held in Aarhus, Denmark, was concerned primarily with husbandry and immunology. The second workshop in Tokyo had but one general paper on husbandry; somewhat over half dealt with various aspects of oncology, and the rest were concerned with immunology, pathology, and infection. As you note from our present program, this third workshop is not limited to the nude mouse but includes presentations on the athymic nude rat and a hairless immunodeficient guinea pig which apparently is athymic. Aside from these interesting additions the rest of the program deals primarily with the nude mouse in studies on infectious diseases, oncologic research, and immunology studies.

In a very real sense, this program is concerned with the potentialities and limitations of the nude mouse as we now understand this highly interesting and useful biological model. For the future, I think that the explosive use of it into new areas of biomedical research is, for the most part, past. New areas probably remain in fields such as endocrinology. Broader use in the study of infectious diseases is bound to occur. It is anticipated that major breakthroughs will come in the technology and the manipulation of the nude mouse.

The World Health Organization has listed six diseases as major targets for study. All but one of these six diseases is vector borne. As you will learn in the next few days, the nude mouse is of value in such diseases, but our concepts are not much more than in the formative stages; much more knowledge will come in the next few years. The one bacterial disease on the WHO list, leprosy, probably can be studied best in the nude mouse. In contrast to the armadillo, which has not been induced to breed in the laboratory to date, the nude mouse is now bred in numbers not thought practicable 3 years ago. The nude mouse is well defined genetically and microbiologically, perhaps more than other biological models, a distinct contrast to such animals as the armadillo.

The sophistication in the technology of mouse rearing is a major accomplishment. The manipulation of the athymic nude factor on to various strains and in combination with other factors bring us close to the realization of the "living test tube" concept. We now know that the spontaneous tumors appear in nude mice as our management practices have increased the longevity of these mice close to the thymic strains and stocks.

Opening Remarks

I would like to make you aware of at least two programs that the International Council for Laboratory Animal Science (ICLAS) is developing that have direct bearing upon the nude mouse. To cope with the increased volume of literature that continues to increase concerning the nude mouse, an information center, with a newsletter under the sponsorship of ICLAS through the guidance and editorship of Jørgen Rygaard, is nearly ready. A genetic and microbiological monitoring system is being developed at the Central Institute for Experimental Animals, Tokyo, and the Veterinary Resources Branch, National Institutes of Health, Bethesda, Maryland, under the direction of T. Nomura and H. E. Hoffman, respectively.

On behalf of our host and ICLAS, I welcome you to the Third International Workshop on Nude Mice. I am confident you will find the exchanges of information among you will be as exciting as the setting you have been placed in. If this be true, this should be one of the most productive meetings in the history of laboratory animal science.

Cluff E. Hopla
President
International Council for
Laboratory Animal Science

CONTENTS

VOLUME 1

Invited Lectures, Infection, Immunology

I.	Nude Mice in the Study of Susceptibility and Responses to Infection with Metazoan and Protozoan Parasites Graham F. Mitchell and Margaret C. Holmes	1
II.	Importance of Microbiological Control in Using Nude Mice Tatsuji Nomura and Naoko Kagiyama	11
III.	Genetic Quality Control of the Nude Mouse Tatsuji Nomura and Takeshi Tomita	23
IV.	The Nude Mouse as a Tool in Cancer Chemotherapy Studies David P. Houchens and Artemio A. Ovejera	27
V.	The Congenitally Athymic Streaker Mouse Leonard D. Schultz, Hendrick G. Bedigian, Hans-Jorg Heiniger, and Eva M. Eicher	33
VI.	Characteristics of Nude Rats Michael F. W. Festing	41
VII.	The Hairless Athymic Guinea Pig Carolyn Reed and John L. O'Donoghue	51

Chapter

1.	Nude Mice as a Model for Chemotherapy of Leprosy Kenji Kohsaka, Kazuo Yoneda, Yumiko Arimochi, Masanao Makino, Tatsuo Mori, and Tonetaro Ito	59
2.	Mechanisms of Resistance of Nude Mice to *Neisseria gonorrhoeae* Philip S. Lamborn, Julia Cauthen, and David J. Drutz	67
3.	Resistance and Immune Response of Nude Mice to Experimental Fungal Infections Tadashi Arai, Akia Shiraishi, Koji Yokoyama, and Kazuhide Nakagaki	77
4.	T-independent Antibody Production in Nude Mice Immunized with a Rodent Malaria Parasite (*Plasmodium berghei*) Seiji Waki and Mamoru Suzuki	91
5.	The Use of CBA/N and Nude Mice to Study the Regulation of B-Cell Activation Following Infection with *Trypanosoma rhodesiense* John F. Finerty, Yvonne J. Rosenberg, Louise Kendrick, Russell P. McKelvin and Carl T. Hansen	103
6.	*Trypanosoma musculi* infection of Nude Mice Bradford O. Brooks and Norman D. Reed	111
7.	Experimental *Pneumocystis carinii* Infection in Nude and Steroid-Treated Normal Mice Peter D. Walzer and Ralph D. Powell, Jr.	123

Contents

8. Characteristic Responses of Nude Mice in Angiostrongyliasis and Echinococcosis
   Masao Kamiya, Yuzaburo Oka, Haruo Kamiya and Tatsuji Nomura ... 133
9. Thymus Dependence of Specific Homocytotropic Antibody Production and Intestinal Mast Cell Accumulation
   Norman D. Reed, Patricia K. Crowle, Rheta S. Booth, and John J. Munoz ... 147
10. Inhibition of Encephalomyocarditis Virus Replication by Macrophages from Athymic (Nude) Mice Sensitized with Nonviable *Mycobacterium tuberculosis*
    Donald L. Lodmell, Robert R. Cent, Jr., Anne M. Pusateri, and Larry C. Ewalt ... 157
11. Friend Leukemia Virus Infection of C57BL/10 Nude Mice
    Radmila B. Raikow, James P. OKunewick, Ruby F. Meredith, and Kathleen C. Magliere ... 169
12. Characterization of a Rat Mutant (rnu^{nz}) Showing Similarities to the Nude Mouse
    Michael V. Berridge, L. Jane McNeilage, Barbara F. Heslop, and Tom E. Miller ... 179
13. Immunologic Studies and Growth of Human Tumors in the Athymic Rat
    M. Joseph Colston, A. Howard Fieldsteel, R. Denise Lancaster, and Peter J. Dawson ... 189
14. Induction of Lymphatic Tissue in the Nude Mouse Dysgenetic Thymus
    Miroslav Holub, Z. Rychter, and A. Machoninová ... 197
15. Induction of a T-Cell-like Response in Athymic Mice
    Gillian Beattie, Joseph Lipsick, Robert A. Lannom, Steven Baird, Nathan O. Kaplan, and Abraham G. Osler ... 207
16. Clearance of Enzymes by the Reticuloendothelial System in Euthymic and Athymic Mice
    J. Gabriel Michael, Linda DiPersio, Andreas P. Kyriazis, and Amadeo J. Pesce ... 217
17. T Helper Cells That Differentiate in an Allogeneic Thymus and the Requirement for H-2 Compatibility for T-B Help
    Berenice Kindred ... 225
18. Tolerance Induction for Alloantigens by Thymus Epithelium
    Richard Hong, Roger Klopp, Robert Struble, and Judith K. Manning ... 231
19. IgE Responses in Nude Mice Reconstituted with Cultured Thymic Fragments
    Judith K. Manning and Richard Hong ... 239
20. Detection of IgE in Congenitally Athymic (Nude) Mice
    Dean Roberts, Toru Takenaka, and Joe M. Jones ... 243
21. Induced Oophoritis and Gastritis in Nude Mice: A New Approach to the Localized Type of Autoimmunity
    Akinori Kojima, Osamu Taguchi, and Yasuaki Nishizuka ... 245

Contents

22. The Nude Mouse in Autoimmune Disease
    Karsten Buschard, Sten Madsbad, Erik Dabelsteen,
    and Jørgen Rygaard ... 255
23. The Immunopathology of Congenitally Athymic (Nude)
    New Zealand Mice
    M. Eric Gershwin and Yoshiyuki Ohsugi ... 263
24. Athymic Mice: An Experimental Animal for the Isolation of
    "Crohn's Disease Agent"
    Kiron M. Das, Isabel Valenzuela and Rachel Morecki ... 271
25. Antibody Response to Allogeneic and Xenogeneic Skin Grafts
    in Nude Mice
    Peter J. A. Capel, Simon P. M. Lems,
    and Robert A. P. Koene ... 275
26. Preliminary Studies of Normal Untreated and/or
    Carcinogen-Treated Adult Human Breast, Prostate,
    and Esophagus as Xenografts in Nude Mice
    Marion G. Valerio, Elliot L. Fineman,
    Ronald L. Bowman, Curtis C. Harris, Benjamin F. Trump,
    Elizabeth A. Hillman, and Barry M. Heatfield ... 283
27. Heterotransplantation of Embryonic Human Organs into
    Athymic (Nude) Mice
    Hideo Nishimura, K. Arishima, C. Uwabe, and K. Shiota ... 297
28. Athymic Nude Mice: Ex Vivo In Vivo Models to Study the
    Development, Growth, and Differentiation of the Normal and
    Neoplastic Xenogeneic Mammary Gland
    Clifford W. Welsch and M. Jean McManus ... 309
29. Endocrine Morphology and Reproductive Function in
    Athymic Nude Mice
    Kowetha A. Davidson, J. Michael Holland, Jerry W. Hall,
    and Lawrence C. Gipson ... 317
    Index ... 327

VOLUME 2

Oncology

30. Spontaneous Neoplasms in Aged Athymic (Nude) Mice
    John W. Parker, Jeanne Joyce, and Paul Pattengale ... 347
31. Spontaneous Tumors of Long-Lived, Reconstituted Nude Mice
    Gordon J. Eaton, R. Philip Custer, and A. Reynolds Crane ... 359
32. Immunologic Mechanisms of Selective Graft Resistance to
    Certain Malignant Tumors and Prevention of Metastases by
    Athymic Nude Mice
    Untae Kim, Tin Han, Swapan Ghosh, Victoria H. Freedman,
    Seung-il Shin, and David Pressman ... 367
33. Thymus-Independent Host Resistance against the Growth and
    Metastases of Allogeneic and Xenogeneic Tumors in Nude Mice
    Rachel S. Simon, Abraham S. Klein, Untae Kim,
    Victoria H. Freedman, and Seung-il Shin ... 379

34. Study of Metastases of Human Malignant Cells in Nude Mice
   Emilio A. Machado, Bismark B. Lozzio, Carmen B. Lozzio,
   Stephen V. Lair, and Patricia A. Maxwell ... 391
35. Detection of Early Malignant Changes in Tissue-Cultured
   Cells Using a Novel Tumorigenicity Assay in Nude Mice
   Robert S. Wells, Evelyn W. Campbell,
   Lawrence M. Holland, Douglas E. Swartzendruber,
   and Paul M. Kraemer ... 403
36. Characterization and Strain Difference of Natural Killer Cells
   in Nude Mice
   Norikazu Tamaoki, Sonoko Habu, Ko Okumura,
   Masataka Kasai, Akira Akatsuka, Tsuyoshi Sato,
   Kazu Shimamura, and Muneo Saito ... 413
37. Resistance of Germ-free Athymic Nude Mice to Two-Stage
   Skin Carcinogenesis
   J. Michael Holland and Eugene H. Perkins ... 423
38. Comparative Effects of Proliferative Agents on Nude Mouse
   Skin, Pig Skin, and Pig Skin Xenogeneic on Nudes
   Gerald G. Krueger, Donald A. Chambers,
   and N. Jane Shelby ... 435
39. Analysis of Human Tumor Growth in Nude Mice
   Jørgen Fogh, John Tiso, Thomas Orfeo, Jens M. Fogh,
   Walter P. Daniels, and Francis E. Sharkey ... 447
40. Increased Mitotic Activity in Human Tumors Transplanted to
   Athymic Nude Mice
   Francis E. Sharkey and Marcia Bains-Grebner ... 457
41. Human Neuroblastoma Ocular and Bone Heterotransplants
   Lawrence Helson and Christiane Helson ... 463
42. Growth of Human Tumors in Three Strains of Nude Mice
   with Additional Immune Deficiencies and Immunologic
   Characterization of the Strains
   Beppino C. Giovanella, Kenneth J. McCormick,
   Carl Hansen, and John S. Stehlin ... 471
43. Tumorigenicity of Epstein-Barr Virus-Transformed
   Human Lymphocytes in Conditioned Nude Mice
   Shaw Watanabe, Yukio Shimosato, Masahito Kuroki,
   Setsuo Hirohashi, Hidechika Okada, Shigeo Suzuki,
   and Yutaka Tsutsumi ... 481
44. Effects of Thymus Grafts in Nude Mice Transplanted with
   Human Malignant Tumors
   Grete Krag Jacobsen, Carl O. Povlsen and Jørgen Rygaard ... 493
45. Rejection of Virus Persistently Infected Tumor Cells and
   Its Implications for Regulation of Tumor Growth and
   Metastasis in Athymic Nude Mice
   Lola M. Reid, Nagahiro Minato, Charlotte Jones,
   Barry Bloom, and John Holland ... 505
46. Isolation of Human Tumor Cell Lines from Xenotransplanted
   Human Tumors in Nude Mice by Immune Elimination of
   Host Fibroblasts

Contents

	Tetsuro Okabe, Atsuko Suzuki, Nakaaki Ohsawa, and Toyozo Terasima	527
47.	The Behavior of Co115 Human Colon Carcinoma in Nude Mice Bernard Sordat, Rosemary K. Lees, E. Bogenmann, and Geronimo Terres	543
48.	Usefulness of Nude Mice in Detecting Function of Human Tumors Jun-ichi Hata, Norikazu Tamaoki, and Yoshito Ueyama	557
49.	Xenotransplanted Human Tumors Producing Colony-Stimulating Factor (CSF) in Nude Mice as a Source for Large-Scale Production of Human CSF Nakaaki Ohsawa, Noriharu Sato, Sigetaka Asano, Masayoshi Ono, Hitoshi Nomura, Tomoko Nakanishi, Asako Shibukawa, Yuko Miyazono, and Yoshito Ueyama	573
50.	Biochemical Markers of Human Tumors Grown in Athymic (Nude) Mice Amadeo J. Pesce, S. Dingle, Linda DiPersio, Andreas P. Kyriazis, and J. Gabriel Michael	579
51.	Phenotypic Character of Serially Heterotransplanted Tumors in Athymic Mice Donald E. Slagel, William B. Bevins, and John J. Beasley	589
52.	Human Tumor-Induced Expression of Endogenous Murine Leukemia Viruses of Athymic Mice James W. Gautsch, Aileen F. Knowles, Fred C. Jensen, and Nathan O. Kaplan	601
53.	Some Studies on the Characterization of a Transplantable Androgen-Dependent Human Prostatic Carcinoma (PC 82) Johan C. Romijn, K. Oishi, G. J. van Steenbrugge, J. Bolt-deVries, and F. H. Schröder	611
54.	Comparison between the Chemotherapy of Human Cancer Xenografts in Nude Mice and the Clinical Responses Observed in the Donor Patients Masahide Fujita and Tetsuo Taguchi	621
55.	Studies on Experimental Chemotherapy against Cancer of Digestive Organs Transplanted in ^{60}Co Irradiated Nude Mice E. Kawamura, Tatsuo Suzuki, T. Kurakawa, T. Miyagawa, H. Toyoda, Y. Miura, N. Matsuzaki, S. Katagiri, Y. Kawakubo, I. Umezawa, Y. Suzuki, and A. Ghoda	631
56.	Retinoblastoma in the Eyes of Nude Mice: Quantitative Assessment of Therapy Brenda L. Gallie, E. Y. Chew, M. Chang, and Robert A. Phillips	641
57.	Evaluation of the Response of a Panel of Human Melanoma Tissue-Cultured Cell Lines Xenografted in Nude Mice to Four Anticancer Drugs of Known Clinical Activity Robert E. Bellet, Victoria Danna, Michael J. Mastrangelo, Gordon J. Eaton, and David Berd	649
58.	Experimental Therapy of Human Tumors Heterotransplanted in Nude Mice by Continuous Infusion of Short-Acting Chemotherapeutic Agents	

	Shih-shun Lee, Beppino C. Giovanella, John S. Stehlin, Jr., and Jan C. Brunn	657
59.	Effect of Single-Dose X Irradiation on Growth Delay and Flow Cytometric DNA Distribution of a Human Colonic Carcinoma Transplanted to Nude Mice	
	Mogens Spang-Thomsen, Lars Vindeløv, I. J. Christensen, Jakob Visfeldt, and Arne Nielsen	665
60.	Primary and Acquired Resistance to Alkylating Agents in Heterotransplants of Human Melanomas and Colon Carcinomas	
	Rainhardt Osieka and Carl G. Schmidt	675
	Index	325 & 685

I

Nude Mice in the Study of Susceptibility and Responses to Infection with Metazoan and Protozoan Parasites

Graham F. Mitchell* and Margaret C. Holmes

Laboratory of Immunoparasitology, The Walter and Eliza Hall Institute of Medical Research, Royal Melbourne Hospital P.O., Victoria 3050, Australia.

Introduction

Hypothymic nude mice are proving to be as important to the experimental immunoparasitologist as they are to the experimental cellular immunologist; such mice greatly facilitate the analysis of the immunologic aspects of natural mouse–parasite relationships and of T-cell-dependent reactions to natural and unnatural mouse parasites. Although precise quantitative and anatomic details of induction, expression, and suppression of *specific* antiparasite immune effector mechanisms cannot be determined without isolated and purified parasite antigens, nude mice are nevertheless a powerful research tool for rapid dissection of the more immunobiologic aspects of mouse–parasite interactions. From analyses of metazoan, protozoan, and ectoparasitic infections in nude mice, information can be obtained readily on the following at the very least:

1. The consequences of a gross T-cell defect on susceptibility to infection and on genetically based variation in susceptibility to infection
2. The general nature of T-cell-dependent responses to infection
3. The broad immunoparasitological activities of isolated T-cell subpopulations and isolated antibody isotypes in vivo

In this discussion, each of these research outputs from studies on parasitic infections in nude mice is examined, emphasis being on natural mouse–parasite systems. A more comprehensive listing of references than that given here can be found in articles by Mitchell (29,30).

---

* To whom correspondence should be addressed.
© 1982 Gustav Fischer New York, Inc.
Proceedings of the Third International Workshop on Nude Mice.

In this laboratory, BALB/c *nu/nu* (N12), CBA/H *nu/nu* (N7), and C57BL/6 *nu/nu* (N14) mice have been used. Details on their derivation, production, and maintenance are as follows: A noninbred stock bearing the *nu* gene was obtained from Dr. M. Naughton, CSIRO Division of Genetics, Australia, in 1970. A line of C57BL/6 *nu* (N10) was kindly provided by Dr. I. Lefkovits in 1974. Lines congenic with the major strains in use were produced by repeated backcrossing with selection at early stages at the H-2 and Ig1 loci. Backcrossing in successive generations has been continued to N20 but lines were taken out earlier for brother × sister mating when mice were required for special purposes. Inbred partners used for backcrossing were CBA/CaHWehi F? + F16fF26, BALB/cAnBradleyWehi F? + F13fF19, and C58BL/6JWehi F105 + F1fF17. Backcrossing was carried out under conventional conditions but inbred lines were transferred to specific pathogen-free units through caesarian derivation into Trexler type isolators. Lines were maintained germ free for 6 months and tested to confirm absence of virus antibody before transfer to the specific pathogen-free breeding unit. Mice for use in experiments are bred from matings between *nu/nu* × *nu/+* or *nu/+* × *nu/+*. Mice under experiment are maintained under filter hats in conventional mouse rooms, or in negative pressure quarantine rooms.

Susceptibility to Infection in Nude Mice

It is usually a relatively simple matter to determine parasite burdens and variations in infection characteristics in nude versus intact versus T-cell-reconstituted nude mice. Several studies have been performed on the susceptibility of nude mice to parasitic infection and the results are of two basic types, bearing in mind that the term "susceptibility" is rather gross when applied to diverse parasite systems:

Category 1: Increased susceptibility of nude mice to infection (i.e., increased parasite burdens, increased proliferative rate of parasites, prolonged duration of infection, increased host mortality, or failure to develop resistance to reinfection),

Category 2: Comparable susceptibility or even marginally greater resistance of nude mice.

As far as we are aware, it is accurate to say that all natural murine parasites (or, at least, natural rodent parasites) are within the first category; the second category contains various abnormal parasites—abnormal because the organism is not normally a mouse parasite or because of laboratory modification of what was at one time a natural murine parasite.

Examples of category 1 parasites are the pinworms, *Aspicularis tetraptera* and *Syphacia obvelata* (19), *Nematospiroides dubius* (47), *Taenia taeniaeformis* (33), *Mesocestoides corti* (34,46), *Hymenolepis nana* (18,50), *Giardia muris* (53,58), *Hexamita muris* (3,25), some isolates of *Plasmodium yoelii* (8,35,51, 63), *Babesia microti* (8), *Trypanosoma musculi* (4,48), and *Leishmania tropica* (15), the "rat parasites" *Nippostrongylus brasiliensis* (20,21,29), *Hymenolepis diminuta* (1,2,17,50), and *Strongyloides ratii* (H.S. Dawkins, D.I. Grove, and

G.F. Mitchell, unpublished observations), as well as *Trichinella spiralis* (56,57) and *Trypanosoma cruzi* (59). In all these mouse–parasite systems (but again, more particularly in the natural mouse–parasite systems), infection in nude mice is persistent or uncontrolled, spontaneous rejection is absent or delayed, parasites are present in higher numbers, or the life of the host is threatened.

Examples of category 2 parasites are *Ascaris suum* (36), *Fasciola hepatica* (49), *Schistosoma mansoni* (44), various extracellular *Trypanosoma* spp. (6, 23,39; B.M. Ogilvie, personal communication), *Plasmodium berghei* (60,62), and *Babesia rodhaini* (28). Many of the parasites in this category are lethal in intact mice and, as is well known, any parasite that kills, within a matter of weeks, the majority of immunocompetent individuals of a particular host species used in the laboratory can be considered highly abnormal.

The fact that natural murine parasites frequently overwhelm nude mice (an event prevented by T-cell reconstitution) provides compelling evidence that T-cell-dependent, host-protective responses have been of major importance in the evolutionary development of balanced host–parasite relationships. Of course, the restoration of a particular outcome of infection in nude mice by injection of T cells simply indicates that the parameter in question is T cell dependent and not that cell-mediated immunities predominate over humoral immunities in effecting this outcome. In reconstitution experiments, T-cell effects may be quite indirect even when it is shown that specific antiparasite-reactive T cells are involved (the latter being a most difficult exercise). There has been no unequivocal demonstration of "killer" T cells with specificity for, and a direct killing action on, parasitized cells or extracellular parasites akin to the cytotoxic T cells of *in vitro* tumor systems.

Principal findings from studies on metazoan and protozoan parasitic infections in SPF-derived nude mice in this and collaborating laboratories are as follows, the more artificial rat and veterinary parasites not being included:

1. In *Giardia muris* infection (initiated by oral administration of cysts), cyst excretion in feces and intestinal trophozoite numbers remain elevated in nude mice long after the infection has apparently resolved in intact or T-cell-injected nude mice of the genetically resistant BALB/c mouse strain. Intact C3H/He, and in particular males, resemble BALB/c *nu/nu* mice in developing a chronic giardiasis and protective vaccination with trophozoites in Freund's complete adjuvant can be achieved readily in BALB/c but not in C3H/He mice (54). No information is yet available on the nature of T-cell-dependent antitrophozoite immune responses which presumably contribute to the differences in infection characteristics between infected BALB/c *nu/+* and BALB/c *nu/nu* mice.

2. In *Leishmania tropica* infection (initiated by intradermal injection of promastigotes), striking mouse strain variation in susceptibility is apparent and nude mice of the relatively resistant genotypes C57BL/6 and CBA/H are as susceptible as nude mice of the genetically susceptible mouse strain, BALB/c. As few as 1×10^6 syngeneic cortisone-resistant thymocytes will protect CBA/H *nu/nu* mice from infection. [The *L. tropica* organism used in these studies was isolated from an infected human; however, it is generally believed that cutaneous leishmaniasis is a zoonosis, the usual host for the parasite being a desert rodent].

3. In *Taenia taeniaeformis* infection (initiated by oral administration of eggs), mouse strain variation in susceptibility is again very obvious and nude mice of relatively resistant (C57BL/6 and BALB/c) and relatively susceptible (CBA/H) genotypes are highly susceptible. Cysts in the liver are present in greater numbers and are of greater size than in intact $nu/+$ mice of the appropriate genotype. Host-protective antibodies, which are responsible for the expression of impressive concomitant immunity, are absent in the sera of infected nude mice, although such mice readily support the host-protective activities of serum harvested from infected intact mice (38, and see below).
4. In *Nematospiroides dubius* infection (initiated by oral administration of infective third stage larvae, L3), BALB/c nu/nu mice fail to develop the solid resistance to reinfection and expulsion of adults seen in female BALB/c $nu/+$ mice (cf. intact CBA/H male mice) after two or three challenges with L3. However, infected BALB/c nu/nu mice partially inhibit establishment of larvae in the intestinal wall following an additional L3 administration.
5. In *Mesocestoides corti* infection (initiated by oral or intraperitoneal administration of peritoneal larvae), the proliferative rate of larvae in the liver is markedly increased over that in intact mice and nudes succumb to the infection after a few weeks. Such mice can be protected from early death by injection of T cells and parasite proliferation is more restrained in T-cell-reconstituted versus unreconstituted mice.

Responses to Infection in Nude Mice

The list of T-cell-dependent responses to metazoan and protozoan parasitic infection is very long. Responses that are defective or absent in infected nude mice and that can be restored wholly or in part to such mice by injection of T cells include:

1. Peripheral and tissue eosinophilia in various metazoan infections (e.g. 14,16,24,36,42,45)
2. Egg granulomas in *Schistosoma mansoni* infections (5,16,44)
3. Fibrotic encapsulation of the larval cestode, *Mesocestoides corti* in the liver (46)
4. Peritoneal malabsorption in infection with *M. corti* (30,34)
5. Intestinal wall granulomas in *Nematospiroides dubius* infection (47)
6. Splenomegaly in some murine plasmodial infections (52; cf. G.F. Mitchell, unpublished observations)
7. Antihapten antibody responses to haptenated *M. corti* larvae (37)
8. IgG_1 hypergammaglobulinemia in *M. corti* infection (7)
9. Various antiparasite antibody responses involving the IgG and IgE isotypes (and probably IgA, ± IgM isotypes)
10. Production of an anti-*M. corti* hybridoma antibody specificity (32)
11. IgM antiphosphorylcholine responses in *Ascaris suum* infections (36)

12. Splenic direct (IgM?) plaque-forming cell increases to bromelain-treated mouse erythrocytes in murine hemoprotozoal infections (10,55)
13. Mast cell accumulations in the intestinal tract in *Trichinella spiralis* infection (57)
14. Minor villus changes in the intestinal tract in *Giardia muris* infection (13,53)

There are numerous other references to several of these phenomena in T-cell-deficient mice other than nudes. The above list is likely to be expanded to include such diverse responses as goblet cell increases and mucus secretion in intestinal nematode infections (27), mediator-dependent parasitocidal macrophage activation in *Trypanosoma cruzi* infection (43), potentiated reagin responses (22), chronic inflammatory responses other than those already mentioned, and various aspects of architectural disruption of tissues (30), detoxification of noxious parasite substances (61), and *Schistosoma mansoni* parasite egg excretion in feces (11).

It is not known which of the more inflammatory responses in the above listing are the direct results of the activities of T-cell products and which responses result from the activities of T-cell-dependent products such as antiparasite antibodies of various isotypes. The list illustrates the multitude of host responses that are influenced either directly or indirectly by activated T cells. Triggering of T cells may be effected through parasite antigens [the efficiency of this triggering is presumably increased by complexing of parasite antigen with H-2 major histocompatibility complex (MHC) associative recognition molecules of host cell surfaces] or by parasite mitogens. Parasites exhibit great structural and life cycle complexity (9) and it would be surprising if parasites did not contain mitogenic molecules capable of initiating a triggering perturbation of host cell membranes. In order to differentiate between activities of T cells with or without parasite antigen specificity, techniques to deplete reconstitutive T cell inocula (using parasite antigens as solid phase absorbents, for example) of specifically reactive cells will need to be developed. Great difficulties can be anticipated if these antigens need to be complexed with self-MHC antigens (e.g., as an antigen-pulsed macrophage monolayer) in order to achieve sufficient binding strength and thus efficient depletion. The use of the reconstituted nude mouse in immunoparasitological studies is further examined in the next section.

Reconstitution of Nude Mice

One means of investigating the immunologic activities of T cells and T-cell-dependent cellular and molecular accompaniments of parasitic infection is to use infected nude mice as recipients of isolated T-cell subpopulations, enriched parasite antigen-reactive T cells, other cell populations such as eosinophils, and various antibody isotypes. Readouts will include susceptibility to infection and the initiation or modification of the responses to infection mentioned in the preceding section. A difference between intact and T-cell-reconstituted nudes may be apparent if the parasite is located in a site (e.g., skin or intestinal tract) where the generalized epithelial defect in the nude may influence various outcomes and events.

The length of time required for expression of many accompaniments of chronic parasitic infection poses a serious problem in the analysis of immunoparasitological events in reconstituted nude mice. Although techniques for isolation of T-cell subtypes, accessory cells such as eosinophils, and antibody isotypes have improved markedly over the past few years, the methods for isolation of large quantities of pure populations of cells and molecules are often far from ideal. Moreover, methods for assessing the degree of contamination are quite crude in many instances. Because of the long readout time in many nude reconstitution experiments, a minority contaminant lymphocyte population, for example, often has adequate time to expand. Interpretation of findings in cell-injected nude mice must therefore be cautious. This difficulty has already been encountered in analysis of the eosinophilia-promoting activities of Ly1- and Ly2-enriched T-cell populations in nude mice infected with *Mesocestoides corti* (24) and in similar studies on induction of resistance of nude mice (of appropriate genotype) to *Leishmania tropica* infection (G.F. Mitchell and I.F.C. McKenzie, unpublished observations). Other difficulties, some of which are presumably imposed by the generalized epithelial defect in thymus grafted or reconstituted nude mice versus intact mice, have been mentioned earlier and discussed previously (29).

A potent effect of "immune serum" in infection of nude mice with *Taenia taeniaeformis* has been demonstrated (33) and host-protective antibodies can be readily demonstrated in the *T. taeniaeformis*–mouse (e.g., 26,40) and *T. taeniaeformis*–rat systems (e.g., 41). Provided the injection is given at around the time of oral administration of eggs, complete protection can be achieved in nude mice with serum harvested from infected intact (but not infected nude) mice. By fractioning the serum into IgG1-enriched and IgG2-enriched fractions using differential pH elution from protein A–Sepharose columns (12), evidence has been obtained that both fractions are better than either alone at transferring resistance to infection (38). Thus, Ig antibody isotypes may interact in mediating protection, an interpretation in keeping with much evidence that various immunological effector mechanisms interact in mediating host protection against natural parasites (discussed by G.F. Mitchell, 30,31). This example serves to make the point that the infected nude mouse, as a recipient of various inocula, is a powerful means by which "minimal" antiparasite immunological effector mechanisms can be studied in vivo. There is no doubt that nude mice of various genotypes, together with other models such as anti-μ-treated (B-cell deficient) and Biozzi mice, will readily provide important new information on the immunologic aspects of murine host–parasite relationships.

References

1. Andreassen, J., O. Hinsbo, and E.J. Ruitenberg. *Hymenolepis diminuta* infections in congenitally athymic (nude) mice: Worm kinetics and intestinal histopathology. Immunology 34:105–113, 1978.
2. Bland, P.W. Immunity to *Hymenolepis diminuta*: Unresponsiveness of the athymic mouse to infection. Parasitology 72:93–97, 1977.
3. Boorman, G.A., P.H.C. Lina, C. Zurcher, and H.T.M. Nieuwerkerk. *Hexamita* and

Giardia as a cause of mortality in congenitally thymus-less (nude) mice. Clin. Exptl. Immunol. 15:623–627, 1973.
4. Brooks, B.O. and N.D. Reed. Thymus dependency of *Trypanosoma musculi* elimination from mice. J. Reticuloendothel. Soc. 22:605–608, 1977.
5. Byram J.E., and G. von Lichtenberg. Altered schistosome granuloma formation in nude mice. Am. J. Trop. Med. Hyg. 26:944–956, 1977.
6. Campbell, G.H., K.M. Esser, and S.M. Phillips. *Trypanosoma rhodesiense* infection in congenitally athymic (nude) mice. Infect. Immun. 20:714–720, 1978.
7. Chapman, C.B., P.M. Knopf, R.F. Anders, and G.F. Mitchell. IgG_1 hypergammaglobulinaemia in chronic parasitic infections in mice. Evidence that the response reflects chronicity of antigen exposure. Aust. J. Exptl. Biol. Med. Sci. 57:389–400, 1979.
8. Clark, I.A., and A.C. Allison. *Babesia microti* and *Plasmodium yoelii* infections in nude mice. Nature (London) 252:328–329, 1974.
9. Cohen, S., and E.G. Sadun (eds.). Immunology of parasitic infections. Oxford: Blackwell Scientific, 1976.
10. Cox, K.O., R.J. Howard, and G.F. Mitchell. Studies on immune responses to parasite antigens in mice. VII. Cells secreting antibodies to modified mouse erythrocytes in *Babesia rodhaini*-infected mice. Cell. Immunol. 32:223–227, 1977.
11. Doenhoff, M., R. Musallam, J. Bain, and A. McGregor. Studies on the host–parasite relationship in *Schistosoma mansoni*-infected mice: The immunological dependence of parasite egg excretion. Immunology 35:771–778, 1978.
12. Ey, P.L., S.J. Prowse, and C.R. Jenkin. Isolation of pure IgG_1, IgG_{2a}, and IgG_{2b} immunoglobulin from mouse serum using protein A-Sepharose. Immunochemistry 15:429–436, 1978.
13. Ferguson, A., and T.T. MacDonald. Effects of local delayed hypersensitivity on the small intestine, p. 305. *In* R. Porter and J. Knight (eds.), Immunology of the gut. Amsterdam: Elsevier, 1977.
14. Fine, D.P., R.D. Buchanan, and D.G. Colley. *Schistosoma mansoni* infection in mice depleted of thymus-dependent lymphocytes. Eosinophils and immunologic responses to a schistosomal egg preparation. Am. J. Pathol. 71:193–206, 1973.
15. Handman, E., R. Ceredig, and G.F. Mitchell. Murine cutaneous leishmaniasis: Disease patterns in intact and nude mice of various genotypes and examination of some differences between normal and infected macrophages. Aust. J. Exptl. Biol. Med. Sci. 57:9–29, 1979.
16. Hsu, C-K, S.H. Hsu, R.A. Whitney, and C.T. Hansen. Immunopathology of schistosomiasis in athymic mice. Nature (London) 262:397–399, 1976.
17. Isaak, D.D., R.H. Jacobson, and N.D. Reed. Thymus dependence of tapeworm (*Hymenolepis diminuta*) elimination from mice. Infect. Immun. 12:1478–1479, 1975.
18. Isaak, D.D., R.H. Jacobson, and N.D. Reed. The course of *Hymenolepis nana* infections in thymus-deficient mice. Intl. Arch. Allergy Appl. Immunol. 55:504–513, 1977.
19. Jacobson, R.H., and N.D. Reed. The thymus dependency of resistance to pinworm infection in mice. J. Parasitol. 60:976–979, 1974.
20. Jacobson, R.H., and N.D. Reed. The immune response of congenitally athymic (nude) mice to the intestinal nematode *Nippostrongylus brasiliensis*. Proc. Soc. Exptl. Biol. Med. 147:667–670, 1974.
21. Jacobson, R.H., and N.D. Reed. The requirement of thymic competence for both humoral and cell-mediated steps in expulsion of *Nippostrongylus brasiliensis* from mice. Intl. Arch. Allergy Appl. Immunol. 52:160–168, 1976.
22. Jarrett, E., and A. Ferguson. Effect of T cell depletion on the potentiated reagin response. Nature (London) 250:420–422, 1974.
23. Jayawardena, A.N., and B.H. Waksman. Suppressor cells in experimental trypanosomiasis. Nature (London) 265:539–541, 1977.
24. Johnson, G.R., W.L. Nicholas, I.F.C. McKenzie, D. Metcalf, and G.F. Mitchell. Peritoneal cell population of mice infected with *Mesocestoides* as a source of eosinophils. Intl. Arch. Allergy Appl. Immunol. 59:315–322, 1979.
25. Kunstyr, I., B. Meyer, and E. Ammerpohl. Spironucleosis in nude mice; an animal model for immuno-parasitologic studies, pp. 17–27. *In* T. Nomura, N. Ohsawa, N. Tamaoki, and K. Fujiwara (eds.), Proceedings of the second international workshop on nude mice. Tokyo: University of Tokyo Press, 1977.

26. Lloyd, S., and E.J.L. Soulsby. The role of IgA immunoglobulin in the passive transfer of protection to *Taenia taeniaeformis* in the mouse. Immunology 34: 939–945, 1978.
27. Miller, H.R.P., Y. Nawa, and C.R. Parish. Intestinal goblet cell differentiation in *Nippostrongylus*-infected rats after transfer of fractionated thoracic duct lymphocytes. Intl. Arch. Allergy Appl. Immunol. 59:281–285, 1979.
28. Mitchell, G.F. Studies on immune responses to parasite antigens in mice. V. Different susceptibilities of hypothymic and intact mice to *Babesia rodhaini*. Intl. Arch. Allergy Appl. Immunol. 53:385–388, 1977.
29. Mitchell, G.F. Metazoan and protozoan parasitic infections in hypothymic nu/nu ("nude") mice. Contemp. Top. Immunobiol. 8:55–67, 1978.
30. Mitchell, G.F. Responses to infection with metazoan and protozoan parasites in mice. Adv. Immunol. 28:451–511, 1979.
31. Mitchell, G.F. Effector cells, molecules and mechanisms in host-protective immunity to parasites. Immunology 38:209–223, 1979.
32. Mitchell, G.F., K.M. Cruise, C.B. Chapman, R.F. Anders, and M.C. Howard. Hybridoma antibody immunoassays for the detection of parasitic infection: Development of a model system using a larval cestode infection in mice. Aust. J. Exptl. Biol. Med. Sci. 57:287–302, 1979.
33. Mitchell, G.F., J.W. Goding, and M.D. Rickard. Studies on immune responses to larval cestodes in mice. Increased susceptibility of certain mouse strains and hypothymic mice to *Taenia taeniaeformis* and analysis of passive transfer of resistance with serum. Aust. J. Exptl. Biol. Med. Sci. 55:165–186, 1977.
34. Mitchell, G.F., and E. Handman. Studies on immune responses to larvel cestodes in mice. A simple mechanism of nonspecific immunosuppression in *Mesocestoide corti*-infected mice. Aust. J. Exptl. Biol. Med. Sci. 55:615–622, 1977.
35. Mitchell, G.F., E. Handman, and R.J. Howard. Protection of mice against *Plasmodium* and *Babesia* infections: Attempts to raise host-protective sera. Aust. J. Exptl. Biol. Med. Sci. 56: 553–559, 1978.
36. Mitchell, G.F., R.S. Hogarth-Scott, R.D. Edwards, H.M. Lewers, G. Cousins, and T. Moore. Studies on immune responses to parasite antigens in mice. I. *Ascaris suum* larvae numbers and antiphosphorylcholine response in mice of various strains and in hypothoymic nu/nu mice. Intl. Arch. Allergy Appl. Immunol. 52:64–78, 1976.
37. Mitchell, G.F., J.J. Marchalonis, P.M. Smith, W.L. Nicholas, and N.L. Warner. Studies on immune responses to larval cestodes in mice. Immunoglobulins associated with the larvae of *Mesocestoides corti*. Aust. J. Exptl. Biol. Med. Sci. 55:187–211, 1977.
38. Mitchell, G.F., G.R. Rajasekariah, and M.D. Rickard. A mechanism to account for mouse strain variation in resistance to the larval cestode, *Taenia taeniaeformis*. Immunology 39:481–489, 1980.
39. Morrison, W.I., G.E. Roelants, K.S. Mayor-Withey, and M. Murray. Susceptibility of inbred strains of mice to *Trypanosoma congolense*: Correlation with changes in spleen lymphocyte populations. Clin. Exptl. Immunol. 32:25–40, 1978.
40. Musoke, A., and J.F. Williams. Immunoglobulins associated with passive transfer of resistance to *Taenia taeniaeformis* in the mouse. Immunology 28:97–101, 1975.
41. Musoke, A., J.F. Williams. The immunological response of the rat to infection with *Taenia taeniaeformis*. V. Sequence of appearance of protective immunoglobulins and the mechanism of action of 7Sγ2a antibodies. Immunology 29:855–866, 1975.
42. Nielsen, L.F., L. Fogh, and A. Andersen. Eosinophil response to migrating *Ascaris suum* larvae in normal and congenitally thymusless mice. Acta Pathol. Microbiol. Scand. (B) 82:919–920, 1974.
43. Nogueira N., and Z.A. Cohn. *Trypanosoma cruzi*: in vitro induction of macrophage microbioidal activity. J. Exptl. Med. 148:288–300, 1978.
44. Phillips, S.M., and D.G. Colley. Immunologic aspects of host responses to schistosomiasis: Resistance, immunopathology and eosinophil involvement. Prog. Allergy 24:49–182, 1978.
45. Phillips, S.M., J.J. Diconza, J.A. Gold, and W.A. Reed. Schistosomiasis in the congenitally athymic (nude) mouse. I. Thymic dependency of eosinophilia, granuloma formation, and host morbidity. J. Immunol. 118:594–599, 1977.
46. Pollacco, S., W.L. Nicholas, G.F. Mitchell, and A.C. Stewart. T cell-dependent collagenous encapsulating response in the mouse liver to *Mesocestoides corti* (Cestoda). Intl. J. Parasitol. 8:457–462, 1978.

47. Prowse, S.J., G.F. Mitchell, P.L. Ey, and C.R. Jenkin. *Nematospiroides dubius*: Susceptibility to infection and the development of resistance in hypothymic (nude) BALB/c mice. Aust. J. Exptl. Biol. Med. Sci. 56:561–570, 1978.
48. Rank, R.G., D.W. Roberts, and W.P. Weidanz. Chronic infection with *Trypanosoma musculi* in congenitally athymic nude mice. Infect. Immun. 16:715–716, 1977.
49. Rajasekariah, G.R., C.B. Chapman, P.E. Montague, and G.F. Mitchell. *Fasciola hepatica*: attempts to induce protection against infection in rats and mice by injection of excretory/secretory products of immature worms. Parasitology 79: 393–400, 1979.
50. Reed, N.D., D.D. Isaak, and R.H. Jacobson. The use of nude mice in model systems for studies on acquired immunity to parasitic helminths, pp. 3–16. *In* T. Nomura, N. Ohsawa, N. Tamaoki, and K. Fujiwara (eds.), Proceedings of the second international workshop on nude mice. Tokyo: University of Tokyo Press, 1977.
51. Roberts, D.W., R.G. Rank, W.P. Weidanz, and J.F. Finerty. Prevention of recrudescent malaria in nude mice by thymic grafting or by treatment with hyperimmune serum. Infect. Immun. 16:821–826, 1977.
52. Roberts, D.W., and W.P. Weidanz. Splenomegaly, enhanced phagocytosis and anemia are thymus-dependent responses to malaria. Infect. Immun. 20:728–731, 1978.
53. Roberts-Thomson, I.C., and G.F. Mitchell. Giardiasis in mice. I. Prolonged infections in certain mouse strains and hypothymic (nude) mice. Gastroenterology 75:42–46, 1978.
54. Roberts-Thompson, I.C., and G.F. Mitchell. Protection of mice against *Giardia muris* infection. Infect. Immun. 24:971–973, 1979.
55. Rosenberg, Y. Autoimmune and polyclonal B cell responses during murine malaria. Nature (London) 274:170–172, 1978.
56. Ruitenberg, E.J., A. Elgersma, W. Druizinga, and F. Leenstra. *Trichinella spiralis* infection in congenitally athymic (nude) mice. Parasitology, serological and haematological studies with observations on intestinal pathology. Immunology 33:581–587, 1977.
57. Ruitenberg, E. J., and P.A. Steerenberg. Intestinal phase of *Trichinella spiralis* in congenitally athymic (nude) mice. J. Parasitol, 60:1056–1057, 1974.
58. Stevens, D.F., D.M. Frank, and A.A.F. Mahmoud. Thymus dependency of host resistance to *Giardia muris* infection: studies in nude mice. J. Immunol. 120:680–682, 1978.
59. Trischman, T., H. Tanowitz, M. Wittner, and B. Bloom. *Trypanosoma cruzi*: Role of the immune response in the natural resistance of inbred strains of mice. Exptl. Parasitol. 45:160–168, 1978.
60. van Zon, A., W. Eling, and C. Jerusalem. Histo- and immunopathology of malaria (*Plasmodium berghei*) infection in mice. Isr. J. Med. Sci. 14:659–672, 1978.
61. von Lichtenberg, F. Experimental approaches to human schistosomiasis. Am. J. Trop. Med. Hyg. 26:79–87, 1977.
62. Waki, S., and M. Suzuki. A study of malaria immunobiology using nude mice, pp. 37–44. *In* T. Nomura, N. Ohsawa, N. Tamaoki, and K. Fujiwara (eds.), Proceedings of the second international workshop on nude mice. Tokyo: University of Tokyo Press, 1977.
63. Weinbaum, F.I., C.B. Evans, and R.E. Tigelaar. Immunity to *Plasmodium berghei yoelli* in mice. I. The course of infection in T cell and B cell deficient mice. J. Immunol. 117:1999–2005, 1976.

II

Importance of Microbiological Control in Using Nude Mice

Tatsuji Nomura* and Naoko Kagiyama

Central Institute for Experimental Animals, 1430 Nogawa, Takatsu, Kawasaki 213, Japan.

Introduction

If the microbiological quality of nude mice and the quality of care of such animals are not of a high level, reliable experimental results cannot be expected. This contribution offers several basic reasons that microbiological control is indispensable in experiments using nude mice.

Most importantly, various infectious diseases often interfere with the performance of experiments using nude mice. The most prominent feature of these diseases is the wasting syndrome, which has been reported to be closely associated with several infectious agents as listed in Table II-1.

In 1966, Flanagan (4) showed that *Toxoplasma* played a role in the wasting syndrome of nude mice. Since that time, there have been many reports on this syndrome, showing that infections by *Giardia* (1), *Hexamita* (1), and *Pneumo-*

Table II-1. Infectious agents reported to be associated with the wasting syndrome of nude mice

Agents	Reports
Toxoplasma	(4)
Giardia and *Hexamita*	(1)
Pneumocystis carinii	(12)
Mouse hepatitis virus	(3,5,10,14)
Sendai virus	(3,7,13,15)
Mouse adenovirus	(2)

* To whom correspondence should be addressed.
© 1982 Gustav Fischer New York, Inc.
Proceedings of the Third International Workshop on Nude Mice.

cystis (12), as well as viral infections by mouse hepatitis virus (MHV) (3,5, 10,14), Sendia virus (3,7,13,15), and mouse adenovirus (2), can cause the syndrome. In addition, infections with weak pathogens, which are carried by euthymic conventional animals and which cause the wasting syndrome only in nude mice, often disrupt experiments using nude mice. Furthermore, certain kinds of bacteria, such as *Staphylococcus aureus*, which exist on the human body surface and are usually harmless, may also cause severe abscesses in nude mice.

Examples of Microbiological Contamination and Its Causes

In the following paragraphs, some examples of accidental infections of nude mice in Japan are given.

In 1978, the number of nude mice used in Japan exceeded 40,000. During the last 3 years, various accidents owing to infections have occurred in experiments using nude mice. Among these accidents, five typical cases (A, B, C, D, and E) have been selected for discussion here. The causes of the accidents in all five cases were clarified by microbiological checking.

Case A. In the facility shown in Figure II-1, nude mice were used for experiments on the transfer of tumor cells and for the screening of anticancer drugs. The transfer of tumor cells was performed in vinyl isolators, whereas the screening of anticancer drugs was carried out in the open animal room. The wasting syndrome appeared only in those nude mice which were placed in the open

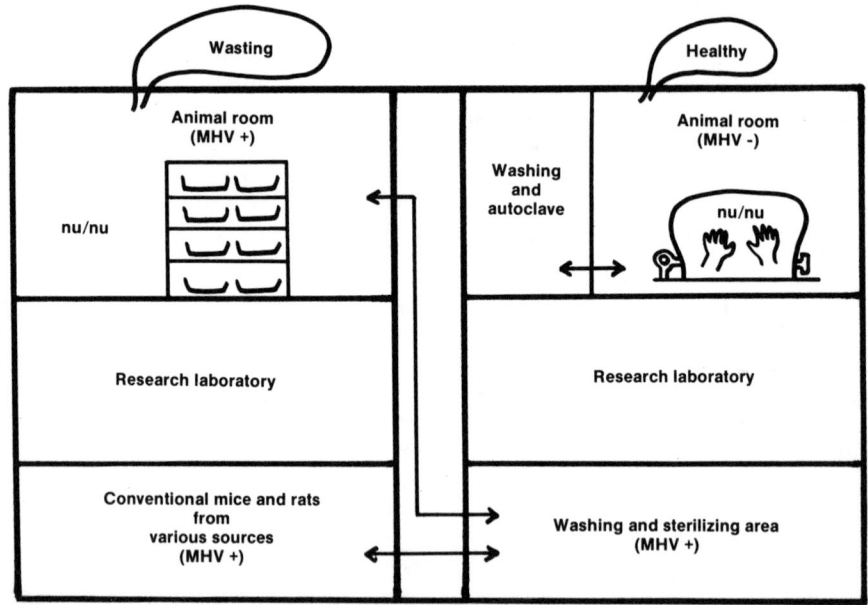

Figure II-1. Accidental infections of nude mice, case A. Cause, contamination of cages in washing and sterilizing area. Problem, layout of facilities.

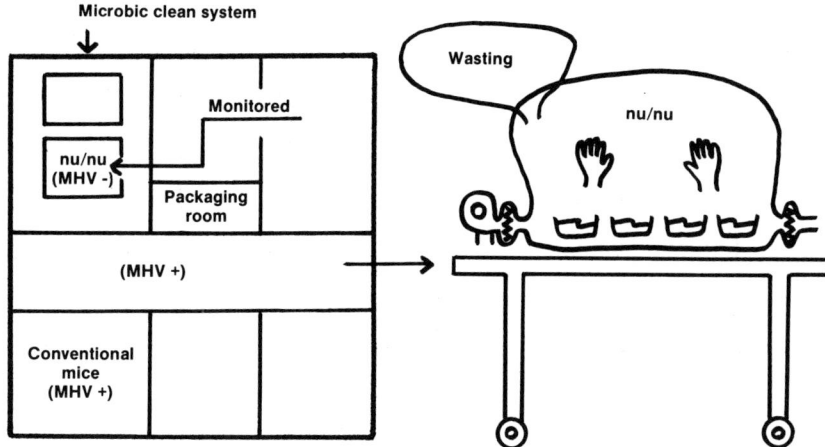

Figure II-2. Accidental infections of nude mice, case B. Cause, contamination during packaging. Problem, layout of facilities.

room, and MHV was isolated from such mice. The cages for these mice were supplied from the washing and sterilizing area on the first floor of the facility. The cages for euthymic conventional mice and rats, latently infected with MHV and used for other experiments, had also been treated in the same washing and sterilizing area. Under such conditions, the route of the MHV infection was suspected to be via the cages supplied from the washing and sterilizing area. The cages for the nude mice should have been supplied from a special washing and sterilizing area for nude mice only. The room in which experiments on transfer of tumor cells were carried out in vinyl isolators was provided with a special adjacent sterilizing room, and the mice tested in this room were all healthy.

Case B. In the facility shown in Figure II-2, nude mice were produced in a microbic clean-room system and supplied to separate experimental facilities, where they were housed in vinyl isolators for the transfer of tumor cells. Nevertheless, the wasting syndrome appeared among the mice in isolators and MHV was isolated. The nude mice kept in the microbic clean-room system had previously been proved MHV-free by monthly serologic monitoring performed on $nu/+$ retired breeders. The supply route of these infected nude mice was traced and it was found that they had been transferred to a packaging room adjacent to a room where conventional mice, latently infected with MHV, were reared. Therefore, MHV from these conventional mice presumably contaminated the packaging room and eventually infected the nude mice during packaging.

Case C. In the facility shown in Figure II-3, nude mice were also used for experiments on the transfer of tumor cells. Because of the small space in the facility, the vinyl isolator had to be placed in the same room with conventional rats infected latently with MHV. The nude mice in the isolator developed the wasting syndrome and MHV infection was detected by examination of their liver tissues. Inspection of the isolator soon revealed leakage. Microbiological checking

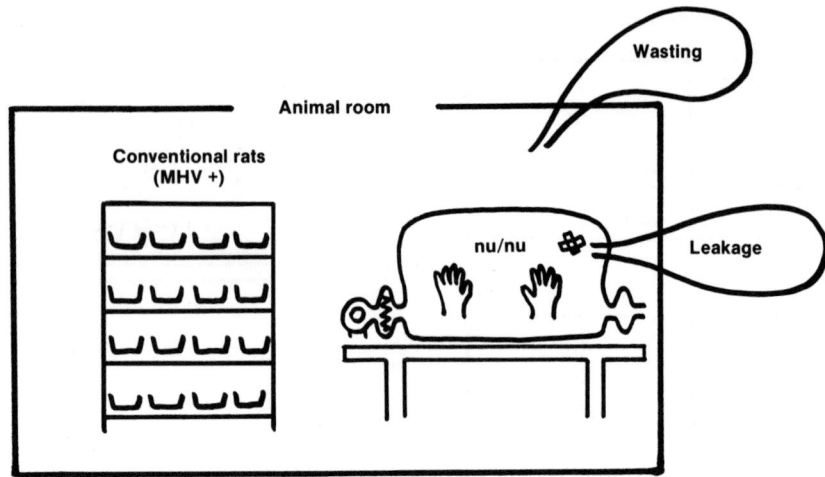

Figure II-3. Accidental infections of nude mice, case C. Cause, isolator leakage during experiment. Problem, maintenance of isolator.

showed that the conventional rats reared in the same room had MHV antibodies. This case indicates that microbiologically controlled rearing systems are of no use unless they are carefully maintained.

Case D. In the facility shown in Figure II-4, nude mice were produced in a special barrier system. All supplies were autoclaved. The caretaker took a shower, wore a clean uniform, and put clean gloves and boots on. Nevertheless, a high incidence of severe abscesses appeared on the faces of the nude mice. Examina-

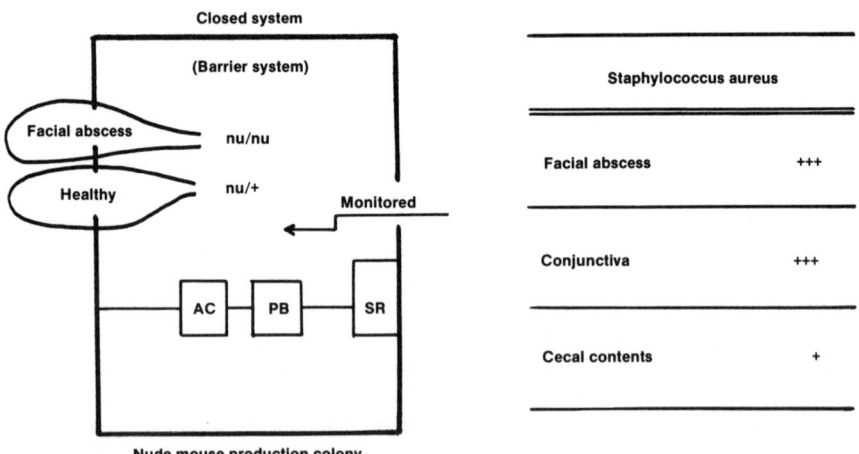

Figure II-4. Accidental infections of nude mice, case D. Cause, contamination from caretaker(?) Problem, limitation of barrier system. AC, autoclave; PB, pass box; SR, shower room.

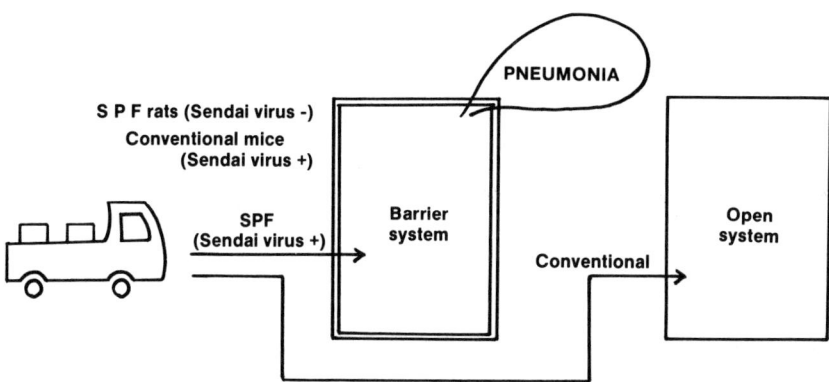

Figure II-5. Accidental infections of SPF rats, case E. Cause, contamination during transportation. Problem, placing SPF rats together with conventional animals during transportation.

tion of the abscesses, conjunctiva, cecal contents, etc., resulted in the isolation *Staphylococcus aureus*. These bacteria were also detected on the skin of the caretaker, suggesting that he was the origin of the infection. This is an example of the problem of the sterilization limits in the barrier system.

Case E. This case involved not nude mice but SPF rats. The SPF rats were transported to the experimental facility where a barrier system was employed (Figure II-5). A few days after arrival, most of the rats showed rough coats with decreased body weights. When some rats were autopsied, hepatization was found in their lungs and after several weeks, Sendai virus antibodies were detected in the remaining rats. However, previous microbiological monitoring had shown that the rats reared in the facility had been free from Sendai virus until that time. Investigations revealed that, during transportation, these SPF rats had been placed together with conventional mice latently infected by the Sendai virus.

Summary. Table II-2 summarizes the causes of the foregoing five accidents and the basic problems involved. In cases A to C, the nude mice suffered from

Table II-2. Causes and problems of accidental infections in nude mice

Case	Disease/Microbe	Cause	Main source of problem
A	Wasting/MHV	Contamination of cages in washing area	Facilities
B	Wasting/MHV	Contamination during packaging	Facilities
C	Wasting/MHV	Contamination from isolator leakage during experiment	Equipment
D	Facial abscess/ S. aureus	Infection from caretaker?	Animal care
E[a]	Pneumonia/ Sendai Virus	Contamination during transportation	Animals

[a] Euthymic SPF rats.

the wasting syndrome owing to MHV infection. In case D, the nude mice had facial abscesses caused by *Staphylococcus aureus*. In case E, the euthymic SPF rats exhibited pneumonia induced by Sendai virus infection.

In case A, the cause was contamination of the cages in the washing and sterilizing area and in case B, the cause was contamination during packaging. In both cases, the problem was improper layout of the facilities. In case C, the contamination occurred because of leakage in the vinyl isolator. The problem in this case was unsuitable maintenance. In case D, the cause appeared to be infection from the caretaker. The problem in this case was the humans themselves who handle animals directly, suggesting the sterilization limits in such a barrier system. In case E, the cause was the contact of SPF rats with conventional mice during transportation. The problem in this case involved the transportation.

Measures to Prevent Contamination

Care and Management

Consideration of the foregoing problems indicates the great importance of microbiological control in rearing systems. Infection from other animals should be avoided by establishing special areas and equipment only for nude mice within the facilities. Since immunodeficient animals, such as nude mice, are very susceptible to infections, their maintenance requires strict microbiological control. The aims in experiments using such immunodeficient animals can be achieved only when the rearing systems are kept completely sterilized under pertinent microbiological controls. Personnel who enter the animal room to handle nude mice should be careful not to touch any other animals or equipment, and if they must handle other animals they should start their work with nude mice before other daily routine work. This is probably the best way to avoid infections carried by humans.

The layout of facilities requires particular attention. Figure II-6 shows an

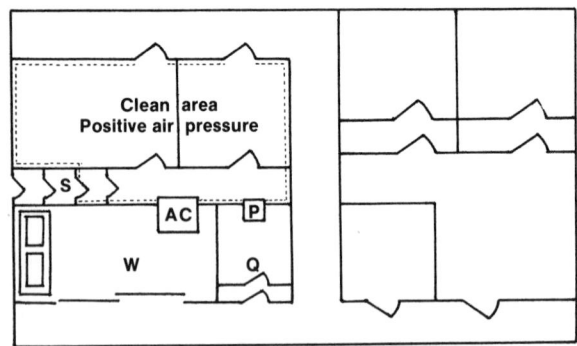

Figure II-6. An example of desirable layout of facilities. AC, autoclave; S, shower room; P, pass box; W, washing room; Q, quarantine room.

Figure II-7. Filter rack system with laminar flow racks and filter caps.

example of facilities based on the necessary principles. The left part of this layout is the clean area, a barrier system which must have a positive air pressure and must be equipped with an autoclave, a pass box, and a shower room. This part should be completely isolated from the contaminated area (the right side in Figure II-6), which is kept under negative air pressure. Although such a barrier system is desirable for rearing euthymic animals, it alone does not achieve perfect microbiological control for nude mice. The barrier system should be provided with special rearing systems or units. The following three systems can be applied according to the aim of the experiments involved.

The first example is the filter rack system shown in Figure II-7, which was introduced by Lane-Petter (8) in 1970. Although this system is simple, its use in barrier facilities makes it possible to prevent cross-contamination between cages. This system is suitable for performing short-term experiments, and it is employed widely in many countries. We are now using this system in our institute for the transfer of tumor cells and the mass production of human tumors using nude mice.

The second example is the microbic clean-room system shown in Figure II-8. The principle of this system was introduced by McGarrity et al. (9). Personnel can enter this room to work. It is possible to keep the room permanently clean by means of refiltered circulating air so that reliable sterilization of air can also be performed. This system is suitable for relatively long-term experiments and can be applied to rooms of any size. We are now using this system in our institute for screening of anticancer drugs using human tumors transplanted in nude mice. We are also trying to improve the system for further investigations.

The third example is the isolator system shown in Figure II-9. This is a

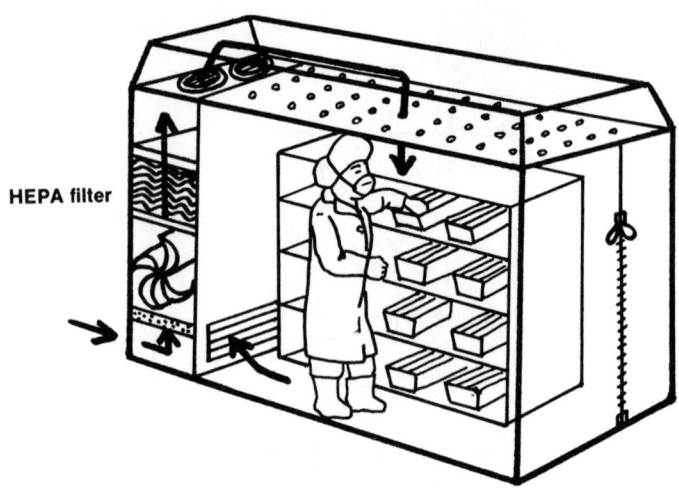

Figure II-8. Microbic clean-room system.

large-scale isolator which was first introduced by Trexler (11) in 1975. This system was developed originally for the isolation of patients with infectious diseases. In this system, personnel can work without touching animals directly and complete microbiological control can be achieved. Complete sterilization of the system and equipment is possible. We are using this system for high-level microbiological control in immunologic experiments. When a negative pressure is applied, this system can be used for experiments on various types of biohazardous agents.

Figure II-9. Large isolator system (P.C. Trexler).

The foregoing three systems have the following common advantages:
1. Sterilized air is supplied by means of high-efficiency particulate air (HEPA) filters.
2. Cross-contamination by animals is prevented.
3. Sterilization of equipment or units can be performed easily.

Microbiological Checking and Monitoring

The basic concept of microbiological monitoring is that animals used in experiments must be quarantined just before they are brought into the facilities and that they must be periodically checked during the experiments. The rearing systems, which include the facilities, equipment, and care, must be examined often to determine whether they are clean. Once any accidental infection has occurred, its cause must be clarified by microbiological checking as soon as possible.

In 1977, the International Committee on Laboratory Animals (ICLA) designated the Central Institute for Experimental Animals as an ICLA Microbiological Nude Mouse Reference Centre. Table II-3 indicates the items and checking methods used for microbiological monitoring of nude mice. These items have been selected with the following conditions in mind: (a) Diseases or inapparent infections are commonly detected in euthymic conventional mouse colonies in Japan and (b) effective checking techniques are available. Other considerations may be added according to local needs. The standard items and techniques for nude mice given in Table II-3 should be adopted in all experimental facilities.

Since nude mice are athymic animals, serologic monitoring is not applicable

Table II-3. Items and checking methods for microbiological monitoring of nude mice (ICLA Microbial Nude Mouse Reference Centre 1977)

Items	Techniques[a]		nu/+	nu/nu
MHV	S		×	
Mouse adenovirus	S		×	
Sendai virus	S		×	
Vaccinia virus	S		×	
Mycoplasma pulmonis	S	E	×	×
Corynebacterium kutscheri	S	E	×	×
Escherichia coli 0115 a,c:K(B)	S	E	×	×
Pasteurella pneumotropica	S	E	×	×
Pseudomonas aeruginosa		E		×
Salmonella spp.	S	E	×	×
Staphylococcus aureus		E		×
Tyzzer's organism	S		×	
Enteric protozoa		E		×
Syphacia obvelata		E		×

[a] S, serological; E, etiological.

Table II-4. Diagnostic methods for clarification of accidental infections in nude mice

Example	Microbes	Diagnostic methods
A	MHV	Serologic testing of mice in the same facilities as nu/nu mice are kept
B	MHV	Serologic testing of specially prepared "decoy" mice
C	MHV	Serologic testing of rats in the same room as nu/nu mice are kept
D	Staphylococcus aureus	Cultivation of samples from both nu/nu mice and caretaker
E[a]	Sendai virus	Serologic testing of recovered rats

[a] Euthymic SPF rats.

to the animals themselves. Instead, serologic monitoring has to be performed on euthymic mice kept in the same room as the nude mice. A characteristic method of microbiological monitoring of nude mice is the periodic testing of specially prepared euthymic "decoy" mice. Table II-4 indicates the diagnostic methods for clarification of accidental infections. The importance of serologic monitoring with decoy mice can be realized from the foregoing explanations of the five accidents. In cases A through C, serologic methods were used to clarify the causes of infections, as shown in the right hand column in Table II-4. In cases A and C, serologic monitoring was performed on euthymic mice or rats which were reared in the same facilities as nude mice. In case B, similar serologic monitoring of specially prepared "decoy" mice solved the problem.

Recently, it has been reported that Sendai virus infection in nude mice can easily be detected by the fluorescent antibody method (6). The method should be useful for monitoring Sendai virus in nude mice themselves because nude mice show more persistent viral infections than euthymic mice.

Conclusion

For success in experiments using nude mice, it is indispensable that microbiologically controlled animals be used. Animals from colonies produced under strict microbiological monitoring must be introduced by means of careful transportation to prevent microbiological contamination. The point at which microbiological control should be applied in using nude mice is the rearing system, including facilities, equipment, and care. The facilities that receive these animals should be provided with special areas for nude mice only together with rearing systems that are perfectly controlled microbiologically to match the aims of the experiments performed. All workers should be careful to avoid needless handling of anything other than the nude mice. It is desirable to start work on nude mice prior to any other daily routine work. Such precautions should lessen or eliminate cross-contamination within the facilities. Finally, it is necessary to check the animals themselves. It is highly desirable to perform microbiological monitoring periodically, no matter whether the nude mice used for experiments are infection-free or not. When any accidental infection is detected, its cause should be clarified and eliminated as soon as possible.

References

1. Boorman, G.A., P.H.C. Lina, C. Zurcher, and H.T.M. Nieuwerker. *Hexamita* and *Giardia* as a cause of mortality in congenitally thymus-less (nude) mice. Clin. Exptl. Immunol. 15:623–627, 1973.
2. Cohen, B.J., and F.G. de Groot. Adenoviral infection in athymic (nude) mice. Lab. Anim. Sci. 26:955–956, 1976.
3. Eaton, G.J., H.C. Outzen, R.P. Custer, and F.N. Johnson. Husbandry of the "nude" mouse in conventional and germfree environment. Lab. Anim Sci 25:309–314, 1975.
4. Flanagan, S.P. 'Nude', a new hairless gene with pleiotropic effects in the mouse. Genet. Res. 8:295–309, 1966.
5. Hirano, N., T. Tamura, F. Taguchi, K. Ueda, and K. Fujiwara. Isolation of low-virulent mouse hepatitis virus from mice with wasting syndrome and hepatitis. Jpn. J. Exptl. Med. 45:429–432, 1975.
6. Iwai, H., Y. Goto, and K. Ueda. Response of athymic nude mice to Sendai virus. Jpn. J. Exptl. Med. 49:123–130, 1979.
7. Iwasaki, Y. Experimental virus infection in nude mice, pp. 457–475. *In* J. Fogh and B.C. Giovanella (eds.), The nude mouse in experimental and clinical research. New York, London: Academic Press, 1978.
8. Lane-Petter, W. A ventilation barrier to the spread of infection in laboratory animal colonies. Lab. Anim. 4:125–134, 1970.
9. McGarrity, G.J., L.L. Coriell, R.W. Schaedler, R.J. Mandle, and A.E. Greene. Medical application of dust-free rooms: III. Use in an animal care laboratory. Appl. Microbiol. 18:142–146, 1969.
10. Sebesteny, A., and A.C. Hill. Hepatitis and brain lesions due to mouse hepatitis virus accompanied by wasting in nude mice. Lab. Anim. 8:317–326, 1974.
11. Trexler, P.C., A.S.D. Spiers, and H. Gaya. Plastic isolators for treatment of acute leukaemia patients under "germ-free" conditions. Br. Med. J. 6:549–552, 1975.
12. Ueda, K., Y. Goto, S. Yamazaki, and K. Fujiwara. Chronic fatal pneumocytosis in nude mice. Jpn. J. Exptl. Med. 47:475–482, 1977.
13. Ueda, K., T. Tamura, H. Machii, and K. Fujiwara. An outbreak of Sendai virus infection in an nude mouse colony, pp. 61–69. *In* T. Nomura, N. Ohsawa, N. Tamaoki, and K. Fujiwara (eds.), Proceedings of the second international workshop on nude mice. Tokyo: University of Tokyo Press, 1977.
14. Ward, J.M., M.J. Collins, and J.C. Parker. Naturally occurring mouse hepatitis virus infection in the nude mouse. Lab. Anim. Sci. 27:372–376, 1977.
15. Ward, J.M., D.P. Houchens, M.J. Collins, D.M. Young, and R.L. Reagan. Naturally-occurring Sendai virus infection in athymic nude mice. Vet. Pathol. 13:36–46, 1976.

III

Genetic Quality Control of the Nude Mouse

Tatsuji Nomura* and Takeshi Tomita

Central Institute for Experimental Animals, 1430 Nogawa, Takatsu, Kawasaki 213, Japan, and Department of Genetics, Faculty of Agriculture, Nagoya University, Furo-cho, Chikusa-ku, Nagoya 464, Japan.

The nude mouse is now used widely in biomedical experiments as a new model animal, and the *nu* gene is being introduced into a number of inbred mouse strains to produce congenic nude strains.

Genetic monitoring of inbred mouse strains was started in Japan based on a plan prepared in cooperation with Dr. H. Hoffman of the National Institutes of Health, USA. The results for 10 inbred strains (a total of 124 strain samples) maintained in 29 institutions have shown that genetic contamination and subline variations are not rare among highly inbred mouse strains commonly employed in biomedical research.

Table III-1 shows that the first four strains, namely A, BALB/c, C3H/He, and NC, from all the institutions showed uniformity in genetic profiles of biochemical and immunological markers, but several cases of subline variations were found in the CBA, AKR, and NZB strains.

It was striking that genetic contamination was undoubtedly confirmed in some breeding of C57BL/6, DBA/2, and KK strains. One case of possible mutation was suggested in a NZB subline.

Similar results were obtained by Hoffman (3,4), Hedrich (2), and Festing (1), and the importance of genetic quality control is now recognized.

Congenic nude mouse strains already established have a variable genetic background, and the genetic profiles need to be monitored.

It is very urgent that genetic quality control be applied to these strains because they are now used extensively for studies of immunology, oncology, and pathology. It is also advisable that whenever "nude" genes are to be introduced into well-established inbred strains, genetic checks be performed on that strain.

---

* To whom correspondence should be addressed.
© 1982 Gustav Fischer New York, Inc.
Proceedings of the Third International Workshop on Nude Mice.

Table III-1. Genetic monitoring of inbred mouse strains

		Chromosome:	1		2	3	4		5	6	7		8		9	11	17	
	Strain	No.[a]	Idh	Pep	Hc	Car	Mup	Gpd	Pgm	Ldr	Gpi	Hbb	Es1	Es2	Mod	Es3	Ce2	H-2
Standard profile	A	8	a	b	0	b	a	b	a	a	a	d	b	b	a	c	a	a
	BALB/c	7	a	a	1	b	a	b	a	a	a	d	b	b	a	a	a	d
	C3H/He	15	a	b	1	b	a	b	b	a	b	d	b	b	a	b	b	k
	NC	2	b	b	0	a	b	b	a	a	a	s	b	b	a	c	b	—
Subline variation	CBA	3	b	b	1	b	a	b	a	a	b	d	b	b	b	c	b	d
		2	b	b	1	a	a	b	a	a	b	d	b	b	a	c	b	?
		2	b	b	1	a	a	b	a	a	b	d	b	b	b	c	b	?
	AKR	2	b	b	0	a	a	b	a	a	a	d	b	b	b	c	b	k
		1	b	b	0	b	a	b	a	a	a	s	b	b	b	c	b	—
		1	b	b	0	a	a	b	a	a	a	d	b	b	a	c	b	—
		1	b	b	0	a	a	b	a	a	a	d	b	b	a	c	b	—
		1	b	a	0	a	a	b	a	a	a	d	b	b	b	c	a?	—
		1	b	b	1	a	a	b	a	a	a	d	b	b	b	c	b	—
		1	ab	b	1	b	a	b	a	a	a	d	b	b	a	c	b	—
	NZB	1	a	c	0	a	a	b	b	a	a	d	b	b	b	c	b	—
		1	a	b	0	a	a	b	ab	a	a	d	a	b	b	c	b	d
		1	a	c	1	a	b	b	b	a	a	s	a	b	b	c	b	—
		1	a	bc	01	a	b	b	b	a	a	d	b	b	a	c	b	—
Contamination	C57BL/6	14	a	a	1	a	b	a	a	a	b	s	a	b	b	a	a	b
		1	ab	ab	1	ab	b	b	ab	a	ab	s	ab	b	b	ac	a	?
	DBA/2	6	b	b	0	b	a	b	b	a	a	d	b	b	a	c	a	d
		1	ab	b	01	b	a	b	b	a	ab	d	b	b	a	c	a	?
	KK	2	a	b	0	a	b	a	a	a	b	s	b	a	a	c	a	—
		1	a	ab	01	a	b	a	a	a	b	sd	b	Not a	a	c	a	—

[a] Number of institutes participating.

Genetic quality control will result in considerable savings of both money and time in avoiding problems arising from the use of genetically contaminated animals.

References

1. Festing, M.F.W. Genetic reliability of commercially-bred laboratory mice. Lab. Anim. 8:265–270, 1974.
2. Hedrich, H.J. Genetic monitoring of the mouse. American College of Laboratory Animal Medicine (ACLAM), American Society of Laboratory Animal Practitioners (ASLAP) Symposium on the mouse in biomedical research. Seattle, Washington, 24 July, 1979.
3. Hoffman, H.A. Genetic quality control of the laboratory mouse (*Mus musculus*), pp. 217–234. *In* H.C. Morse, III (ed.), Origins of inbred mice. New York, London: Academic Press, 1978.
4. Hoffman, H.A., K.T. Smith, J.S. Crowell, T. Nomura, and T. Tomita. Genetic quality control of laboratory animals with emphasis on genetic monitoring, pp. 307–317. *In* A. Spiegel, S. Erichsen, and H.A. Solleveld (eds.), Animal quality and models in biomedical research. Stuttgart: Gustav Fischer Verlag, 1980.

IV

The Nude Mouse as a Tool in Cancer Chemotherapy Studies

David P. Houchens* and Artemio A. Ovejera

Battelle Memorial Institute, 505 King Avenue, Columbus, Ohio 43201.

There have been many reports of the development of the nude mouse as a model for the growth and treatment of human tumor xenografts (2,5,9,10,14) and it can truly be said that in the 10 years since Rygaard and Povlsen described their work (12), the nude mouse has found a home in the area of cancer research.

The use of the nude mouse as a tool in cancer chemotherapy studies has only occurred in the past few years. Therefore, we would like to present some of the ideas that have gone into the development of the drug-testing programs of the National Cancer Institute (NCI) of the United States and describe for you what the status of the nude mice is in this program.

A number of excellent reviews of the historical aspects of the development of drug screening have been written, starting with reviews, of systems prior to the 1950s (15) and continuing through more comprehensive reviews to cover studies up to the present (3,6,17). The earliest attempts to develop a drug-screening program were made in 1934 and since then various tumor systems have been used through the years in both primary and secondary screening programs and in both mice and rats to find drugs that can predict for the clinic (17). Various research institutions worked on the development of these systems. These include Sloan-Kettering in New York, Chester Beatty Institute in England, the University of Tokyo, and the U.S. National Cancer Institute (NCI). In 1955, the Cancer Chemotherapy National Service Center (CCNSC) was started at the NCI to develop a large-scale screening program to test compounds submitted by pharmaceutical companies and universities throughout the world. Such a program ideally should select agents that give good correlation between experimental and clinical studies. The systems selected should select as many compounds as possible that have activity in a clinical situation (true positive compounds) and should also

---

* To whom correspondence should be addressed.
© 1982 Gustav Fischer New York, Inc.
Proceedings of the Third International Workshop on Nude Mice.

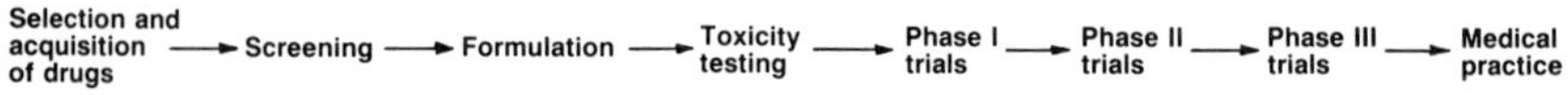

Selection and acquisition of drugs → Screening → Formulation → Toxicity testing → Phase I trials → Phase II trials → Phase III trials → Medical practice

Figure IV-1. New drug development by NCI. (Adapted from ref. 11.)

categorize compounds that are inactive in the clinic (true negative compounds). Additionally, they should have as few false positives and false negatives as possible. The initial program has developed through the years into a comprehensive chemotherapy program (Figure IV-1) (11) that not only provides antitumor information about drugs but also produces genetically defined animals, establishes frozen tumor banks, and has developed the world's largest computerized data base on experimental antitumor activity.

Through the years, 30–40 drugs have been approved for clinical treatment of cancer in the United States. A great deal of improvement has been made in the treatment of certain types of cancer, but other types have been refractory to treatment. Colon, breast, and lung tumors account for approximately 35% of the cancer cases in the United States (7) and the cure rate has not been high. Additionally, major undesirable side effects are seen with many antineoplastic compounds, so the goal of the drug development program is to design drugs that not only are more specific for various tumor types but also are less toxic.

Presently, the flow of drugs by testing in the Division of Cancer Treatment (DCT) of NCI is as shown in Figure IV-2 (16). This tumor panel has been used for the past 3 years and data are still being assessed to determine predictability for the clinic. As can be seen, the human tumor xenografts appear in this panel for the first time as a part of the NCI drug evaluation. Based on pilot studies from the United States, Europe, and Japan, tumors have been tested in this system using the subcutaneous implantation route and drugs have generally been administered by the intraperitoneal route. The reporting of testing by this method has appeared in a number of publications (2,5,9,10,14). Laboratories performing these types of studies for NCI include the Stehlin Cancer Foundation, Houston, Texas; Mason Research Institute, Worcester, Massachusetts; Illinois Institute of Technology Research Institute, Chicago, Illinois; Southern Research Institute, Birmingham, Alabama; and Battelle Memorial Institute, Columbus, Ohio. Presently, about 500–800 compounds a year are tested in the xenograft systems. Table IV-1 shows the number of drugs tested to the present time and Table IV-2 shows the name of drugs that actually produced regression in the

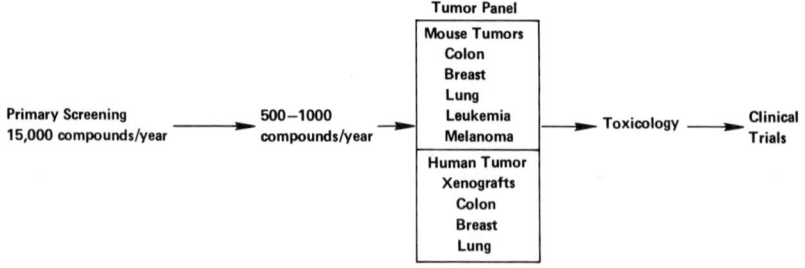

Figure IV-2. Division of Cancer Treatment tumor panel. (Adapted from ref. 16.)

Table IV-1. Human tumor xenograft testing

	Colon			Lung	Breast
	CX-1[a]	CX-2	CX-5	LX-1[a]	MX-1[a]
Number of drugs tested	103	32	45	97	96
Retards tumor growth	0	9	0	4	9
Produces tumor regression	0	2	1	3	13

[a] In the present NCI tumor panel.

three tumor types. However, only one representative tumor of each type (breast, colon, and lung) is used in this system at this time. There are some serious questions that can be raised regarding the use of only one tumor to represent a type. Studies (8) have shown that different tumors of the same type respond differently to the same drugs. This is also reported by Bellet in this workshop (R.E. Bellet et al., Chapter 57, volume 2). Additional considerations in using this system for evaluation of compounds include: (a) slower tumor growth than murine systems, (b) somewhat more elaborate facilities than conventional animals require, (c) greater expense than required for studies in conventional mice, and (d) potential loss of experiments because of infections in the colony.

During the past 2 years, a new protocol using a subrenal capsule implant of tumors in nude mice (1) has been assessed. The goal of such a model would be to test drugs in a shorter period (11 days per assay as compared to 40 days in subcutaneous model). At present, this system is used as a prescreen for priority drugs in the tumor panel. Any compound found to be active in this system is then tested against the subcutaneously implanted xenografts. Other methods could be developed for testing xenografts in nude mice. For instance, implantation of brain tumors or other types by the intracranial route produces reproducible death times (4,13). Of course, no matter what system might be used in growing and testing human tumors, other considerations, such as drug treatment schedule, dose, and route of administration are taken into account.

Over the last 3–4 years, the total number of tests under screening programs

Table IV-2. Drugs inducing tumor regression in human tumor xenografts

MX-1 (Breast)	LX-1 (Lung)	CX-1 (Colon)
DON	DON	None
Melphalan	Azotomycin	
Hexamethylmelamine	Procarbazine	
Cyclophosphamide		
Mitomycin C		
Azotomycin		
Vincristine		
Methyl CCNU		
Dibromodulcitol		
Pentamethylemelamine		
Piperazinedione		
AT 125		
Trimethyltrimethylomelamine		

in all tumor systems has been reduced from approximately 300,000 to 200,000. With this reduction and with increased effort to determine predictability for the clinic, an obvious need is present for valid correlations between animal and clinical studies. The nude mouse seems to be a tool that can be used effectively in establishing these correlations.

References

1. Bogden, A., D. Kelton, W. Cobb, and H. Esber. A rapid screening method for testing chemotherapeutic agents against human tumor xenografts, pp. 231–250. In D. Houchens and A. Ovejera (eds.), Proceedings of the symposium on the use of athymic (nude) mice in cancer research. New York, Stuttgart: Gustav Fischer, 1978.
2. Giovanella, B., J. Stehlin, and R. Shepard. Experimental chemotherapy of human breast carcinomas heterotransplanted in nude mice, pp. 475–481. In T. Nomura, N. Ohsawa, N. Tamaoki, and K. Fujiwara (eds.), Proceedings of the second international workshop on nude mice. Tokyo: University of Tokyo Press, 1977.
3. Goldin, A., S.A. Schepartz, J.M. Venditti, and V.T. DeVita, Jr. Historical development and current strategy of the National Cancer Institute drug development program, pp. 165–245. In V.T. DeVita, Jr., and H. Busch (eds.), Methods in cancer research. New York, San Francisco, London: Academic Press, 1979.
4. Houchens, D.P., A.A. Ovejera, and S.M. Balthaser. Growth and treatment of a human medulloblastoma in athymic (nude) mice. Workshop on Brain Tumor Immunology. Durham, North Carolina, 1977.
5. Houchens, D.P., A.A. Ovejera, and A.D. Barker. The therapy of human tumor in athymic (nude) mice, pp. 267–289. In D.P. Houchens, and A.A. Ovejera (eds.), Proceedings of the symposium on the use of athymic (nude) mice in cancer research. New York, Stuttgart: Gustav Fischer, 1978.
6. Johnson, R.K., and A. Goldin. The clinical impact of screening and other experimental tumor studies. Cancer Treat. Rev. 2:1–31, 1975.
7. Myers, M., and L. Axtell. Introduction, pp. 1–10. In L. Axtell, A. Asine, and M. Myers (eds.), Cancer patient survival report No. 5. Washington, D.C.: U.S. Government Printing Office, 1976.
8. Osieka, R., D. Houchens, A. Goldin, and R.K. Johnson. Chemotherapy of human colon cancer xenografts in athymic (nude) mice. Cancer 40:2640–2650, 1977.
9. Ovejera, A., D. Houchens, and A. Barker. Chemotherapy of human tumor xenografts in genetically athymic mice. Ann. Clin. Lab. Sci. 8:50–56, 1978.
10. Povlsen, C., and G. Jacobson. Chemotherapy of a human malignant melanoma transplanted in the nude mouse. Cancer Res. 35:2790–2796, 1975.
11. Rothenberg, L., and R. Terselic. Management of the National Cancer Institutes drug research program through application of the linear array concept. Cancer Chemother. Rep. 54:303–310, 1970.
12. Rygaard, J., and C.O. Povlsen. Heterotransplantation of a human malignant tumor to "nude" mice. Acta. Pathol. Microbiol. Scand. 77:758–760, 1969.
13. Shapiro, W.R., G.A. Basler, N.L. Chernik, and JB. Posner. Human brain tumor transplantation into nude mice. J. Natl. Cancer Inst. 62:447–453, 1979.
14. Shimosato, Y., T. Kameya, K. Nagai, S. Hirohashi, T. Koide, H. Hayashi, and T. Nomura. Transplantation of human tumor in nude mice. J. Natl. Cancer Inst. 56:1251–1260, 1976.
15. Spencer, M.C., D.H. Algire, and N.G. Smith. Cancer Chemotherapy Supplement XXXVI: A bibliography of agents supplement 1955–1959. Cancer Res. 25 (suppl., part 2) No. 4:637–1075, 1965.
16. Venditti, J.M. Foreword, pp. ix–xii. In D.P. Houchens and A.A. Ovejera (eds.), Proceedings of the symposium on the use of athymic (nude) mice in cancer research. New York, Stuttgart: Gustav Fischer, 1978.
17. Zubrod, C. G., S. Schepartz, J. Leiter, J. Endicott, L. Carrese, and C. Baker. The chemotherapy program of the National Cancer Institute: history, analysis, and plans. Cancer Chemother. Rep. 50:349–540, 1966.

General Discussion

CROWLE: I am intrigued by your use of the brain as an injection site of tumors and death as an end point in studies on the effects of drugs on tumor growth. Do you think that the blood–brain barrier restricts the entry of drugs enough to decrease the efficacy of the drugs?

HOUCHENS: Neurosurgeons assure me that after injection of tumor cells the blood–brain barrier ceases to exist within a few days.

SPANG-THOMSEN: You clearly demonstrated that because of great individual variation in response, tumor panels are not representative for, e.g., the biological tumor types involved. How can your results be used by clinicians? How can your data be conclusive for tumors other than the few tumors examined?

HOUCHENS: At the present, NCI is using only one tumor of each type for evaluation. We know that one tumor cannot predict for all tumors of that type. However, a drug need not be active against all tumors in the NCI panel to move on to toxicologic and clinical testing. It might be assumed that the ideal way for testing would be to use a panel of several xenografts of the same type. This, of course, would be extremely costly and require a large amount of space in a laboratory for the maintenance of nude mice. Although once the development and assessment of nude mice as predictors of antineoplastic therapy is completed, this may prove to be a possibility.

V

The Congenitally Athymic Streaker Mouse

Leonard D. Shultz,* Hendrick G. Bedigian, Hans-Jorg Heiniger, and Eva M. Eicher

The Jackson Laboratory, Bar Harbor, Maine 04609.

Introduction

Although the congenitally athymic nude mouse has proved to be an extremely powerful tool in biomedical research, investigations with nude mice are often subject to a great deal of variability because of heterogeneity of genetic background. While the recent backcrossing of the nude mutation onto several well-characterized inbred strains should greatly reduce this variability, we cannot exclude the possibility that observed phenotypic differences between nude mice and littermate heterozygote or $+/+$ controls are caused by one or more genes closely linked to the nude locus rather than to nu itself. The recent occurrence of a remutation at the nude locus called "streaker" (nu^{str}) in the inbred AKR/J strain now provides us with a model of congenital thymic aplasia in which the athymic animals differ from their intact controls by a single gene. In this report, we review the development of the streaker mutation and compare its phenotypic expression with that of the nude mutation.

Linkage and Allelism Tests

A spontaneous recessive mutation characterized by hairlessness was found in 1974 in a colony of AKR/J mice in the Animal Resources Department at the

---

* To whom correspondence should be addressed.
© 1982 Gustav Fischer New York, Inc.
Proceedings of the Third International Workshop on Nude Mice.

Jackson Laboratory. The hairless deviants were found to be athymic and linkage tests showed that the new mutation mapped on chromosome 11, 20.6 ± 4.9 recombination units from Rex (*Re*). Subsequently, crosses between AKR/J mice carrying the new mutation and noninbred mice (imported from the National Institutes of Health) carrying the nude mutation showed that this new mutation was allelic with nude (*nu*) (6). Even though these genes are found to be allelic they are not necessarily identical since there are many known instances of multiple alleles at the same locus with significant differences in phenotypic expression (17,19). Thus, the remutation at the nude locus was given a new name (streaker) (6) with the genetic symbol (nu^{str}) indicating both the independent occurrence of the streaker mutation and its allelism with *nu*.

Lifespan of Streaker Mice

Death from intercurrent infection early in life limits investigations with nude mice to short-term experiments. Although this can be circumvented by raising the mice in sterile germ-free isolators (12), the technique is both costly and cumbersome. In a study of survival under conventional mouse room conditions as maintained at the Jackson Laboratory, we found that the mean lifespan of 37 streaker mice observed from weaning to death was 236.3 ± 13.6 days, compared with 307.4 ± 16.9 days for 35 phenotypically normal littermate controls (18). Mice of the AKR/J strain show nearly a 100% incidence of thymic lymphoma and the phenotypically normal controls in this study died with gross symptoms of thymoma. However, the streaker mice were refractory to thymoma development and instead showed symptoms of infection and wasting disease at death.

Lymphoid System

Lymphoid architectural abnormalities seen in streaker mice do not differ markedly from those described for nude mice. In streaker mice, the deep cortical areas of lymph nodes, periarteriolar regions of the spleen, and the corridor of tissue between the primary nodules of Peyer's patches are severely depleted of cells (18). Similar changes have been described in detail for nude mice on a variety of inbred backgrounds (4,15). In addition, Peyer's patches of both streaker (18) and nude (15) mice lack well-developed germinal centers and have decreased numbers of phagocytic macrophages compared with intact control mice. Numbers of Thy-1-positive lymphocytes (T cells), as determined by indirect immunofluorescence with anti-Thy-1.1 antisera, varied ranging from undetectable levels to 6.8% in lymphoid tissues of streaker mice (18). This is in agreement with similar findings in nude mice (13,14). All of the streaker mice so far studied for T-cell numbers were born from heterozygous ($+/nu^{str}$) mothers. Placental transfer of T cells or thymic factors to streaker mice in utero may contribute to the finding of Thy-1-positive cells in this mutant.

Total and Differential Leukocyte Counts

The peripheral blood leukocyte counts of streaker mice do not differ significantly from those of AKR/J +/+ controls. However, streaker mice show a significant decrease in percentage of lymphocytes and a concomitant increase in percentage of granulocytes (18). Ten-week-old streaker mice showed only 44.2% peripheral blood lymphocytes compared with 78.0% lymphocytes in AKR/J +/+ controls. Likewise, the percentage of neutrophils was increased to 52.2% in streaker mice compared with 20.4% for the +/+ controls. The percentage of eosinophils was slightly increased in streaker mice at 10 weeks of age and there was no significant effect of the streaker mutation on the level of blood monocytes (18). In contrast to the finding of a normal total leukocyte count in streaker mice, nude mice have been reported to have a marked leukopenia (15).

Immune Function

Streaker mice have been tested for a variety of humoral and cell-mediated immune responses. In a study to determine the ability of streaker mice to reject H-2-incompatible split thickness skin grafts, it was found that whereas streaker mice retained their allografts throughout their lifespan (24–34 weeks after grafting), control AKR/J +/+ (H-2^k) mice rejected similar allografts from C57BL/6J (H-2b) donors within 2 weeks after grafting (18). Streaker mice also showed a marked decrease in humoral response to thymic-dependent antigens. Compared with AKR/J +/+ controls, streaker mice had a 20-fold reduction in the IgM

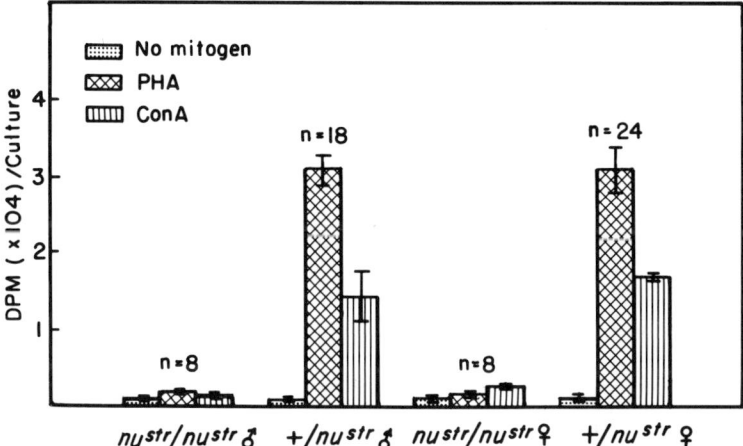

Figure V-1. Response of AKR/J nu^{str}/nu^{str} and $+/nu^{str}$ mice to T-cell mitogens. Whole blood cultures were prepared as described previously (9). The mice were 3–5 months old. Numbers of mice in each group indicated by n.

Table V-1. Serum immunoglobulin levels in streaker mice

Genotype[b]	Immunoglobulin levels (mg/100 ml)[a]				
	IgG1	IgG2A	IgG2B	IgA	IgM
$+/+$	45.5 ± 1.50	250.4 ± 22.39	209.6 ± 14.96	19.2 ± 2.54	21.6 ± 1.12
nu^{str}/nu^{str}	5.5 ± 1.20	154.2 ± 26.75	68.6 ± 8.27	11.0 ± 0.79	15.2 ± 0.91

[a] Arithmetic mean ± standard error.
[b] Sera from six 10-week-old female AKR/J $+/+$ and five 10-week-old female AKR/J nu^{str}/nu^{str} mice were tested by single radial immunodiffusion as previously described (18).

plaque-forming cell response to sheep erythrocytes (18). Proliferative response of streaker mice to T-cell mitogens is also severely depressed. As shown in Figure V-1, peripheral blood lymphocytes from streaker mice were nonresponsive to the mitogenic effect of phytohemagglutinin and cocanavalin A in a blood microculture assay (9). In contrast, AKR/J $+/+$ mice showed high reactivity to both these mitogens. Impaired humoral response of streaker mice is also reflected in a quantitative decrease in serum immunoglobulin levels. At 10 weeks of age, streaker mice showed decreases in levels of all major classes of serum immunoglobulins compared with those of AKR/J $+/+$ controls (18). As shown in Table V-1, IgG1 and IgG2B are most severely decreased in streaker mice, whereas IgM expression was least affected by the streaker mutation. There is some discrepancy among investigators concerning the effect of the nude mutation on serum immunoglobulin levels (2,3,16). This lack of agreement may result in part from the marked effect of strain background (11) and age (7) on immunoglobulin expression.

Endogenous Murine Leukemia Virus Expression

AKR/J mice are chronically infected with murine leukemia viruses and show a high incidence of thymic lymphoma (8). In a preliminary study, which compared levels of ecotropic C-type RNA virus in tail homogenates from streaker and littermate control mice, it appeared that streaker mice had increased virus titers (18). However, in a more recent study, which compared ecotropic virus expression from spleen extracts of streaker and littermate control mice, it was found that there was no significant effect of the streaker mutation on ecotropic virus expression (1). In contrast, the streaker mutation had a pronounced effect on the expression of xenotropic virus. At 5 months of age, streaker mice showed significant decreases in levels of spleen-associated xenotropic virus compared with normal littermate controls. However, by 8 months of age, streaker and control mice showed no significant difference in xenotropic virus titers. Attempts to isolate MCF virus from spleens or tumor tissue from streaker mice were unsuccessful, whereas dual tropic mink cell focus-inducing (MCF) virus was readily isolated from thymuses of AKR/J $+/+$ mice (1). These findings suggest that presence of the thymus or high persistent titers of both xenotropic and ecotropic virus or both are necessary for expression of MCF virus.

Tumorigenesis

Mice of the AKR/J strain develop thymic lymphoma at nearly 100% incidence by 1 year of age (8). Although thymectomy of AKR/J mice prevents the development of thymomas, approximately 10% of AKR/J mice thymectomized between 35 and 100 days of age develop either nonthymic lymphomas or myeloid leukemia (10). We found that streaker mice are, as expected, refractory to the development of thymic lymphoma (18). However in our research colonies, we have observed to date nine cases of spontaneous neoplasms in streaker mice. As shown in Table V-2, these were all lymphoreticular neoplasms with the majority defined according to Dunn's morphological criteria (5) as reticulum cell sarcomas. Preliminary data shows that thymus reconstitution of streaker mice results in an increased incidence of lymphomas compared with untreated streakers. These tumors include both reticulum cell neoplasms and thymic lymphomas. Analysis for murine leukemia virus expression of one of the neoplasms (reticulum cell sarcoma) from a thymus-reconstituted streaker mouse showed low levels of ecotropic and xenotropic virus but no expression of dual tropic MCF virus (1).

Potentialities and Limitations of the Streaker Mouse

From the limited range of investigations so far with streaker mice we cannot at this time determine whether the nude and streaker mutations are identical or represent different alleles at the same locus. Such a determination would require comparison of many diverse phenotypic effects of these genes on the same genetic background and ultimately would entail comparison of the gene products of nu and nu^{str}. At this time, the observed differences in phenotypic expression between nu and nu^{str} are no greater than the reported variations among different experiments with nude mice. Of greater importance is the potential use of streaker mice in biomedical research. The spontaneous occurrence of the streaker mutation in an inbred mouse strain provides the scientific community with a coisogenic

Table V-2. Spontaneous neoplasms in AKR/J nu^{str}/nu^{str} mice

Age (days)	Sex	Tumor
152	M	Reticulum cell sarcoma type A
166	F	Reticulum cell sarcoma type B
215	F	Reticulum cell sarcoma type A
229	M	Reticulum cell sarcoma type B
256	M	Reticulum cell sarcoma type B
268	M	Granulocytic leukemia
399	M	Reticulum cell sarcoma type B
420	M	Lymphocytic lymphoma
465	M	Granulocytic leukemia

line for the nude locus. Furthermore, the AKR/J strain background enables us to examine the association between thymus function, murine leukemia virus expression, and development of lymphomas. It is hoped that future experimentation with streaker mice will contribute to our understanding of thymic function and will elucidate the role of the immune system in tumorigenesis.

Acknowledgments

This work was supported by U.S. National Institutes of Health Grants CA 20408 from the National Cancer Institute and RR 01183 from the Division of Research Resources and by contract N01 CP33255 from the National Cancer Institute.

References

1. Bedigian, H.G., L.D. Shultz, and H. Meier. Expression of endogenous murine leukemia viruses in AKR/J streaker mice. Nature (London) 279:434–436, 1979.
2. Bloemmen, J., and H. Eyssen. Immunoglobulin levels of sera of genetically thymusless (nude) mice. Eu. J. Immunol. 3:117–118, 1973.
3. Crewther, P., and N.L. Warner. Serum immunoglobulins and antibodies in congenitally athymic (nude) mice. Aus. J. Exptl. Biol. Med. Sci. 50:625–635, 1972.
4. DeSousa, M.A.B., D.M.V. Parrott, and E.M. Pantelouris. The lymphoid tissue in mice with congenital aplasia of the thymus. Clin. Exptl. Immunol. 4:637–644, 1969.
5. Dunn, T.B. Normal and pathologic anatomy of the reticular tissue in laboratory mice, with a classification and discussion of neoplasms. J. Natl. Cancer Inst. (USA) 14:1281–1433, 1969.
6. Eicher, E.M. Remutations. Mouse News Lett. 54:90, 1976.
7. Fahey, J.L., and W.F. Barth. The immunoglobulins of mice. 4. Serum immunoglobulin changes following birth. Proc. Soc. Exptl. Biol Med 118:596–600, 1965.
8. Hartley, J.W., N.K. Wolford, L.J. Old, and W.P. Rowe. A new class of murine leukemia virus associated with development of spontaneous neoplasms. Proc. Natl. Acad. Sci. USA 74:789–792, 1977.
9. Heiniger, H.J., J.M. Wolf, H.W. Chen, and H. Meier. A micromethod for lymphoblastic transformation of mouse lymphocytes from peripheral blood. Proc. Soc. Exptl Biol. Med. 143:6–11, 1973.
10. Nakakuki, K., H. Shisa, and Y. Nishizuka. Prevention of AKR leukemia by thymectomy at various ages. Acta Haematol. 38:317–323, 1967.
11. Natsuume-Sakai, S., K. Motonishi, and S. Migita. Quantitative estimations of five classes of immunoglobulin in inbred mouse strains. Immunology 32:861–866, 1977.
12. Outzen, H.C., R.P. Custer, G. Eaton, and R.T. Prehn. Spontaneous and induced tumor incidence in germfree "nude" mice. J. Reticuloendothel. Soc. 17:1–9, 1975.
13. Raff, M.C. Theta-bearing lymphocytes in nude mice. Nature (London) 246:350–351, 1973.
14. Roelants, G.E., F. Loor, H. Von Boehmer, J. Sprent, L.B. Hagg, K.S. Mayor, and A. Ryden. Five types of lymphocytes characterized by double immunofluorescence and electrophoretic mobility. Organ distribution in normal and nude mice. Eur. J. Immunol. 5:127–131, 1975.
15. Rygaard, J., and C.O. Povlsen. Effects of homozygosity of the nude (nu) gene in three inbred strains of mice. Acta Pathol. Microbiol. Scand. (A) 82:48–92, 1974.
16. Salomon, J.C., and H. Bazin. Low levels of some serum immunoglobulin classes in nude mice. Rev. Eu. Etudes Clin. Biol. 17:880–882, 1972.

17. Searle, A.G. Mouse gene list. Mouse News Lett. 60:2–26, 1979.
18. Shultz, L.D., H.J. Heiniger, and E.M. Eicher. Immunopathology of streaker mice: A remutation to nude in the AKR/J strain, pp. 211–222. *In* M.E. Gershwin and E.L. Cooper (eds.), Comparative and developmental aspects of immunity and disease. New York: Pergamon Press, 1978.
19. Silvers, W.K. (ed.). The coat colors of mice. A model for mammalian gene action and interaction. New York: Springer Verlag, 1979.

General Discussion

FINERTY: Do the isotype levels change in nu^{str}/nu^{str} mice after immunization?

SHULTZ: We have not yet quantitated the levels of serum immunoglobulins after immunization of streaker mice.

OVEJERA: What are the phenotypic differences among nu/nu, nu^{str}/nu^{str}, and nu/nu^{str}? Between $nu/+$ and $nu^{str}/+$?

SHULTZ: Although the nude and streaker mutations are allelic, phenotypic characteristics of nude and streaker mice cannot be considered to be identical without extensive comparative testing. It is difficult to compare our findings with the myriad of apparent discrepancies in reported characteristics of nude mice. At present, the differences in phenotypic expression between the streaker and nude mutations do not appear any greater than reported variations among different experiments with nude mice.

MICHAEL: Are these mice now commercially available?

SHULTZ: Streaker mice are available in small numbers from individual research colonies through the Animal Resources Department of the Jackson Laboratory.

B. LOZZIO: You said the nu^{str}/nu^{str} mice do not develop thymoma. Absolutely none? How many have you examined?

SHULTZ: Although we have observed to date nine cases of spontaneous lymphoreticular neoplasms in streaker mice in approximately 100 aging animals autopsied, we have not yet seen a streaker mouse with a neoplasm involving thymic lymphocytes. However, several streaker mice that have received subcutaneous thymus tissue from AKR/J $+/+$ donors have subsequently developed lymphomas that appear to be of thymic origin.

VI

Characteristics of Nude Rats

Michael F.W. Festing

Medical Research Council Laboratory Animals Centre, Woodmansterne Road, Carshalton, Surrey SM5 4EF, England.

Introduction

The nude mouse is probably the most important single mutant used in biomedical research. A considerable amount is now known about its biology, and there is little doubt that it is destined to play an increasingly important role in research in immunology, cancer, and the study of infectious disease. The rediscovery in 1977 of an apparently similar mutation in the rat, which first occurred some time prior to 1953 (9), is clearly an event of some importance as the rat is numerically the second most important laboratory animal, after the mouse. Apart from the fact that the nude rat is larger than the nude mouse, and therefore more convenient for experiments involving surgery, the differences and similarities between the two mutants may well provide additional information on the role of the cell-mediated immune function, as well as being of direct importance in the study of the immunology of the rat.

Fortunately, studies of the biology of the nude mouse provide a framework for the study of the nude rat. It was immediately apparent that preliminary studies of the new mutation should be directed at the cell-mediated and humoral immune responses, the response to skin and tumor allografts and xenografts, and the response of the nude rat to infectious organisms, including parasites, viruses, and bacteria. Unfortunately, there is usually a considerable delay between the completion of research and the appearance of the paper in the open literature, and for this reason, although a number of projects on the nude rat have now been completed, few have yet been published. The aim of this review is to report on the current status of knowledge of the nude rat, using both published and unpublished information.

© 1982 Gustav Fischer New York, Inc.
Proceedings of the Third International Workshop on Nude Mice.

History and Genetics

The history of the Rowett nude rat mutation has been described by Festing et al. (9). Briefly, hairless mutant animals were first observed in an outbred colony of hooded rats maintained at the Rowett Research Institute, Aberdeen, Scotland in 1953. A breeding colony of the rats was maintained with difficulty until the early 1960s, when it died out (12). Apparently, it was also established that the homozygous mutant animals were athymic, but the importance of the thymus in cell-mediated immunity was not established at that time.

Homozygous mutant animals were again found in a mating established in 1975. It therefore appears as though the mutation was maintained in the outbred stock at a low gene frequency for more than 22 years. The alternative explanation, namely that there was a second mutation in the same colony, cannot be ruled out, but it seems highly improbable. By 1975 the importance of the thymus was clearly understood, and the nude mouse was well known, so the potential value of the nude rat was quickly recognized. The colony was "cleaned up" to a germ-free status, and the breeding stock was transferred to the Medical Research Council Laboratory Animals Centre for further characterization and distribution to research workers.

It was soon established that the nude characteristic results from an autosomal recessive mutation with full penetrance, although so far its linkage group is not known. The gene symbol *rnu* has been assigned to the Rowett nude mutation.

In 1976 another entirely independent nude rat mutation was discovered at the Victoria University of Wellington (3). Homozygotes are hairless and athymic with associated immunodeficiencies. However, the homozygotes are phenotypically different from the Rowett nude rat in that the latter has more hair, with a completely normal tail, whereas the tail of the New Zealand nude is completely hairless.

Recently, a cross has been made between the Rowett and the New Zealand nudes, (the New Zealand nudes being supplied by Dr. Barbara Heslop) and it has been shown that the F1 hybrid is also nude, and therefore the genes are allelic. It appears as though the Rowett phenotype is dominant, although only small numbers have been studied so far (M.F.W. Festing, B.F. Heslop, and M.V. Berridge, in preparation). The New Zealand nude has been tentatively assigned the gene symbol rnu^N.

Morphology, Growth, and Reproduction

Homozygous *rnu/rnu* rats may be recognized soon after birth by their bent vibressae. Subsequently they never grow a complete coat of hair. At weaning (about 3 weeks of age) they are usually completely hairless, apart from their bent vibressae, and at this age they frequently have a brown scaley substance on the surface of their skin. Eventually these scales disappear, and for brief periods animals may grow a substantial coat of hair. This is never as dense as in the normal rat, and is gradually lost again. Eventually, older homozygotes are largely hairless, although they nearly always have some long hairs on their body, and the tails appear to be normal.

In contrast, New Zealand nudes are much more hairless, with most, although not all of their vibressae missing, and few if any body hairs at any time. Nor do the New Zealand nude homozygotes have any hair on their tails, and in this respect they resemble the nude mouse. However, it is not fully established whether this is a genetic background effect or a true difference between the two alleles.

Growth rate is reduced to about 70% of normal (8) in both sexes, in the Rowett nude homozygotes.

Homozygous *rnu/rnu* rats of both sexes are fully fertile, but as with the nude mouse, the female nudes have difficulty under normal conditions in rearing their young (8). Whether or not survival of the young could be improved by changes in husbandry and by selection is not known, although this has been found to be the case in nude mice.

The usual breeding method is to use homozygous nude males and heterozygous females. Under these conditions, about 36% of the pups at 4 days and 34% of the weaned young were found to be homozygous nude, rather than the 50% that would be expected if the viability of the nude pups were equal to that of the heterozygotes. However, from these figures, it appears as though most of the losses of homozygotes were in utero, or immediately postpartum.

Estimates of total productivity were based on 33 pairs of an outbred genetically variable stock. Average litter size was 9.5 young, with 1.17 total young, and 0.48 nude young weaned per breeding female per week.

Lifespan and Pathology

No detailed studies of the lifespan of nude rats have yet been carried out, and virtually all the available data are of an anecdotal nature. Although the Rowett nude rat is undoubtedly susceptible to both natural and experimental infections, it appears to live longer in conventional conditions than the nude mouse. When homozygous nude rats bred in "SPF" conditions are brought into a conventional animal house, there may be some deaths within the first 2–3 weeks, and at a later date the nude rats may develop some respiratory disease, but most animals will usually survive for 6–24 months. It is not clear why the nude rat is so much more robust than the nude mouse, but it may be that there are fewer common viruses that affect rat than mouse colonies. In the nude mouse most deaths among conventional colonies can be attributed to virus infections such as murine hepatitis virus. However, the nude rat is highly susceptible to the bacterium causing Tyzzer's disease, and on two occasions colonies maintained at the Laboratory Animals Centre have had to be killed out because of this disease, which is difficult to eliminate.

Conjunctivitis leading to periorbital abscesses may be a problem in some cases and may lead to blindness unless treated. Mildly affected individuals may be treated by bathing the eye with warm water, but more severely affected individuals may need antibiotic therapy (13).

Anatomic Features

No detailed studies of the anatomy of the skin have yet been published. The most detailed histological studies carried out so far have been on the lymphoid organs. Parallel studies of the thymus area, the lymph nodes, the spleen, and the Peyer's patches by Vos et al. (18), Brooks et al. (4), and Pritchard and Eady (14) are essentially all in agreement. An accumulation of brown fat in the thorax replaces normal thymus tissue, although there is a small nonfunctional thymic rudiment. Insufficient numbers of nude rats have been studied so far to be certain that there is never any functional thymus tissue.

The lymph nodes are of normal size but are depleted of lymphocytes in the paracortical areas, as in the nude mouse. Brooks et al. (4) found only 15×10^6 mononuclear cells per lymph node in nudes as compared with 98×10^6 cells per lymph node in heterozygotes. In spleen the ratio of red to white pulp appears to be normal. According to Brooks et al. (4),

> In the red pulp, the only noticeable difference was an increase in the number of megakaryocytes in rnu/rnu animals. In the white pulp the periarteriolar sheaths were much reduced in size in rnu/rnu rats, possessing at most only five to six layers of small lymphocytes; in some sheaths only pyroninophilic medium lymphocytes were present round the arteriole. The marginal zone surrounding the white pulp, however, was often more developed than that surrounding white pulp in control animals and germinal centers in the splenic follicles were more distinct, larger, contained more mitotic figures and tingible body macrophages, and possessed wider mantle layers than those of controls.

The Peyer's patches were studied by Pritchard and Eady (14) and by Brooks et al. (4), who concluded that

> Peyer's patches in $rnu/+$ rats contained four or more lymphoid nodules most of which comprised a germinal centre and cuff of small lymphocytes. The region between adjacent nodules also contained numerous small lymphocytes. Peyer's patches in rnu/rnu rats were much smaller with fewer lymphoid nodules. Germinal centres in the nodules were less developed, and the areas between adjacent nodules were almost totally devoid of small lymphocytes.

Histological studies of the pituitary, thyroid, adrenals, ovaries, and testes of rnu/rnu compared with $rnu/+$ rats showed no abnormalities in the homozygous nude rats. Neither were there any important differences in the submandibulary salivary gland, although there was some evidence that the normal sexual dimorphism was less pronounced (18).

The proportion of lymphocytes carrying T- and B-cell surface markers in nude rats was studied by Brooks et al. (4), using the W3/13 monoclonal antiserum, which reacts specifically with thymic-derived lymphocytes, and rabbit anti-rat-Ig antiserum. An indirect fluorescent antibody technique was used. In the spleen 17% of the lymphocytes of rnu/rnu rats were W3/13 positive, compared with 49% in heterozygotes. There was also an increase in the null cells from 9% in heterozygotes to 19% in rnu/rnu rats. The percentage of Ig-positive cells was increased from 42% in heterozygotes to 64% in nudes, but the absolute numbers of B cells were within the normal range in the nudes. A similar, although more extreme pattern was observed with lymph node cells. In heterozygotes there were

Table VI-1. Immune response of nude rats to T- and B-cell mitogens

Function[a]	Observation	Reference
PHA (T)	No response	4
		19
ConA (T)	Occasional small response	4
	No response	19
Pokeweed (T?)	No response	19
LPS (B)[b]	Normal	19
Staphylococcus aureas (B)	Normal	4

[a] Type of mitogen given in parentheses.
[b] LPS, lipopolysaccharides.

23% Ig-positive, 68% W3/13-positive, and 9% null cells, whereas in nudes there were 70% Ig-positive, 4% W3/13-positive, and 26% null cells.

It is worth noting that although as many as 17% of the spleen cells had surface antigens of apparent thymic origin, this did not necessarily imply that these cells were functional T lymphocytes.

According to Ford and Smith (10) output of thoracic duct lymphocytes (TDL) was only 15–20% of normal in nude rats 24 h after cannulation. When normal TDL were injected intravenously into nude rats they circulated normally, with the same number recovered from the thoracic duct in nude and normal rats. They suggested that the distribution of TDL in nude rats was as would have been expected if the nude TDL were entirely B lymphocytes.

Immune Functions

There have now been several studies on various aspects of the immune function on nude rats. The response to a number of mitogens and antigens is summarized in Tables VI-1 and VI-2. Apart from a very slight response to concanavalin A in some individuals (4), there was no response to T-cell mitogens, but the response to B-cell mitogens was within the normal range. Similarly, no response was observed to T-cell-dependent antigens, such as ovalbumen, tetanus toxoid, tuberculin (19), sheep erythrocytes, or bovine gammaglobulin (14).

Table VI-2. Response to antigens assumed to be T-cell dependent in nude rats

Antigen	Result	Reference
Ovalbumen	NR[a]	19
Tetanus toxoid	NR	19
Sheep erythrocytes	NR	14
Tuberculin hypersensitivity	NR	19
Bovine gammaglobulin	No eosinophilia or bronchoconstriction	14

[a] NR, no response.

Table VI-3. Response of nude rats to experimental infections

Infective agent	Result	Reference
Nippostrongylus brasiliensis	Prolongation of egg production	14
		15
Eimeria nieschulzi	No "memory" of previous infection	15
Taenia crassiceps	Death of nudes	Chernin, personal communication
Sendai virus	Virus persists—no deaths	5

The immune response of nude rats to experimental infections with parasites and Sendai virus is summarized in Table VI-3. Both Rose et al. (15) and Pritchard and Eady (14) found that experimental infections with *Nippostrongylus brasiliensis* resulted in a prolongation of egg production in homozygous nudes compared with heterozygous or normal controls, although egg production in the nudes was not prolonged indefinitely.

More clear-cut results were obtained following experimental infections with *Eimeria nieschulzi*. Not only was the peak output of oocysts greater in nude than in normal rats, but both a second and a third challenge infection resulted in a high oocysts output comparable with that found in the primary infection. In heterozygotes the second challenge infection had a considerably reduced oocysts output, and it was impossible to establish a third challenge infection. Thus it is clear that the response of nude rats to experimental parasitic infections is highly abnormal, and it seems highly probable that this is because of a defective cell-mediated immune response.

The study by Carthew and Sparrow (5) on the course of experimental infection with Sendai virus is of some practical importance, as this virus is common in animal houses. In nude rats, the course of infection was abnormal, with a delay in the establishment of the virus in the nude rats compared with heterozygotes. However, there were no deaths from this experimental virus infection, although whether nude rats would show equal resistance to a natural infection has yet to be determined.

Response to Grafts

There have now been several studies of the fate of skin and tumor allografts and xenografts in nude rats. These are summarized in Table VI-4. Skin allografts of the types tried so far have been permanently accepted, but in view of the fate of some skin xenografts reported below, further research will be needed to determine whether such grafts are always accepted.

An interesting study of the fate of a fibrosarcoma tumor allograft has been reported by Eccles et al. (6). They found that in the isogenic host, 28–33% of the tumor mass consisted of host cells that had infiltrated into the tumor. In nude rats, however, the degree of host infiltration was substantially less, amount-

Table VI-4. Response of nude rats to allografts and xenografts

Tissue	Species	Result	Reference
Allografts			19
Skin		Accept	
			7
Fibrosarcoma		Increased metastasis; reduced cellular infiltration	6
Xenografts			
Skin	Mouse	Some accepted	8
	Hamster, Gerbil	Some rejected c50 days	
Tumors	Mouse (5)[a]	All accepted	6
	Mouse (1)	Accepted	16
	Mouse (4)	Two accepted	16
	Human (4)	Accepted	2
	Human (20)	Six accepted; did not grow as well as in nude mice	1
Pituitary	Human	Accepted; Secreted hormones	7

[a] Number of tumors transplanted.

ing to only about 4–7%. These differences were associated with a complete lack of antibody response to the tumor in the nudes, compared with a low, although easily detectable response in the isogenic host, presumably to tumor-specific antigens. Tumors were rejected by the nonisogenic heterozygous nudes, and this was also accompanied by a strong antibody response.

The most interesting observation, however, was that the transplanted tumors failed to metastasize in eight isogenic hosts, whereas in all of seven nude hosts the tumors metastasized to various sites including lungs, lymph nodes, the kidney, or to subcutaneous regions. They suggest from these findings that metastatic spread may be controlled to some extent by the cell-mediated immune system. It is particularly interesting to note this high level of metastatic spread in a tumor allograft, in view of the usual lack of such spread in tumor xenografts.

The fate of tumor and skin xenografts is less clear-cut. Mouse, hamster, and gerbil skin xenografts have been attempted, and some have healed in well and been accepted for long periods. In particular, xenografts of gerbil tail skin on one nude rat were accepted for the life of the rat (several months). Not all grafts were successfully established, although this may have been because of technical problems associated with healing of the grafts, rather than acute rejection. However, one graft of hamster skin and three grafts of mouse skin to the back of five nude rats were accepted and healed in well for about 50 days, but these subsequently became scabby and were rejected between about the 45th and 120th day after grafting (7). Whether this implies some residual cell-mediated immune response or whether this is an antibody-mediated rejection as described by Koene et al. (11) is not known.

The response of nude rats to tumor xenografts has been somewhat variable (Table VI-4). Eccles et al. (6) found that five different mouse tumors were accepted. Festing et al. (9) established one mouse lymphoma but had less success with some other tumors. Salomon et al. (16) successfully established two mouse and four human tumors. Similarly, Bastert et al. (2) attempted to grow 20 human tumors in nude rats and successfully established six of these. However, they found that these human tumors did not grow as well in nude rats as in nude mice. Finally, Althoff et al. (1) successfully established in nude rats a human pituitary graft which secreted hormones.

Discussion

Enough is now known about the nude rat to make it clear that in many ways it is the rat equivalent of the nude mouse. There is no evidence for any cell-mediated immune response in the relatively small numbers of rats that have been examined so far, and there are many examples of allografts and xenografts being accepted, with no sign of immunologic rejection. Rejection of some mouse and hamster skin xenografts has been recorded, but this may have resulted from an antibody-mediated rejection process rather than from any residual cell-mediated immune response. Obviously, this phenomenon needs to be studied in more detail, and in particular it will be interesting to discover whether skin xenograft rejection depends on factors such as the size of the graft and the genetic background of the nude host. Such studies will depend on the availability of inbred strains to which the nude gene has been backcrossed. Obviously, such strains will take several years to develop, although some indication may be given from studies of partially backcrossed stocks.

The value and place of the nude rat in immunology, cancer research, and studies of parasitic diseases will only become clear in a few years time. The nude rat appears to have one important advantage over the nude mouse, and that is that it is apparently more robust when maintained in conventional conditions. Whether this robustness is caused by a species difference in the types of infection commonly found in the animal house or whether it results from some residual cell-mediated immunity is not yet known. The observation that in some cases skin xenografts may be rejected again may be because of a better immune response in nude rats than in nude mice (which do not reject xenografts), or it may be because of the greater efficiency of the complement system in rats than in mice.

The nude rat is also likely to be of value in those studies involving surgery. An interesting example is that of the "epigastric pouch" method of tumor xenografting developed by Steinau et al. (17), in which the tumor is transplanted into a subcutaneous patch supplied by the superficial epigastric artery and vein, and the whole pedicle is then covered in a polythene bag. In this way, it is possible to grow a tumor that has its own separate blood supply. This blood supply can then be sampled or perfused separately with drugs.

Of course, the size of nude rats, while being an advantage for some studies, may well be a disadvantage for others in that rats are less economical than mice. For this reason alone, it is clear that the nude rat will in no way become a rival of the nude mouse; rather, it will become complementary to the nude mouse in much the same way that the two species are themselves complementary in biomedical research.

References

1. Althoff, P.H., G. Bastert, K.H. Usadel, U. Schwedes, U. Steinau, H. Eichholz, H.P. Fortmeyer, and I. Klempa. Xenografts of benign and malignant endocrine tissues in thymusaplastic nude mice and rats: Development and function, pp. 383–407. In G. Bastert, H.P. Fortmeyer, and H. Schmidt-Matthiesen (eds.), Thymusaplastic nude mice and rats in clinical oncology. Stuttgart: Gustav Fischer Verlag, 1981.
2. Bastert, G., H. Eichholz, H.P. Fortmeyer, R.-Th. Michel, R. Huck, and H. Schmidt-Matthiesen. Comparison of human breast cancer xenotransplantation into nu/nu mice and rnu/rnu rats, pp. 229–233. In G. Bastert, H.P. Fortmeyer, and H. Schmidt-Matthiesen (eds.), Thymusaplastic nude mice and rats in clinical oncology. Stuttgart: Gustav Fischer Verlag, 1981.
3. Berridge, M.V., N. O'Kech, L.J. NcNeilage, B.F. Heslop, and R. Moore. Rat mutation ($NZNU$) showing "nude" characteristics. Transplantation 27:410–413, 1979.
4. Brooks, C.G., P.J. Webb, R.A. Robins, G. Robinson, R.W. Baldwin, and M.F.W. Festing. Studies on the immunobiology of rnu/rnu "nude" rats with congenital aplasia of the thymus. Eur. J. Immunol. 10:58–65, 1980.
5. Carthew, P. and S. Sparrow. Sendai virus in rnu/rnu rats and germ-free AGUS rats. Res. Vet. Sci. 29:289–292, 1980.
6. Eccles, S.A., J.M. Styles, S.M. Hobbs, and C.J. Dean. Metastasis in the nude rate associated with lack of immune response. Br. J. Cancer 40:802–805, 1979.
7. Festing, M.F.W. The Rowett athymic nude rat, pp. 15–23. In G. Bastert, H.P. Fortmeyer, and H. Schmidt-Matthiesen (eds.), Thymusaplastic nude mice and rats in clinical oncology. Stuttgart: Gustav Fischer Verlag, 1981.
8. Festing, M.F.W. and D.P. Lovell. The breeding of athymic nude rats, pp. 81–84. In A. Spiegel, S. Erichsen, and H.A. Solleveld (eds.), Animal quality and models in biomedical research. Stuttgart: Gustav Fischer Verlag, 1980.
9. Festing, M.F.W., D. May, T.A. Connors, D. Lovell, and S. Sparrow. An athymic nude mutation in the rat. Nature (London) 274:365–366, 1978.
10. Ford, W.L. and M.E. Smith. More on the nude rat. Rat News Lett. 2:15–16, 1977.
11. Koene, R.A.P., P.G.G. Gerlag, J.J. Jansen, J.F.H. Hagemann, and P.G.A.B. Wijdeveld. Rejection of skin grafts in the nude mouse. Nature (London) 251:69–70, 1974.
12. May, D. More on the nude rat. Rat News Lett. 2:14, 1977.
13. Moore, J.G. Conjunctivitis in the nude rat (rnu/rnu). Lab. Anim. 13:35, 1979.
14. Pritchard, D.A. and R.P. Eady. Some aspects of immunology in the nude rat, pp. 67–79. In S. Sparrow (ed.), Immunodeficient animals in cancer research, New York: Macmillan, 1980.
15. Rose, E.M., B.M. Ogilvie, P. Hesketh, and M.F.W. Festing. Failure of nude (athymic) rats to become resistant to reinfection with the intestinal coccidian parasite Eimeria nieschulzi or the nematode Nippostrongylus brasiliensis. Parasite Immunol. 1:125–132, 1979.
16. Salomon, J.-C., N. Lynch, and J. Prin. Graft susceptibility of nude rats and mice to animal and human tumors and to hybrid cell lines, pp. 159–165. In S. Sparrow (ed.), Immunodeficient animals in cancer research. New York: Macmillan, 1980.
17. Steinau, H.U., G. Bastert, H. Eichholz, H.P. Fortmeyer, and H. Schmidt-Matthiesen. Epigastric pouching technique: human xenografts in rnu/rnu rats, pp. 531–542. In G. Bastert, H.P. Fortmeyer, and H. Schmidt-Matthiesen (eds.), Thymusaplastic nude mice and rats in clinical oncology. Stuttgart: Gustav Fischer Verlag, 1981.
18. Vos, J.G., J.M. Berkvens, and B.C. Kruijt. The athymic nude rat I. Morphology of lymphoid and endocrine organs. Clin. Immunol. Immunopathol. 15:213–228, 1980.
19. Vos, J.G., J.G. Kreeftenberg, and B.C. Kruijt. The athymic nude rate II. Immunological characteristics. Clin. Immunol. Immunopathol. 15:229–237, 1980.

VII

The Hairless Athymic Guinea Pig

Carolyn Reed* and John L. O'Donoghue

University of Rochester School of Medicine and Dentistry, Division of Laboratory Animal Medicine, Box 674, Rochester, New York 14642, and Eastman Kodak Company, Health, Safety and Human Factors Laboratory, Toxicology Section, Building 320, Rochester, New York 14650.

Introduction

Genetic hairlessness is a condition that occurs in a variety of animals. At least four mutant strains of hairless mice (6) have been described, and three hairless mutants in rats have been described (8). Hairlessness has also been reported in rabbits (7) and swine (2), as has featherlessness in chickens (11). The histological defects associated with these mutants can be separated into three categories: (a) reduced numbers of hair follicles (rabbits, swine) (b) abnormal keratinization of the hair shaft (naked mouse), and (c) abnormal keratinization combined with interruption in cyclic activity of hair growth (hairless and rhino mice) (5).

In 1966, a mouse mutant was described that, in addition to being hairless, lacked a thymus (5,9). The nude mouse has since been well characterized and extensively used in elucidating the role of the thymus in immunology, oncology, microbiology, and intraspecies tissue grafting studies. The value of nude mice makes the occurrence of a similar mutation in other species of interest from the aspect of comparison to the nude mouse and for use in areas where the mouse may not be suited.

Two nude athymic rat mutants have been described recently. One was originally observed in a colony of hooded rats at the Rowett Research Institute in England (4) and the other was found in a colony of outbred albino rats at Victoria University of Wellington, New Zealand (1). Although both appear to have depressed or absent T-cell function, the New Zealand mutant seems to be less hardy than the hooded mutant and has less hair (1). Both mutations behave as an autosomal recessive.

---

* To whom correspondence should be addressed.
© 1982 Gustav Fischer New York, Inc.
Proceedings of the Third International Workshop on Nude Mice.

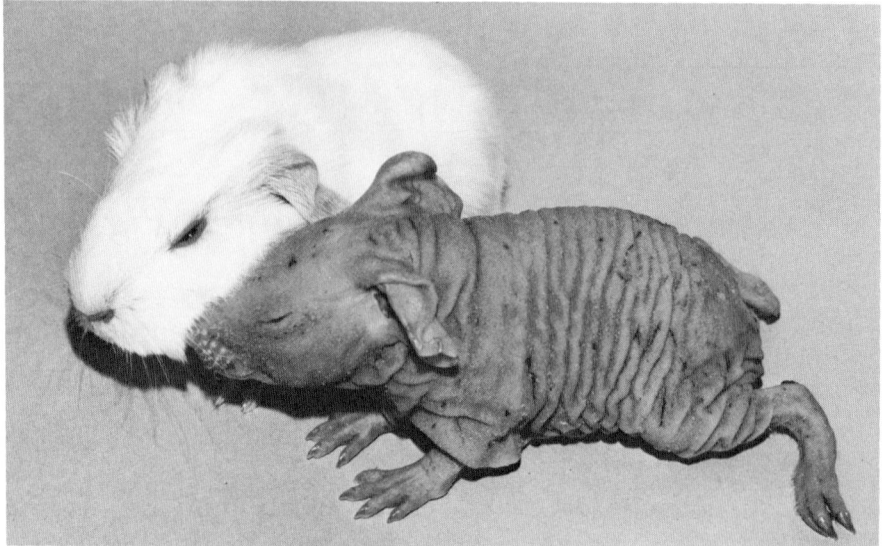

Figure VII-1. Hairless guinea pig with littermate.

In 1975, a hairless athymic guinea pig (Figure VII-1) was born into a colony of Hartley stock guinea pigs used for toxicologic research in Rochester, New York. The colony was started from six pairs of outbred Hartley guinea pigs and has been maintained as a closed colony for the past 10 years. Animals were randomly bred within the colony. As breeding efficiency declined, breeders were replaced with younger animals.

At birth, the hairless guinea pigs were smaller than their littermates; they had tannish colored, wrinkled skin, and stunted vibrissae. A few animals had sparse, short body hair and others had hairs around the nostrils and on the feet at birth. In both cases, these hairs were quickly lost. In some guinea pigs the hair coat grew enough to give a fuzzy appearance, but this was then lost. At each new molt less hair developed.

On gross necropsy examination of the hairless guinea pigs, there was tissue present in the cervical area that consisted of nodules of lymphoid tissue. Histologically, these nodules were not sharply defined into cortical and medullary areas and appeared similar to lymph nodes. No Hassel's corpuscles were observed, but cystic structures lined by respiratory epithelium were rarely present. These structures appeared to represent a rudimentary or hypoplastic thymus. The thymus in normal guinea pigs consists of two finely lobulated structures connected by an isthmus and located entirely within the cervical region; histologically the lobules are well demarcated into cortical and medullary areas with very prominent Hassel's corpuscles. In hairless guinea pigs, germinal follicles were reduced or absent in peripheral lymph nodes and in intestinal lymphoid tissue. One 9-month-old hairless guinea pig showed evidence of paracortical lymphoid depletion.

The hair defect resulted from production of abnormal hair shafts and a small decrease in the total number of hair follicles. The shafts bent and curled as they entered the upper portion of the piliary canal. The canal itself was distended and filled with keratin. Compared to normal guinea pigs, the epidermis of the hairless guinea pig was thickened by one to two cell layer increases in the stratum granulosum and stratum spinosum. When stained with hematoxylin and eosin (H&E), the cortex of the hair shaft of the hairless guinea pig remained acidophilic and under polarized light the shafts were not birefrigent. Both of these factors are indicative of defective hair shaft formation. In contrast, the cortex of the hair shaft of the normal guinea pig developed a pale yellow color on H&E stain and was strongly birefrigent under polarized light. A hairless guinea pig examined at 9 months of age had a marked reduction in the number of hair follicles and a lack of dilated piliary canals.

As initially observed with nude mice, survival of the hairless athymic guinea pigs was poor. None survived when housed under conventional conditions. To improve viability, mutants were isolated from their siblings and housed at temperatures of 24.4–25.5°C. Several groups of guinea pigs were housed under filter top cages with an apparent decrease in neonatal mortality. Once replaced in the colony in conventional housing, these guinea pigs became ill and died after a period of a few months.

A small group of hairless guinea pigs was placed in flexible plastic isolators in an attempt to promote long-term survival. Food, water, and bedding were autoclaved before passage into the isolators and sterile vitamin supplements were added to the food. These animals were 2–6 months of age at the time they entered the isolators. Although they were separated from the guinea pig colony, they were housed with their mothers and had an opportunity to acquire bacterial and viral flora. Three of the four guinea pigs died within 6 weeks of being put in the isolators.

The most successful method of maintaining the hairless guinea pigs was to separate them and their mothers from normal littermates within 24 hours of birth. Both were then placed in chambers of the type used in toxicologic testing. The chambers were equipped with high-efficiency particulate air (HEPA) filters and each held two cages. Food, water, and bedding were not autoclaved. Hairless guinea pigs kept in this manner survived 6 months.

The majority of the hairless guinea pigs died or were killed during the first week of life because of maternal rejection or traumatization by siblings. Animals that died after weaning succumbed to infectious agents that were low-order pathogens or nonpathogenic in animals with an intact immune system. These included cytomegalovirus infections, systemic balantidiasis, *Pneumocystis carinii*, and fungal and pyogenic bacterial infections.

Cytomegalovirus infections are common in guinea pigs but generally infect the salivary gland, producing enlarged nuclei and large intranuclear inclusion bodies. In the hairless guinea pigs, cytomegalovirus encephalitis was found in five animals. It was characterized by large pale areas of malacia with typical inclusion bodies in the nucleus of endothelial cells. Similar inclusions were found in epithelial and endothelial cells of the lung, liver, kidney, gastrointestinal tract, heart, and ovary.

Balantidium caviae is a ciliated protozoan that is found in the lumen of the cecum and colon and occasionally in the mucosa (12). In the hairless guinea

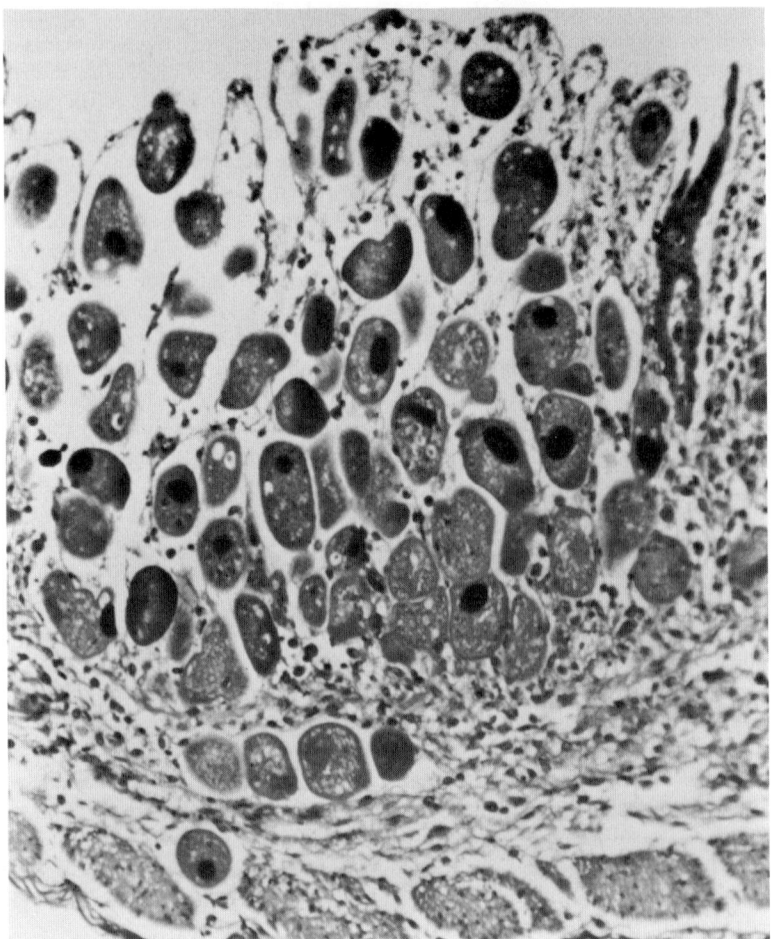

Figure VII-2. *Balantidium caviae* in the lamina propria of the small intestine of a hairless guinea pig. H&E, ×200.

pigs, *Balantidium* heavily infected the lamina propria (Figure VII-2). In one animal the protozoan migrated to the lung (Figure VII-3) and in others it could be found in association with mesenteric lymph nodes.

Pneumocystis carinii infections are usually found in immunosuppressed humans and animals. Two hairless guinea pigs were infected with *Pneumocystis*. Large areas of the alveolar fields were filled with pink foamy material (Figure VII-4). The organism itself was visible when stained with Gomori's methenamine silver stain and periodic acid-Schiff stain (Figure VII-5). These infections along with the abnormal thymic tissue found suggest that in addition to the hair defect, these guinea pigs are immunodeficient.

White blood cell counts and differentials done on six hairless guinea pigs showed white cells in the range described for normal guinea pigs. An interesting

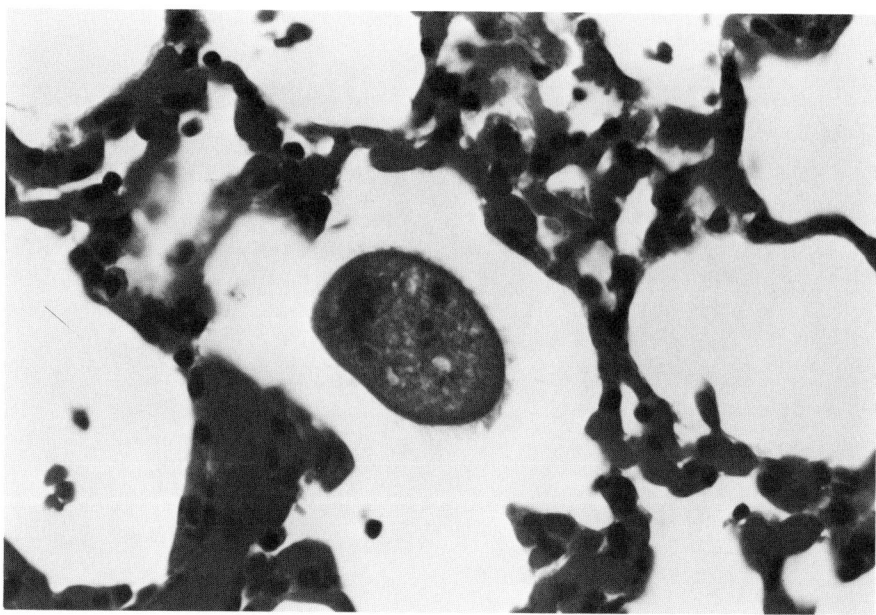

Figure VII-3. *Balantidium caviae* in the lung of a hairless guinea pig. H&E, ×500.

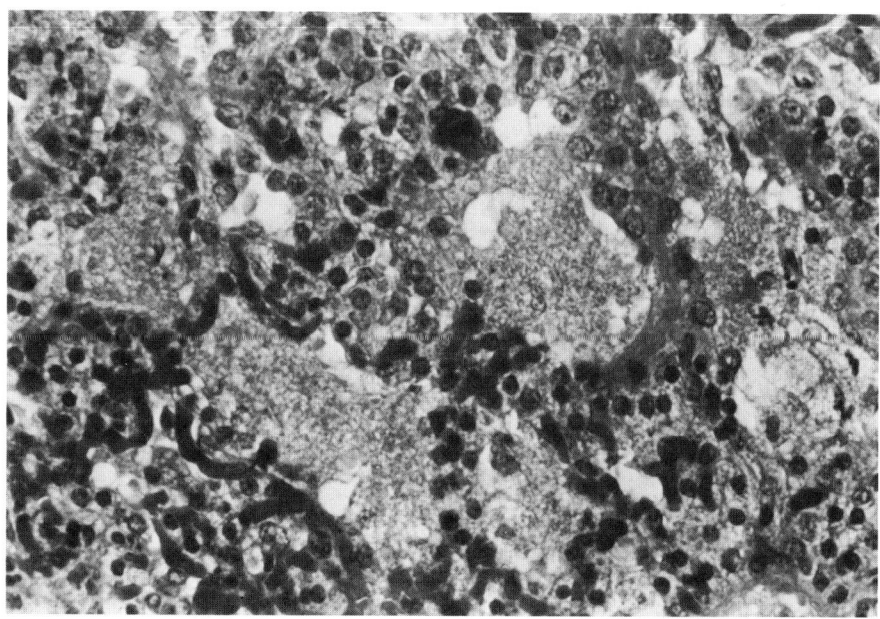

Figure VII-4. Alveoli of a hairless guinea pig filled with foamy material characteristic of *Pneumocystis carinii*. H&E, ×200.

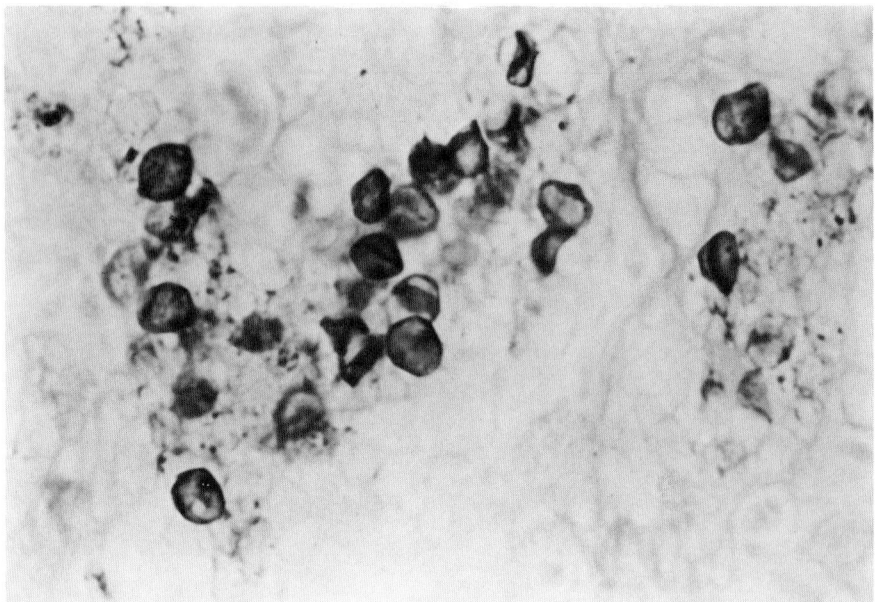

Figure VII-5. *Pneumocystis carinii* organisms in lung of a hairless guinea pig. GMS, ×500.

feature was the absence of Kurloff bodies, which are mononuclear leukocytes containing round or ovoid inclusions (10). Some may originate from thymus, but most come from the spleen.

Gammaglobulin of the hairless guinea pigs was decreased in comparison to haired guinea pigs. Immunoelectrophoresis (done by Dr. Fred Quimby) showed an absence of immunoglobulin G in these hairless guinea pigs.

The mode of inheritance of the hairless defect has not been determined. In the past 4 years, 59 hairless guinea pigs have been born in the Hartley stock colony. There were 30 males and 29 females. The frequency of hairlessness has been 1/1000 live births. The hairless guinea pigs were born in six of the 28 breeding groups. A breeding group was composed of 10 females and 1 male. Four of these breeding groups had more than one litter with hairless pigs born in it while the remaining two groups each had one litter that contained hairless guinea pigs.

Male hairless guinea pigs were fertile. One hairless male sired three litters, and one of the seven offspring was hairless. Hairless female guinea pigs were mated unsuccessfully with hairless male guinea pigs. The breeding method most often tried was to breed a homozygous male to a presumed heterozygous female. The difficulty was to identify the heterozygous females correctly. Six hairless guinea pigs were recently moved from inhalation chambers to isolators to establish breeding pairs. Offspring will be caesarian derived and foster nursed on caesarian-derived guinea pigs.

In summary, the hairless guinea pig has a defect in hair production that parallels that found in the nude mouse (3). No normal thymic tissue has been found, and T-cell areas in the spleen and lymph nodes are depleted of cells. The guinea pigs are hypogammaglobulinemic, and they are susceptible to infections that suggest a defect in cellular immunity.

Still unknown is the mode of inheritance of the defect, whether T cells are actually present, and whether T-cell function is depressed or absent. If these questions can be answered, the hairless guinea pig may be another useful model for research in oncology, immunology, and infectious disease, particularly cytomegalovirus infections.

References

1. Berridge, M.V., N. O'Kech, L.J. McNeilage, B.F. Heslop, and R. Moore. Rat mutant (NZNU) showing "nude" characteristics. Transplantation 27:410–413, 1979.
2. David, L.T. Histology of the skin of the Mexican hairless swine (*Sus scrofa*). J. Anat. 50:283–292, 1932.
3. Eaton, G. Hair growth waves and cycles in nude mice, pp. 89–93. *In* T. Nomura, N. Ohsawa, N. Tamaoki, and K. Fujiwara (eds.), Proceedings of the second international workshop on nude mice. Tokyo: University of Tokyo Press, 1977.
4. Festing, M.F.W., D. May, T.A. Connors, D. Lovell, and S. Sparrow. An athymic nude mutation in the rat. Nature (London) 274:365–366, 1978.
5. Flanagan, S.P. Nude, a hairless gene with pleiotropic effects in the mouse. Genet. Res. Camb. 8:295–309, 1966.
6. Green, M. Mutant genes and linkages, pp. 87–150. *In* E.L. Green (ed.), Biology of the laboratory mouse, 2nd ed. New York: McGraw Hill, 1966.
7. Kislovsky, D.A. Naked, a recessive mutation in the rabbit. J. Hered. 19:438–439, 1928.
8. Palm, J., and F.G. Ferguson. Fuzzy, a hypotrichotic mutant in linkage group I of the Norway rat. J. Hered. 67:284–288, 1976.
9. Pantelouris, E.M. Absence of thymus in a mouse mutant. Nature (London) 217:370–371, 1968.
10. Sisk, D.B. Physiology, pp. 67–69. *In* J.E. Wagner and P.J. Manning (eds.), The biology of the guinea pig. New York: Academic Press, 1976.
11. Soames, R.G., Jr. Delayed feathering, a third allele at the K locus of the domestic fowl. J. Hered. 60:281–286, 1969.
12. Vetterling, J.M. Protozoan parasites, pp. 183–184. *In* J.E. Wagner and P.J. Manning (eds.), Biology of the guinea pig. New York: Academic Press, 1976.

General Discussion

WALZER: Is the species of *Balantidium* you found natural to guinea pigs? It was impressively desimminated. As far as I know, there are no other reports of *Balantidium* infection dissiminating in immunodeficient animals.

C. REED: We have been calling the *Balantidium* found in the hairless guinea pigs *B. caviae*. It is a protozoan found naturally in guinea pigs and is reported to be morphologically distinct from *B. coli*. We did not try to demonstrate that it was *B. caviae* by cultivation or immunological differences.

1

Nude Mice as a Model for Chemotherapy of Leprosy

Kenji Kohsaka,* Kazuo Yoneda, Yumiko Arimochi, Masanao Makino, Tatsuo Mori, and Tonetaro Ito

Department of Leprology, Research Institute for Microbial Diseases, Osaka University, Yamada-oku, Suita-shi, Osaka 565, Japan.

Abstract

Leprosy patients were treated with 450 mg of rifampicin daily, and materials for inoculum were obtained by biopsy before and after treatment. *Mycobacterium leprae* obtained from the patients were inoculated into the right hind foot pads of nude mice. The results of the experiments indicate that rifampicin shows tremendous initial killing effect for *M. leprae*; the bacilli lost infectivity for nude mice after only 2 days of administration with 450 mg of the drug to man. For chemoprophylactic studies, *M. leprae* from experimental leprosy in nude mice were used, and the infected nude mice were given 0.5 mg (once) or 0.2 mg (6 days a week, for 2 weeks) of rifampicin orally. The results suggest that a single administration of 1,500 mg rifampicin or 600 mg daily for 2 weeks of rifampicin may be effective as chemoprophylaxis of human leprosy. Chemoprophylactic administration with a daily dose of 0.03 mg of DDS or 0.2 mg of clindamycin for 1 month or 0.03 mg of minocycline for 3 months was not able to prevent the growth of *M. leprae* in nude mice.

Introduction

It was previously reported by us (1,2) that lepromatoid lesions developed in nude mice inoculated with *Mycobacterium leprae*, and acid-fast bacilli obtained from the lesions were identified as *M. leprae*. Furthermore, we (3) reported that a new model of experimental leprosy was established using nude mice; successive transmission of *M. leprae* which proliferated in the lesions of first infected nude mice to other nude mice was confirmed, and the reproducibility of trans-

---

* To whom correspondence should be addressed.
© 1982 Gustav Fischer New York, Inc.
Proceedings of the Third International Workshop on Nude Mice.

mission to nude mice of *M. leprae* derived from several different patients was established. We have used this new model in studies of the chemotherapy and chemoprophylaxis of leprosy.

Materials and Methods

Animal. Nude mice (BALB/c-*nu/nu*) bred in our laboratory under Specific Pathogen Free (SPF) conditions were used in these experiments. Mice were maintained in plastic isolators or in a well-controlled environment.

Inoculum. In the chemotherapy experiments, bacillary suspensions of *M. leprae* were prepared from lepromata of three patients. Suspensions were made with *M. leprae* from the experimental leprosy in nude mice for chemoprophylactic experiments. Bacillary suspensions were made with F 12 (tissue culture medium), and 0.03 or 0.05 ml of the suspensions was inoculated into the right hind foot pads of the nude mice.

A relapsed patient (lepromatous leprosy) was treated with rifampicin 450 mg daily for 2 months, and materials for inoculum were biopsied before treatment and after 2 months of treatment. A borderline leprosy (BL) patient was treated with rifampicin, 450 mg daily, and biopsy was done before treatment, on day 4 with 2 days intermission after 2 days treatment, on day 9 with 2 days intermission after 1 week's treatment, and after 1 month of treatment. Another BL patient was treated with 450 mg daily of rifampicin and biopsied before treatment and on day 4 with 2 days intermission after 2 days treatment. The reason we set intermission before biopsy was to eliminate rifampicin from the body fluid.

In the experiments of chemoprophylaxis, the nude mice were infected with *M. leprae* from experimentally infected nude mice. The infected mice were given 0.5 mg (once) or 0.2 mg (6 days a week, for 2 weeks) of rifampicin orally; the doses of 0.5 and 0.2 mg per mouse are equivalent to human doses of 1500 mg and 600 mg. Clindamycin, 0.2 mg, or 0.03 mg of diaminodiphenylsulfone (DDS) were given to the infected mice for 1 month from day 20 after infection; the dose of 0.2 mg per mouse is equivalent to 600 mg per human and 0.03 mg is nearly 100 mg per human. Some infected nude mice were given 0.03 mg of minocycline orally 6 days a week for 3 months.

For examination, the number of bacilli in infected foot pads was determined and foot pads were also examined histopathologically.

Results

Effect of Rifampicin Treatment of Human Patient on Subsequent Growth in Nude Mice of M. leprae *from Inocula*

Experiments were carried out three times with materials taken from three patients. Table 1-1 shows the results of experiment 1 in which 3.4×10^5 *M. leprae* from tissue taken before rifampicin treatment of the patient were inoculated into foot pads of the nude mice; 2.0×10^8 bacilli were harvested after 12 months.

Table 1-1. Effect of rifampicin treatment of inoculum donor patient on growth of *M. leprae* in nude mice (Expt. 1)

Material	Inoculum	Harvest of acid-fast bacilli
Taken before rifampicin treatment of patient (6 mice)	$3.4 \times 10^5/0.03$ ml	2.0×10^8/foot pad (12 months)
Taken after treatment of patient with rifampicin 450 mg daily, 2 months (5 mice)	$2.3 \times 10^5/0.03$ ml	$<1.0 \times 10^5$/foot pad (22 months)

In the treated group, 2.3×10^5 bacilli were inoculated after 2 months of patient treatment, and the infected mice were killed after 12 months of infection up to 22 months, but no acid-fast bacilli were found in tissue homogenates of the foot pads. The results suggest that there were no infectious bacilli in 10^4 of *M. leprae* of inoculum after 2 months of rifampicin treatment with 450 mg daily.

Table 1-2 shows the results of experiment 2. Using material taken before treatment of the donor patient, 7.6×10^5 cells of *M. leprae* were inoculated into foot pads and the number of bacilli reached up to 3.0×10^9 after 12 months. Remarkable swelling of infected foot pad was noted macroscopically and lepromatoid lesions were seen at the site of inoculation. However, there was no growth of acid-fast bacilli in the three groups infected with the materials taken after rifampicin treatment of the patient. Even in the group infected with *M. leprae* recovered after only 2 days treatment, no acid-fast organisms were seen in tissue homogenates of the infected foot pads after 20 months of infection, and no tendency of multiplication of acid-fast bacilli was observed by histopathological examination.

Table 1-3 shows the results of experiment 3. As shown in the table, 1.2×10^7 *M. leprae* were inoculated into foot pads of nude mice and, in the untreated group, a similar number of bacilli was recovered from the foot pad in the eighth month after inoculation. In the treated group, 5.9×10^6 of bacilli were inoculated after 2 days treatment with rifampicin; 6.6×10^4 of bacilli were counted in tissue homogenates of infected foot pads after 8 months. The numbers of bacilli gradually decreased in the treated group.

Table 1-2. Effect of rifampicin treatment of inoculum donor patient on growth of *M. leprae* in nude mice (Expt. 2)

Material	Inoculum	Harvest of acid-fast bacilli
Taken before rifampicin treatment of patient (10 mice)	$7.6 \times 10^5/0.03$ ml	3.0×10^9/foot pad (12 months)
Taken after treatment of patient with rifampicin 450 mg daily, 2 days (5 mice)	$3.6 \times 10^6/0.03$ ml	No growth of acid-fast bacilli (20 months)
Taken after treatment of patient with rifampicin 450 mg 1 week 7 (mice)	$3.2 \times 10^7/0.03$ ml	No growth of acid-fast bacilli (20 months)
Taken after treatment of patient with rifampicin 450 mg 1 month (6 mice)	$2.7 \times 10^5/0.03$ ml	No growth of acid-fast bacilli (20 months)

Table 1-3. Effect of rifampicin treatment of inoculum donor patient on growth of *M. leprae* in nude mice (Expt. 3)

Group	Inoculum	Harvest 7 months	8 months
Taken before rifampicin treatment of patient (5 mice)	$1.2 \times 10^7/0.05$ ml	2.0×10^7/foot	2.9×10^7/foot
Taken after treatment of patient with rifampicin 450 mg daily, 2 days (5 mice)	$5.9 \times 10^6/0.05$ ml	3.7×10^5/foot	6.6×10^4/foot

The results of these experiments indicate tremendous initial killing effect of rifampicin to *M. leprae*; the bacilli lost infectivity to nude mice after only 2 days administration of 450 mg of the drug to patients.

Chemoprophylaxis of Experimental Leprosy

Table 1-4 shows the effect of rifampicin in experimental leprosy. *Mycobacterium leprae* (5.2×10^6 cells) was inoculated into foot pads of nude mice, and the mice were divided into three groups. One month after the infection, one group was given 0.5 mg of rifampicin, another group was given 0.2 mg of rifampicin, 6 days a week for 2 weeks, and the third group was not treated with the drug.

In the untreated control group nearly the same amount of bacilli as in the inoculum was recovered from the foot pad 7 months after inoculation, and macroscopic swelling of the foot pad was observed in the 12 months. However, 12 months after inoculation in both groups treated with rifampicin no acid-fast bacilli were seen in tissue homogenates of infected foot pad, and swelling of the foot pad was not observed. The results suggested that single administration of 1500 mg or 2 weeks of 600 mg daily of rifampicin may be effective as chemoprophylaxis of human leprosy.

Table 1-5 shows the use of clindamycin and DDS for prophylaxis of experimental leprosy. *Mycobacterium leprae* (1.2×10^8 cells) was inoculated into the right hind foot pad, and from day 20 after infection 0.2 mg of clindamycin or 0.03 mg of DDS was given daily for 1 month. The infected mice were kept in a

Table 1-4. Prophylactic effect of rifampicin in experimental leprosy in nude mice

Drug treatment of nude mice	Inoculum	Findings of foot pad	
Untreated control (15 mice)	$5.2 \times 10^6/0.03$ ml	2.3×10^9/foot Swelling	(7 months) (12 months)
Rifampicin 0.5 mg once (10 mice)	$5.2 \times 10^6/0.03$ ml	$<8 \times 10^4$/foot No swelling	(11 months) (12 months)
Rifampicin 0.2 mg 2 weeks (10 mice)	$5.2 \times 10^6/0.03$ ml	$<8 \times 10^4$/foot No swelling	(11 months) (12 months)

Table 1-5. Prophylactic effect of clindamycin or DDS in experimental leprosy in nude mice[a]

Drug treatment of nude mice[b]	Acid-fast bacilli harvested at 6 months	
Untreated control (5 mice)	3.7×10^7/foot pad	(killed)
Clindamycin	1.6×10^9/foot pad	(died)
0.2 mg, 1 month	4.9×10^9/foot pad	(killed)
(600 mg/man)	7.5×10^9/foot pad	(killed)
(5 mice)	NT (2 mice)[c]	(died)
DDS	$2.0 \times 10/^7$foot pad	(died)
0.03 mg, 1 month	$1.9 \times 10/^7$foot pad	(killed)
(100 mg/man)	$2.6 \times 10/^7$foot pad	(killed)
(5 mice)	NT (2 mice)	(died)

[a] Inoculum: 1.2×10^8/0.03 ml, injected into right hind foot pad.
[b] Drugs were given 6 days a week for 1 month (orally).
[c] NT, not tested.

conventional but well-controlled animal room without plastic isolators, and drugs were given 6 days a week for 1 month. The results of assays done 6 months after infection indicate that there was no significant difference among the three groups.

The results indicated that chemoprophylaxis with daily dose of 100 mg of DDS or 600 mg of clindamycin for 1 month is not effective enough to depress the growth of *M. leprae* in the individual who has impairment of cell-mediated immunity.

Table 1-6 shows the results of chemoprophylaxis with minocycline of experimental leprosy in nude mice. The experiment was carried out under two different conditions, in SPF plastic isolators or in the well-controlled environment. *Mycobacterium leprae* (3.2×10^6 cells) was inoculated into foot pads, and the mice were divided into two groups, SPF and conventional. Furthermore, each group was divided into two subgroups; one was the untreated control and the other was treated. Starting the day after infection, the infected mice were given 0.03 mg of minocycline 6 days a week for 3 months. Table 1-6 shows the results

Table 1-6. Prophylactic effect of minocycline in experimental leprosy in nude mice

Drug treatment of nude mice[a]	Inoculum[b]	Harvest	
		6 months	8 months
Untreated control	3.2×10^9/0.05 ml	1.2×10^7	6.0×10^{7c}
Minocycline 0.03 mg, 3 months (100 mg/man)	3.2×10^9/0.05 ml	9.2×10^9	7.4×10^{7c}

[a] Nude mice were maintained in well-controlled environment and were given drug orally 6 days a week for 3 months.
[b] *M. leprae* were inoculated into right hind foot pads and harvested at 6 and 8 months after infection.
[c] Slight swelling of infected foot pads was observed in 8 months.

obtained 6 and 8 months after infection from the mice reared in well-controlled environment without plastic isolators. As shown in the table, similar numbers of bacilli were recovered from the foot pads, and slight swelling of the infected foot pad was observed macroscopically in both groups. There was no significant difference between treated and control groups.

Discussion

Earlier (1–3) we described the use of nude mice in a new model of experimental leprosy. We now describe the application of our new experimental system to studies of chemotherapy and chemoprophylaxis of leprosy.

The results of the experiments indicate that rifampicin shows tremendous initial killing effect on *M. leprae*; the bacilli lost infectivity to nude mice after only 2 days administration with 450 mg/day of the drug to man. It was also suggested that single administration of 1500 mg once or 600 mg daily for 2 weeks of rifampicin may be effective as chemoprophylaxis of leprosy. Chemoprophylactic administration with a daily dose of 0.03 mg of DDS or 0.2 mg of clindamycin for 1 month or 0.03 mg of minocycline for 3 months is not effective enough to depress the growth of *M. leprae* in nude mice. It seems that investigations of these drugs should be continued.

Finally, it is hoped that the nude mouse technique will be adopted at many institutes, and studies of chemotherapy and chemoprophylaxis of leprosy will be promoted rapidly by experimental leprosy in nude mice.

Acknowledgments

This investigation received support from the Chemotherapy of Leprosy (THELEP) component of the UNDP/World Bank/WHO Special Programme for Research and Training in Tropical Diseases. This study was also supported by a grant from United States–Japan Cooperative Medical Science Program, Grant-in-Aid for scientific research from the Ministry of Education, Science and Culture, Japan, Sasakawa Memorial Health Foundation, and Osaka Dermatological Institute, for which grateful acknowledgment is made.

References

1. Kohsaka, K., T. Mori, and T. Ito. Lepromatoid lesion developed in nude mouse inoculated with *Mycobacterium leprae*. La Lepro (Jpn. J. Leprosy) 45:177–187, 1976.
2. Kohsaka, K., T. Mori, and T. Ito. 77–84. Successful transmission of *Mycobacterium leprae* to nude mouse, *In* T. Nomura, N. Ohsawa, N. Tamaoki, and K. Fujiwara (eds.), Proceedings of the second international workshop on nude mice. Tokyo: University of Tokyo Press, 1977.
3. Kohsaka, K., M. Makino, T. Mori, and T. Ito. Establishment of experimental leprosy in nude mice. Jpn. J. Bacteriol. 33:389–394, 1978.

General Discussion

KRUEGER: How do you quantify the number of microorganisms per foot pad? Do you correct for weight of tissue?

KOHSAKA: Tissue homogenate was prepared from an infected foot pad, and the number of *M. leprae* in the homogenate was determined microscopically by using our modification of Shepard's technique. Weight of tissue is therefore not important. What is important is total removal of the foot pad and extraction of all bacilli contained within the pad; the results are then based on the number of bacilli per foot pad rather than on a weight basis.

KRUEGER: Many patients with leprosy are resistant to standard therapies. Does such resistance carry over to *M. leprae*-infected nudes? If so, this has important clinical significance.

KOHSAKA: Drug resistance of *M. leprae* is a very important problem in leprosy research and in the clinic. In our experiments, *M. leprae* from a patient treated with DDS or promin over 20 years was shown to be highly resistant to DDS in experimental leprosy in nude mice. The results suggest that the nude mouse is useful in studies of drug-resistant *M. leprae*.

2

Mechanisms of Resistance of Nude Mice to Neisseria gonorrhoeae

Philip S. Lamborn,[*] Julia Cauthen, and David J. Drutz[†]

Infectious Diseases Section, Audie L. Murphy Memorial Veterans' Hospital, and Department of Medicine,[†] University of Texas Health Science Center, 7703 Floyd Curl Drive, San Antonio, Texas 78284.

Abstract

Both nu/nu and $nu/+$ mice were highly resistant to intravenous, intraperitoneal, or mucosal (conjunctival, oral, rectal, genital) infection with *Neisseria gonorrhoeae*. Resistance is atttributable in part to a potent serum bactericidal system which was demonstrable in vitro with either fresh or heat-inactivated (56°C for 1 h) serum; however, not all gonococcal strains were killed by the serum. Resistance is also attributable to the bactericidal activity of phagocytes as demonstrated by in vitro studies with peritoneal exudate cells. Neither iron–dextran nor human semen increased the infectivity of gonococci for the mice, but inoculation of gonococci in the presence of mucin resulted in abrogation of host defenses, presumably by interfering with phagocyte function. Both nu/nu and $nu/+$ mice died of gonococcal septicemia under these circumstances; no foci of suppurative metastatic infection were demonstrable at autopsy. Despite studies by others which suggest that gonococci may survive phagocytosis by human leukocytes in vitro, these microorganisms do not behave in the manner of facultative intracellular pathogens in nude mice.

Introduction

Neisseria gonorrhoeae is currently one of the most common causes of infection in the world (18). As research progresses in an attempt to develop a vaccine against this microorganism, it has become apparent that the immune response to the gonococcus is incompletely understood.

---

[*] Deceased.
[†] To whom correspondence should be addressed.
© 1982 Gustav Fischer New York, Inc.
Proceedings of the Third International Workshop on Nude Mice.

Although gonococci are known to possess highly effective antiphagocytic surface factors (15,23), it has generally been considered that death of the microorganism takes place once phagocytosis is accomplished (4,7). Recent studies by Smith and his colleagues in England suggest that gonococci may survive and even multiply in human polymorphonuclear leukocytes (PMNs) and mononuclear phagocytes (MNs) (25), thus behaving as facultative intracellular microorganisms. However, our own *in vitro* studies (5), and those by others (19), do not support the concept of significant intracellular gonococcal survival.

Unmanipulated laboratory mice are ordinarily resistant to local or systemic gonococcal infection (8). Because nude mice often manifest increased susceptibility to intracellular pathogens, we have designed studies to test the influence of the dysthymic state on the susceptibility of mice to local and systemic experimental gonococcal infection.

Materials and Methods

Mice. The mice used in these studies were of BALB/c lineage, breeding pairs having been obtained originally from Sprague Dawley (Madison, Wisconsin). Both specific pathogen-free (SPF) and germ-free nu/nu and $nu/+$ mice have been bred subsequently in our own laboratories (27). Approximately equal numbers of male and female mice were used in all experiments.

Gonococci. The gonococci employed were strains NRL-1384 (Neisseria Reference Laboratory, Seattle, Washington) and BSDH (provided by Drs. Harry Smith and David Veale, University of Birmingham, Birmingham, England), as well as ATCC-11689 and AMVAH (9). NRL-1384 is an extremely well-characterized bloodstream isolate which possesses all known virulence characteristics for invasive infection including the arginine–hypoxanthine–uracil (AHU$^-$) auxotype, resistance to lysis by human serum, and exquisite susceptibility to penicillin (9). BSDH is a genital isolate which has been shown by Dr. Veale and his colleagues to survive phagocytosis by human leukocytes *in vitro* (25).

The methods for growing and harvesting the gonococci have been described (5,6). The presence of hairlike pili on all gonococcal strains was verified by electron microscopy. Pili are surface structures that are thought to contribute to gonococcal virulence at least partially through an ability to prevent phagocytosis (15). Both transparent and opaque colony variants of the gonococcal strains were studied (22). Transparent colonies produce more severe infections in chick embryo models of gonococcal infection than do opaque colonies (20).

For studies employing mucin (Hog Gastric Mucin; Difco, Detroit, Michigan), the gonococci were suspended in 5% mucin in normal saline immediately prior to their intraperitoneal injection. Preliminary studies established that mucin had no adverse effect upon either the mice or the gonococci.

Susceptibility of gonococci to mouse serum. The technique employed was the plaque assay method which has been described by Corbeil and her co-workers (2). Briefly, 0.05-ml portions of freshly collected serum were deposited on lawns of gonococci spread over the surfaces of plates containing GC agar base–IsoVitaleX (Baltimore Biologic Laboratories; Cockeysville, Maryland) and diethylaminoethyl dextran. Inoculum sizes on individual plates varied by 10-fold and ranged from 10^4 to 10^7 colonies per plate. Gonococcal growth after overnight incubation was graded as no inhibition, partial inhibition, or complete inhibition. All data reported in this study pertain to complete inhibition of gonococcal growth. The technical limits of this procedure are such that killing of 10^3 or fewer gonococci would not be detectable.

Susceptibility of gonococci to mouse peritoneal exudate cells. Peritoneal exudate cells (PECs) were harvested aseptically by saline lavage of the peritoneal cavity of nu/nu and $nu/+$ mice freshly killed by cervical dislocation (uninduced PECs). Cells obtained in this manner were mostly MNs. For induced PECs, mice received intraperitoneal trypticase soy broth, 1–4 hours before sacrifice; the predominant PECs under these circumstances were PMNs.

Cells were washed by centrifugation and transferred in a concentration of 10^6 to 10^7 cells/ml (1-ml/Leighton tube) to Leighton tubes containing glass cover slips. After 1 hour at 37°C to permit attachment, cells were washed and then allowed to ingest NRL-1384 or BSDH in the presence of an opsonic system consisting of 10% heat-inactivated fetal calf serum (FCS; Grand Island Biological Co., Grand Island, New York) in McCoy's 5A medium (Grand Island Biological Co.). The ratio of gonococci to phagocytes was 10:1. After 45 min of incubation, extracellular gonococci were destroyed by 30 min exposure to 30% fresh rat serum (FRS) which lysed both BSDH and NRL-1384 but had no effect on the phagocytes (5). Phagocytes were ruptured by sonication in the cold to release true intracellular gonococci, and viable organisms were enumerated by plate dilution. The proportion of gonococci killed was calculated from the number of true intracellular gonococci identified by direct visual count in a duplicate Leighton tube cover slip preparation treated with FRS.

Results

Inoculation of Gonococci by Local and Systemic Routes

Table 2-1 summarizes the results of attempts to infect SPF nu/nu and $nu/+$ mice by the intravenous, intraperitoneal, conjunctival, oral, rectal, and genital routes. The mice received 10^6 to 10^8 gonococci on all mucosal surfaces simultaneously. Under no circumstances did the mice show evidence of disease. No

Table 2-1. Incidence of infection following intravenous, intraperitoneal, or orifice (eye, mouth, rectum, genital) inoculation with 10^6 to 10^8 gonococci

Gonococcal strains	nu/nu	$nu/+$
BSDH	0/24[a]	0/24
NRL-1384	0/36	0/35
Others	0/12	0/12
Total	0/72	0/71

[a] Infected mice/inoculated mice.

Table 2-2. Attempts to enhance virulence of gonococci (10^6 to 10^8 BSDH/ NRL-1384)

Treatment	nu/nu	$nu/+$
Iron dextran	0/8[a]	0/8
Semen: Normal	0/10	0/10
GC infected	0/4	0/4
Chamber passaged GC	0/4	0/4
Total	0/26	0/26

[a] Infected mice/inoculated mice.

more than transient colonization of mucosal surfaces could be established, and no evidence of hematogenous dissemination of infection could be obtained by bacteriologic or histological criteria. These studies involved not only BSDH and NRL-1384, but two other gonococcal strains (AMVAH and ATCC-11689) used previously in studies of experimental gonococcal endocarditis (9). All microorganisms were piliated and no more than 12–16 h of age; they included pure cultures of both transparent and opaque colony variants.

Attempts to Enhance Gonococcal Virulence

Studies were next conducted with a design to enhance the virulence of strains BSDH and NRL-1384 (Table 2-2). First, gonococci were inoculated intraperitoneally (IP) along with iron dextran (Imferon, Merrell-National Labs, Cincinnati, Ohio) in doses of 5–10 mg/kg (0.1–0.2 ml/mouse). The rationale was that the iron-requiring gonococcus might benefit from the presence of added iron to overcome the host's competitive transferrin iron-binding system (16). The iron–dextran had no effect on the viability of the gonococci *in vitro*. None of the 16 mice so inoculated developed evidence of infection. Next, gonococci were injected IP along with 1.0-ml aliquots of fresh or frozen ($-70°C$) human semen from normal donors with no history of gonorrhea or other venereal disease. The rationale was that gonococci are normally delivered in a milieu of semen during genital transmission of gonorrhea. Further, Stites and his colleagues have shown that semen may possess local immunosuppressive activity within the uterus, which activity blocks the rejection of allogeneic sperm (21). We reasoned that such local immunosuppression might also increase the likelihood of producing infection in the mice. However, none of the 20 mice receiving gonococci in normal semen developed gonococcal infection. (In no case did the semen affect the viability of the gonococci in control experiments). An additional eight mice were therefore inoculated IP with freshly ejaculated semen obtained from four men with active (culture-proved) gonococcal urethritis. In this case the gonococci and semen went within 1–2 h from patient to mouse. However, the mice failed to develop infection even though the gonococci were demonstrably viable at the time of inoculation. Finally, eight nu/nu and $nu/+$ mice were inoculated IP with BSDH gonococci which had been passaged in subcutaneous chambers in a rabbit to insure maximal virulence (26). Again, none of the mice developed evidence of gonococcal infection.

Studies in Germ-free Mice

Nude mice with a normal bowel flora may demonstrate paradoxical resistance to facultative intracellular pathogens—a phenomenon apparently related to the presence of macrophage activation (14). Germ-free mice do not manifest such macrophage activation (12,17). Eighteen nu/nu and 27 $nu/+$ germ-free mice challenged by the intravenous (IV) or IP routes with 10^6–10^8 BSDH or NRL-1384 gonococci failed to develop evidence of infection. These studies suggest that the resistance of nude mice is not a function of the presence of activated macrophages attributable to a normal bowel flora.

Studies of the Mechanisms of Resistance of Nude Mice to Gonococci—Ability of Mouse Serum to Kill Gonococci

The serum of most standard laboratory animals (1,5,9,10) and humans (11) possesses potent antigonococcal activity. The results of studies of the antigonococcal activity of mouse sera are shown in Table 2-3. The data indicate that 10^7 gonococci of strain NRL-1384 were killed, whether the mouse serum was fresh or heat inactiviated (56°C for 1 h). Two control strains of gonococci (ATCC-11689, AMVAH) showed the same results. Identical killing occurred regardless of whether the serum came from germ-free, SPF, nu/nu, or $nu/+$ mice. In striking contrast, strain BSDH was resistant to the serum of any of the animals. These studies were repeated on numerous occasions, always with similar results, and indicate that the resistance of nu/nu and $nu/+$ mice to strain NRL-1384, but not to strain BSDH, could be attributable to innate antigonococcal activity of the mouse sera.

Table 2-3. Bactericidal activity of mouse serum for four gonococcal strains (plaque assay)

Gonococcal strain	Source of serum[a]	Number of gonococci killed
NRL-1384	nu/nu SPF (F)	10^7
ATCC-11689	(HI)	10^7
AMVAH	GF (F)	10^7
	(HI)	10^7
	$nu/+$ SPF (F)	10^7
	(HI)	10^7
	GF (F)	10^7
	(HI)	10^7
BSDH	nu/nu SPF (F)	$\leq 10^{3b}$
	(HI)	$\leq 10^3$
	GF (F)	$\leq 10^3$
	(HI)	$\leq 10^3$
	$nu/+$ SPF (F)	$\leq 10^3$
	(HI)	$\leq 10^3$
	GF (F)	$\leq 10^3$
	(HI)	$\leq 10^3$

[a] SPF, specific pathogen free; GF, germ free; F, fresh; HI, heat inactivated.
[b] Technical limit of the assay.

Antigonococcal Activity of Mouse PECs

Because it seemed apparent that cell-mediated immunity, generalized macrophage activation, or serum antigonococcal activity could not account for the resistance of nude mice to BSDH, studies of the antigonococcal activity of mouse PECs were carried out.

As shown in Figure 2-1, a mean of 97% of ingested gonococci were killed after 30 min of phagocytosis, with individual observations ranging from 90 to 99.05%. By 2 h, about 99% (mean) of the organisms were killed (range 95–99.09%). There were no differences in the ability of PECs to kill NRL-1384 and BSDH. Moreover(data not shown), there were no differences in the ability of induced or uninduced, nu/nu or $nu/+$, SPF or germ-free PECs to kill gonococci *in vitro*.

Effect of Mucin on Intraperitoneal Infection with Gonococci

In order to test the hypothesis that phagocytic cells were the principal host defense mechanism of nu/nu and $nu/+$ mice against strain BSDH, a series of studies was performed in which gonococci were inoculated intraperitoneally in the presence of mucin. Mucin is considered to act, at least in part, by blocking the ingress of inflammatory cells into the peritoneal cavity. Moreover, it interferes with the subsequent process of phagocytosis (10,13,28).

As shown in Table 2-4, BSDH killed 19 or 36 nu/nu and 14 or 22 $nu/+$ mice under these circumstances. This was true in both SPF and germ-free animals (individual data not shown). The mice died of gonococcal septicemia, proved by blood culture, but showed no evidence of acute peritonitis or suppurative metastatic infection at autopsy.

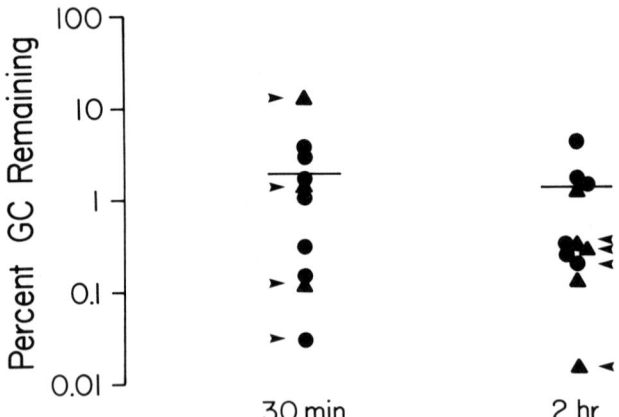

Figure 2-1. Percentage viable gonococci remaining after incubation (30 min and 2 h) with mouse peritoneal exudate cells. BSDH is represented by circles, NRL-1384 by triangles. Arrowheads indicate studies done with cells from nu/nu mice; all other studies employed cells from $nu/+$ mice. The mean value for each column is indicated by a horizontal line.

Table 2-4. Incidence of death following intraperitoneal inoculation of mice with gonococci and mucin

Gonococcal strain	nu/nu	$nu/+$
BSDH	19/36[a]	14/22
NRL-1384	0/12	0/12

[a] Dead mice/inoculated mice.

In striking contrast, 0 of 12 nu/nu and 0 of 12 $nu/+$ mice infected with NRL-1384 in mucin became ill. This was true whether transparent or opaque colony types were employed (data not shown). Sacrificed mice had no evidence of bacteremia, peritonitis, or metastatic infection. Presumably, the bactericidal serum activity for NRL-1384 protected these mice from infection, even when mucin interfered with phagocytic function.

Discussion

These studies neither support nor refute the contention of Smith and his colleagues that gonococci may survive *in vitro* within human leukocytes (25). However, they indicate unequivocally that nude mice, whether SPF or germ-free, possess formidable resistance to the gonococcus, and that such resistance is based upon both phagocytic and serum bactericidal capability. Even with strain BSDH, a strain unaffected by mouse serum, nu/nu and $nu/+$ mice resist infection until phagocytic mechanisms are overwhelmed by mucin.

Most bacteria pursue one of three courses when inoculated into nude mice. First, they may fail to produce infection—a pattern characteristic of the pneumococcus (29). Here, PMNs may be of major importance in protecting the animals from infection since pneumococci are not killed directly by serum. Second, infections may occur, but are paradoxically more severe in $nu/+$ than nu/nu mice. *Listeria monocytogenes*, a facultative intracellular pathogen, behaves in this manner (14). The paradoxical protection of nu/nu mice is thought to be related to spontaneous activation of their macrophages by the presence of a bowel flora; such macrophage activation is not demonstrable in germ-free animals (17). Third, progressive infection resulting in fatality may be encountered with greater frequency in nu/nu mice. This pattern is typical of obligate intracellular pathogens, such as pathogenic mycobacteria (24) or *Histoplasma capsulatum* (27).

Neisseria gonorrhoeae behaves more in the manner of the pneumococcus, a distinct extracellular pathogen. When the contribution of serum bactericidal activity is obviated by using a strain that resists mouse serum *in vitro*, as is the case with BSDH, both nu/nu and $nu/+$ mice, whether SPF or germ-free, are completely resistant to infection. Only when phagocytic function is abrogated by the use of mucin is the gonococcus capable of initiating an infection (3,13). Under these circumstances, both nu/nu and $nu/+$ mice develop fatal infections. Therefore, the presence or absence of a functioning thymus gland is of little apparent significance in the resistance of these mice to experimental gonococcal infection.

These studies suggest that the gonococcus behaves in the manner of a classical extracellular pathogen in nu/nu and $nu/+$ mice.

Acknowledgments

The authors thank Drs. Harry Smith and David Veale of the University of Birmingham, Birmingham, England for providing strain BSDH gonococci. We also gratefully acknowledge the technical assistance of Mr. James F. Legendre.

These studies were supported by grants AI-12469 and AI-14272 from the National Institutes of Health, and by the General Medical Research Service of the Veterans Administration.

References

1. Arko, R.J., K.H. Wong, S.E. Thompson, and F.J. Steurer. Animal models of gonococcal infection: Mouse chamber model for complement-mediated immunity, pp. 303–306. In G.F. Brooks, E.C. Gotschlich, K.K. Holmes, W.D. Sawyer, and F.E. Young (eds.), Immunobiology of *Neisseria gonorrhoeae*. Washington, D.C.: American Society for Microbiology, 1978.
2. Corbeil, L.B., A.C. Wunderlich, J.I. Ito, and J.A. McCutchan. Plaque assay for measuring serum bactericidal activity against gonococci. J. Clin. Microbiol. 8:618–620, 1978.
3. Corbeil, L.B., A.C. Wunderlich, J.A. McCutchan, and A.I. Braude. Murine model of gonococcal bacteremia, pp. 371–381. In G.F. Brooks, F.C. Gotschlich, K.K. Holmes, W.D. Sawyer, and F.E. Young (eds.), Immunobiology of *Neisseria gonorrhoeae*. Washington, D.C.: American Society for Microbiology, 1978.
4. Densen, P., and G.L. Mandell. Gonococcal interactions with polymorphonuclear neutrophils. Importance of the phagosome for bactericidal activity. J. Clin. Invest. 62:1161–1171, 1978.
5. Drutz, D.J. Intracellular fate of *Neisseria gonorrhoeae*, pp. 232–235. In G.F. Brooks, F.C. Gotschlich, K.K. Holmes, W.D. Sawyer, and F.E. Young (eds.), Immunobiology of *Neisseria gonorrhoeae*. Washington, D.C.: American Society for Microbiology, 1978.
6. Drutz, D.J. Hematogenous gonococcal infections in rabbits and dogs, pp. 307–313. In G.F. Brooks, F.C. Gotschlich, K.K. Holmes, W.D. Sawyer, and F.E. Young (eds.), Immunobiology of *Neisseria gonorrhoeae*. Washington, D.C.: American Society for Microbiology, 1978.
7. Gibbs, D.L., and R.B. Roberts. Interaction in vitro between human polymorphonuclear leukocytes and *Neisseria gonorrhoeae* cultivated in the chick embryo. J. Exptl. Med. 141:155–171, 1975.
8. Hill, J.H. Experimental infection with *Neisseria gonorrhoeae*. II. Animal inoculations (2 parts). Am. J. Syph. Gonorr. Vener. Dis. 28:334–378, 471–510, 1944.
9. Kaspar, R.L., and D.J. Drutz. Perihepatitis and hepatitis as complications of experimental endocarditis due to *Neisseria gonorrhoeae* in the rabbit. J. Infect. Dis. 136:37–42, 1977.
10. Keefer, C.S., and W.W. Spink. Studies of gonococcal infection. IV. The effect of mucin on the bacteriolytic power of whole blood and immune serum. J. Clin. Invest. 17:23–30, 1938.
11. McCutchan, J.A., S. Levine, and A.I. Braude. Influence of colony type on susceptibility of gonococci to killing by human serum. J. Immunol. 116:1652–1655, 1976.
12. Meltzer, M.S. Tumoricidal responses in vitro of peritoneal macrophages from conventionally housed and germ-free nude mice. Cell. Immunol. 22:176–181, 1976.

13. Miller, C.P. The enhancement of virulence of the gonococcus for the mouse. Am. J. Syph. Gonorr. Vener. Dis. 28:620–626, 1944.
14. Nikol, A.D., and P.G. Bonventre. Anomalous high native resistance of athymic mice to bacterial pathogens. Infect. Immun. 18: 636–645, 1977.
15. Ofek, I., E.H. Beachey, and A.L. Bisno. Resistance of *Neisseria gonorrhoeae* to phagocytosis: Relationship to colonial morphology and surface pili. J. Infect. Dis. 129:310–316, 1974.
16. Payne, S.M., and R.A. Finkelstein. Pathogenesis and immunology of experimental gonococcal infection: Role of iron in virulence. Infect. Immun. 12:1313–1318, 1975.
17. Rao, G.F., W.E. Rawls, D.Y.E. Perey, and W.A.F. Tompkins. Macrophage activation in conventionally athymic mice raised under conventional or germ-free conditions. J. Reticuloendothel Soc. 21:13–20, 1977.
18. Rein, M.F. Epidemiology of gonococcal infections, pp. 1–31. *In* R.B. Roberts (ed.), The gonococcus. New York: John Wiley and Sons, 1977.
19. Rest, R.F. Killing of *Neisseria gonorrhoeae* by human polymorphonuclear neutrophil granule extracts. Infect. Immun. 25:574–579, 1979.
20. Salit, I.E., and E.C. Gotschlich. Gonococcal color and opacity variants: Virulence for chicken embryos. Infect. Immun. 22:359–364, 1978.
21. Stites, D.P., C.J. Lammel, and G.F. Brooks. Immunosuppressive action of human seminal plasma: Identification of factors which influence cell-mediated immunity or humoral immunity and phagocytosis, pp. 344–345. *In* G.F. Brooks, E.C. Gotschlich, K.K. Holmes, W.D. Sawyer, and F.E. Young (eds.), Immunobiology of *Neisseria gonorrhoeae*. Washington, D.C.: American Society for Microbiology, 1978.
22. Swanson, J. Cell wall outer membrane variants of *Neisseria gonorrhoeae*, pp. 130–137. *In* G.F. Brooks, E.C. Gotschlich, K.K. Holmes, W.D. Sawyer, and F.E. Young (eds.), Immunobiology of *Neisseria gonorrhoeae*. Washington, D.C.: American Society for Microbiology, 1978.
23. Swanson, J., and G. King. *Neisseria gonorrhoeae*—Granulocyte interactions, pp. 221–226. *In* G.F. Brooks, E.C. Gotschlich, K.K. Holmes, W.D. Sawyer, and F.E. Young (eds.), Immunobiology of *Neisseria gonorrhoeae*. Washington, D.C.: American Society for Microbiology, 1978.
24. Ueda, K., S. Yamazaki, and S. Someya. Experimental mycobacterial infection in congenitally athymic "nude" mice. J. Reticuloendothel. Soc. 19:77–90, 1976.
25. Veale, D.R., C.W. Penn, and H. Smith. Capacity of gonococci to survive and grow within human phagocytes, pp. 227–231. *In* G.F. Brooks, E.C. Gotschlich, K.K. Holmes, W.D. Sawyer, and F.E. Young (eds.), Immunobiology of *Neisseria gonorrhoeae*. Washington, D.C.: American Society for Microbiology, 1978.
26. Veale, D.R., H. Smith, K.A. Witt, and R.B. Marshall. Differential ability of colonial types of *Neisseria gonorrhoeae* to produce infection and an inflammatory response in subcutaneous perforated plastic chambers in guinea pigs and rabbits. J. Med. Microbiol. 8:325–335, 1975.
27. Williams, D.M., J.R. Graybill, and D.J. Drutz. Histoplasma capsulatum infection in nude mice. Infect. Immun. 21:973–977, 1978.
28. Wilson, G.S., and A. Miles. The influence of diet, environment, hormones, chemical and chemotherapeutic agents, and other factors on immunity, pp. 1600–1647. *In* Toplcy and Wilson's Principles of Bacteriology, Virology and Immunity. Baltimore: Williams & Wilkins, 1975.
29. Winkelstein, J.A., and A.J. Swift. Host defense against the pneumococcus in T lymphocyte-deficient nude mice. Infect. Immun. 12:1222–1223, 1975.

General Discussion

RAIKOW: Have you tried depleting mice of macrophages by using silica or strontium-89 treatment?

DRUTZ: No.

WEIDANZ: Have you or others used human tumors, of urethral or prostatic origin, for example, in attempts to grow a parasite like this?

DRUTZ: No, we haven't.

UNIDENTIFIED: If you infect mice with the serum-susceptible strain and then inject the same mice with the BSDH strain plus mucin, do you get increased resistance to BSDH?

DRUTZ: We have not tried that.

3

Resistance and Immune Response of Nude Mice to Experimental Fungal Infections

Tadashi Arai,* Akio Shiraishi, Koji Yokoyama, and Kazuhide Nakagaki

Research Institute for Chemobiodynamics, Chiba University, 1-8-1, Inohana, Chiba 280, Japan.

Abstract

Resistance and immune response of nude mice to *Candida albicans*, *Aspergillus fumigatus*, and *Sporothrix schenckii* were studied. Accurate determination of *C. albicans* fungal growth in tissue is difficult; the postinfection survival of athymic nude (nu/nu) mice was shorter than that of their normal littermates ($nu/+$) within a certain inoculum range. When large inocula were used, there was no difference in the survival period between the two groups of mice. Histopathology revealed no essential difference in the development of *A. fumigatus* infection in nu/nu and $nu/+$ mice. On the other hand, nu/nu mice were much less resistant to *S. schenckii* infection than $nu/+$ mice and the importance of T-cell function was evident. In experimental murine mycoses with such opportunistic fungi as *C. albicans* and *A. fumigatus*, nonspecific defense mechanisms seem to play a more important role than specific cellular immunity.

Introduction

Generalized mycoses resulting from superinfection or opportunistic infection of compromised human hosts is an important problem, particularly in view of the fact that, unlike the case for bacterial infections, only a few adequate chemotherapeutic treatments have been developed to date. Furthermore, the mechanism(s) of the immune responses to these mycoses has not been sufficiently elucidated to make successful immunotherapy possible. The lack of appropriate animal models contributes to the difficulties in studying the host defense mechanism to these mycoses.

---
* To whom correspondence should be addressed.
© 1982 Gustav Fischer New York, Inc.
Proceedings of the Third International Workshop on Nude Mice.

Nude (nu/nu) mice proved to be an excellent animal model for studying thymus-dependent immunity against various bacterial infections, particularly those caused by intracellular bacteria. At the Second International Workshop on Nude Mice in Tokyo in 1976, some discussions were created on whether nude mice were more or less resistant to fungal infection than normal mice. Since then, Cutler (6) and Rogers et al (10) reported the enhanced resistance of nu/nu mice to experimental candidiasis infection, whereas Corbel and Eades (5) detected no difference in the resistance of nu/nu and $nu/+$ mice to mucormycosis. Nn/nu mice were found to be more sensitive than $nu/+$ mice to experimental murine coccidioidomycosis (3), to histoplasmosis (18), and to *Cryptococcus neoformans* infection (4,7).

In our laboratory, experimental murine mycoses have long been studied with special reference to drug evaluation, and the natural defense mechanism against *Candida albicans* was also investigated in vitro (2,17). In addition, we studied the natural resistance and immune response of nu/nu mice to pathogenic fungi. Experimental candidiasis (1,8,14), sporotrichosis (15,16), and aspergillosis (12) have been studied to date and our results are summarized in the present report.

Materials and Methods

Animals. Five-week-old, male nude mice of BALB/c background, their normal littermates, and in some experiments BALB/c mice and DDY mice were used. Each group consisted of 10 mice.

Test fungi. *Candida albicans* 7N, grown on Sabounaud's dextrose agar for 18 h or *Sporothrix schenckii* SP17, grown on brain heart infusion agar for 72 h, were suspended in saline and used for the inoculations. Spores of *Aspergillus fumigatus* K-2 grown on potato dextrose agar for 1 week were suspended in M/15 phosphate buffer (pH 7.4) containing 0.01% Tween 80 and used for infection.

Enumeration of viable cells in organs. The animals were sacrificed by cervical dislocation and the pertinent organs were removed, weighed, and homogenized separately with 10 ml of sterile saline. In *C. albicans* and *S. schenckii* studies, each homogenate was diluted with sterile saline and the number of viable fungal cells was determined by the plate dilution method on Sabouraud's dextrose agar containing 0.002% chloramphenicol. In *C. albicans* infection, determination of colony-forming units is, as discussed below, only arbitrary and far from correct reflection of real counts of viable cells. The units are expressed as the number of organisms per gram of wet weight. In the case of *A. fumigatus*, which predominantly grows in mycelial form in tissue, quantitative enumeration was impossible. Three plates each of 1:10 and 1:100 agar dilutions of the organ or lymph node homogenates were prepared. Two pluses denote that fungal growth was observed in all three plates of the 1:100 dilution, one plus that fungal growth was observed in all three plates of the 1:10 dilution, and plus and minus means that growth took place in only one or two plates of the 1:10 dilution.

Spleen cell transfer. Spleens from immunized or nonimmunized mice were minced and gently squeezed between two glass slides in Hank's balanced salt solution (HBSS). The spleen cell suspensions were filtered through sterile gauze, the cell concentration was adjusted to 2.5×10^8 cells/ml, and 0.2 ml of this suspension was injected intravenously 1 day before infection.

Histology. Organs were fixed in 10% buffered neutral formalin and embedded in paraffin. Histological sections were stained with hematoxylin and eosin (H&E) or periodic acid-Schiff (PAS).

Chemicals. OK-432 (Picibanil: streptococcal preparation), which has been developed as an anticancer agent (9) and is known to be an immunopotentiator and macrophage activator (11) was furnished by Chugai Pharmaceutical Co., Tokyo. One KE of the agent corresponds to 0.1 ml dry weight of penicillin-treated *Streptococcus pyogenes*. Carrageenan was purchased from Sigma Chemical Co., USA, and autoclaved.

Intracellular killing of *S. schenckii* by peritoneal macrophages. OK-432-treated, immunized, and normal BALB/c mice were sacrificed by bleeding. Ten milliliters of HBSS containing 100 units of heparin were injected into the peritoneal cavity of these mice and the peritoneal fluids of each group were pooled. After washing the peritoneal cells with HBSS, cells were suspended in RPMI 1640 medium containing 20% calf serum (CS); 5×10^6 cells were inoculated into a petri dish containing a cover slip and incubated for 3 h at 37°C with 5% CO_2 in air. Nonadherent cells were removed with RPMI 1640 medium containing 20% CS. Then 5×10^6 *S. schenckii* cells were added to each dish and after 1 h incubation at 37°C with 5% CO_2 in air. The nonadherent cells were removed with RPMI 1640 medium. RPMI 1640 was added to each dish and incubation continued for an additional 4 h. The number of viable fungal cells was determined by the agar plate dilution method after the macrophages were disrupted with sterile distilled water.

Results

Experimental Candidiasis

Figure 3-1 shows the typical growth of *C. albicans* in mouse kidney. The fungus grows primarily in mycelial form. Therefore, it is important to note that, in a strict sense, the colony-forming units in the tissue homogenate by no means equal the proliferation of viable cells.

Different mouse strains have different resistance to *C. albicans* infection. The age-related sensitivity of DDY mice is shown in Figure 3-2. When 1×10^5 cells were inoculated intravenously, the mice died between 7 and 14 days later and, although there was little age-related difference in the mean survival periods, 4-week-old mice showed a reproducibly higher resistance. Therefore, 5-week-old mice were used in the experimental infections.

Nu/nu mice and their *nu/+* littermates received 1×10^5 cells intravenously or 1×10^8 cells intraperitoneally. In both types of infection, *nu/nu* mice exhibited shorter survival periods (Figure 3-3). This difference was more pronounced following intraperitoneal infection; even when massive doses of *C. albicans* were injected into *nu/+* mice few of the animals died, whereas most of the *nu/nu* mice died within 25 days. On the other hand, no difference in

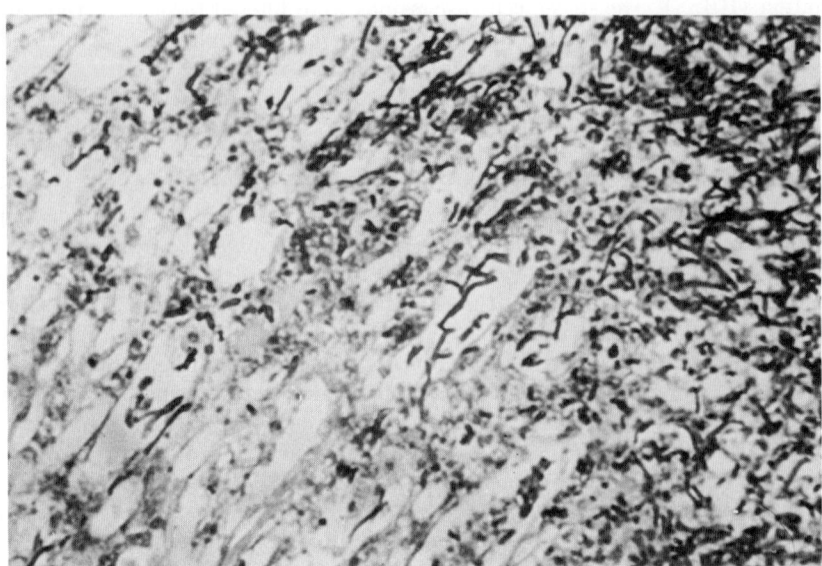

Figure 3-1. Fungal elements in the kidney of a mouse infected with *C. albicans* (1×10^5 cells, IV).

survival was noted when inocula as large as 3×10^6 were injected intravenously (Figure 3-4). These results suggest that within a certain inoculum range, the resistance of *nu/nu* mice to *C. albicans* is impaired. On the other hand, following intravenous infection wtih 1×10^4 cells, there was no significant difference in the growth of *C. albicans* in the organs of *nu/nu* and *nu/+* mice as far as colony-forming unit is concerned (Table 3-1).

To study the effect of OK-432 on the resistance to *C. albicans*, BALB/c and *nu/nu* mice received a daily intraperitoneal dose of 0.5 KE for 4 consecutive days. Two days after the last injection, each animal was intravenously infected with 5×10^4 cells of *C. albicans*. The effect of this treatment is shown in Tables 3-2 and 3-3. Even taking the ambiguity of this enumeration into consideration, the reduction of fungal growth in OK-432-treated mice was remarkable in both cases.

Table 3-1. Colony-forming units in organs of *nu/nu* and *nu/+* mice infected with *C. albicans*

Days after infection	Colony-forming units per organ					
	Liver		Kidney		Spleen	
	nu/nu	*nu/+*	*nu/nu*	*nu/+*	*nu/nu*	*nu/+*
1	200	80	1,900	1,300	150	100
3	60	50	5,000	2,000	100	50
5	40	30	9,000	6,000	0	0
7	10	20	7,000	17,000	0	0
9	5	0	18,000	25,000	0	0

Table 3-2. Effect of OK-432 treatment on the growth of *C. albicans* in BALB/c mice

Mice	Colony-forming units[a] per gram		
	Liver	Kidney	Spleen
OK-432 treated	70	6,000	30
Control	420	81,000	350

[a] Three days after infection.

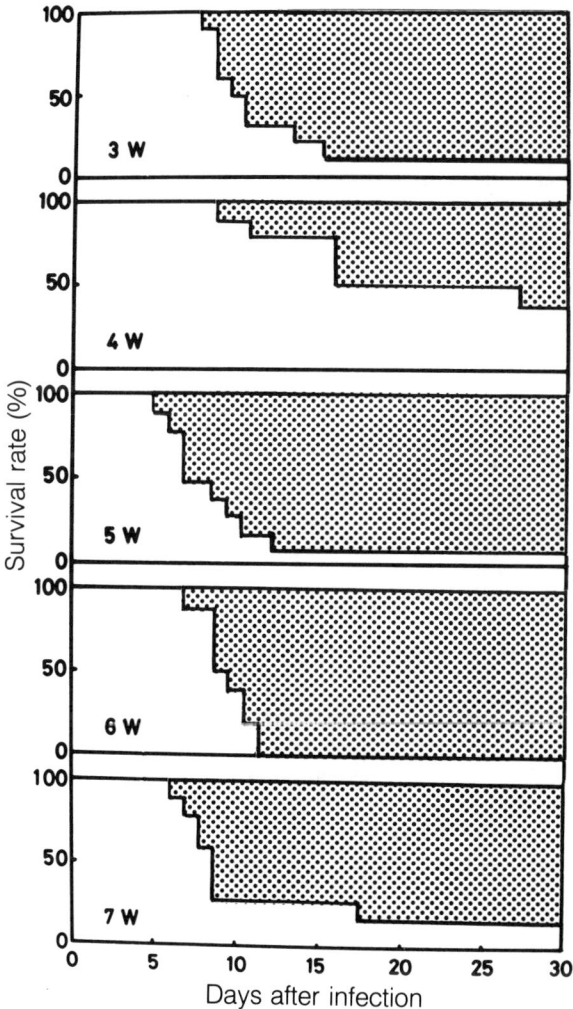

Figure 3-2. Age-related sensitivity of DDY mice to *C. albicans* infection (1×10^5 cells, IV).

Table 3-3. Effect of OK-432 treatment on the growth of *C. albicans* in nude mice

Days after infection	Mice	Colony-forming units per gram	
		Liver	Kidney
4	OK-432 treated	90	6,800
	Control	840	21,000
7	OK-432 treated	30	6,800
	Control	60	40,700

Experimental A. fumigatus *Infection*

The growth of *A. fumigatus* K-2, which forms characteristic microabscesses in mouse brain, can be quantitated histopathologically. Within the first month of intravenous infection with 1×10^6 spores of the K-2 strain, 2 of 10 *nu/nu* and 1 of 10 *nu/+* mice died, indicating that there was no difference in resistance. Furthermore, there was no quantitative or qualitative difference in fungal growth in the mouse brains. Figure 3-5 shows the progression of microabscess formation. At 3 days after infection, the enlargement of microabscesses was evident and at day 7, mononuclear cells were seen to surround aggregated neutrophils. There was no difference between *nu/nu* and *nu/+* mice; in both groups neutrophil infiltration reached its maximum 2–3 days after infection and the number

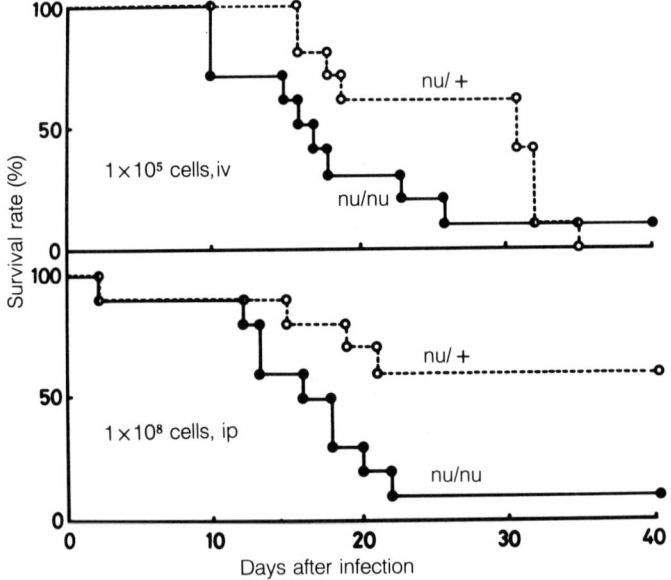

Figure 3-3. Survival of *nu/nu* and *nu/+* mice infected with *C. albicans* (1×10^5, i.v.; 1×10^8, IP).

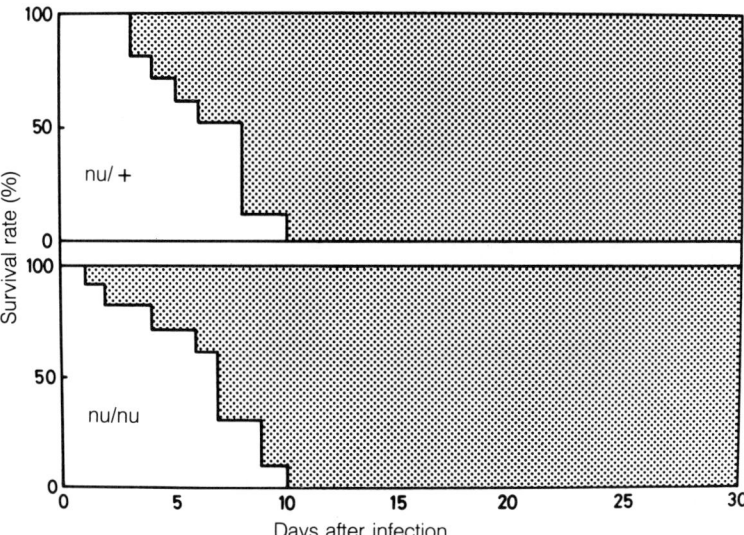

Figure 3-4. Survival of $nu/+$ and nu/nu mice IV infected with 3×10^6 cells of *C. albicans*.

of mononuclear cells peaked 5–7 days after infection. Thereafter, the microabscesses began to disappear and by day 20 they had healed completely (Table 3-4).

Qualitative examination of fungal growth in various organs of nu/nu and $nu/+$ mice also revealed no difference between the two groups (Table 3-5).

To determine the effect of carrageenan in *A. fumigatus* infection, BALB/c mice were intraperitoneally injected with 500μg of the agent 24 h prior to infection with 1×10^6 spores. All carrageenan-treated mice died within 14 days (Figure 3-6).

Table 3-4. Histopathological findings in brains from nu/nu and $nu/+$ mice infected with *A. fumigatus*[a]

Days after infection	Presence of mononuclear cells[a]		Infiltration of neutrophils[a]		Presence of hyphae	
	nu/nu	$nu/+$	nu/nu	$nu/+$	nu/nu	$nu/+$
1	+	−	−	−	+	−
2	+	+	−	−	−	+
3	+++	+++	+	±	+	+
4	+++	+++	++	+	+	−
5	++	++	++	+++	−	−
7	+	+	+++	+++	−	−
20	−	−	−	−	−	−

[a] Abundance of neutrophils or mononuclear cells: none (−); few (+); many (+++).

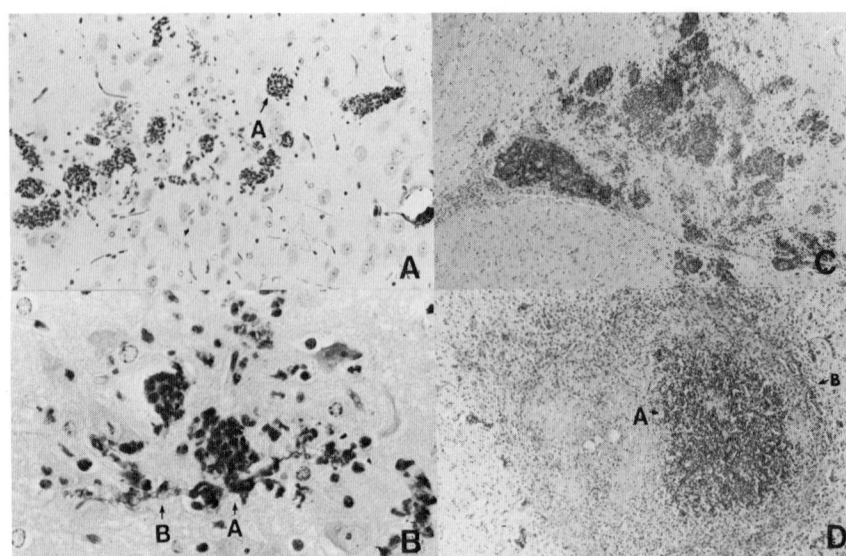

Figure 3-5. Histopathology of brains from *A. fumigatus*-infected mice (1×10^6 spores, IV). **a** H&E-stained brain section of a *nu/nu* mouse 1 day after infection. The formation of microabscesses (A) is noted. ×128. **b** The same section as in **a** stained by PAS. Note formation of microabscesses (A) around fungal hyphae (B). ×320. **c** Brain of a *nu/nu* mouse 3 days after infection. Note microabscess enlargement. PAS, ×50. **d** Brain of a *nu/+* mouse 7 days after infection. Aggregated PMNs (A) are surrounded by mononuclear cells (B). PAS, ×50.

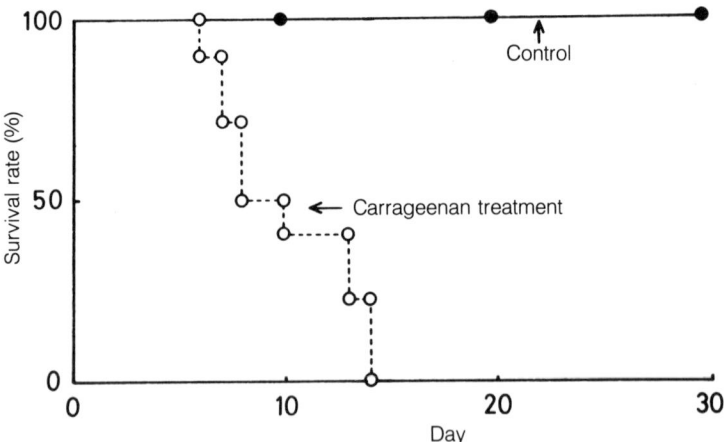

Figure 3-6. Effect of carrageenan treatment (500 µg, IP) in *A. fumigatus*-infected BALB/c mice (1×10^6 spores 24 h after carrageenan administration).

Table 3-5. Recovery of *A. fumigatus* from organs from nu/nu and $nu/+$ mice[a]

Days after infection	Brain nu/nu	$nu/+$	Lung nu/nu	$nu/+$	Liver nu/nu	$nu/+$	Kidney nu/nu	$nu/+$	Spleen nu/nu	$nu/+$
1	−	−	−	−	+	+	−	±	±	+
3	+	+	−	−	+	±	±	±	±	±
5	++	++	±	−	−	−	+	+	±	−
7	+	±	−	−	−	−	+	±	−	±
9	±	±	−	−	±	±	++	±	±	±
12	±	−	N.d.	N.d.	±	−	±	±	±	−

[a] Symbols as in Table 3-4; n.d., Not determined.

Experimental S. schenckii *Infection*

Intravenous or intratesticular inoculation with 1×10^6 or 1×10^5 *S. schenckii* cells, respectively, resulted in progressive chronic infection in both nu/nu and $nu/+$ mice. However, the animals usually survived for approximately 50 days and the survival of nu/nu and $nu/+$ mice was not compared. Fungal growth was determined by the number of colony-forming units in the liver and spleen. As shown in Figure 3-7, colony-forming units in intravenously infected nu/nu mice increased with time, where in $nu/+$ mice a decrease in their number began on the twelfth postinfection day. By day 22, the fungal cells in the liver and spleen of $nu/+$ mice had almost completely disappeared. When *S. schenckii* was infected intratestically, the fungus proved to be more virulent. Although it was not completely eliminated from the organs of $nu/+$ mice, its growth in the liver, kidney, spleen, lymph nodes, and testis of nu/nu mice was markedly higher than in the same organs of $nu/+$ mice (Table 3-6). BALB/c mice were immunized with an intravenous injection of 5×10^5 viable cells of *S. schenckii*, intravenously challenged 18 days later with 1×10^6 cells, and sacrificed 10 days after the challenge. Colony-forming units in the liver and spleen were counted and compared with the number of colony-forming units in nonimmunized mice infected with an identical dose. As shown in Figure 3-8, fungal growth in both organs was markedly lower in the immunized mice.

Table 3-6. Growth of *S. schenckii* in organs of nu/nu and $nu/+$ mice

| Days after infection | Mice | Colony-forming units per gram | | | | | DTH[a] |
		Liver	Kidney	Spleen	Lymph nodes	Testis	
16	nu/nu	70	0	260	250	375,000	−
	$nu/+$	0	0	90	0	160,000	+
29	nu/nu	445	29,000	6,800	710	16,700,000	−
	$nu/+$	45	820	2,300	500	790,000	+
50	nu/nu	13,500	35,200	52,400	88,000	248,000,000	−
	$nu/+$	1,370	6,290	7,900	17,000	15,300,000	+

[a] DTH, delayed hypersensitivity to *S. schenckii* antigen.

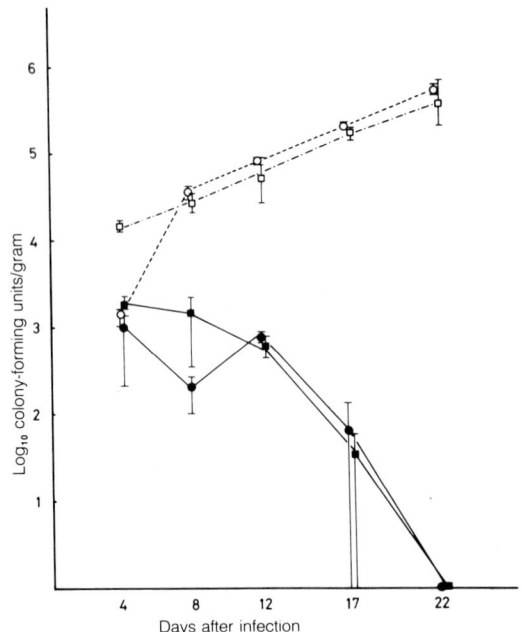

Figure 3-7. Viable S. schenckii organisms in the livers of nu/nu (--O--) and nu/+ mice (—●—), and in the spleens of nu/nu (--□--) and nu/+ mice (—■—), after intravenous infection with 1×10^6 yeast cells.

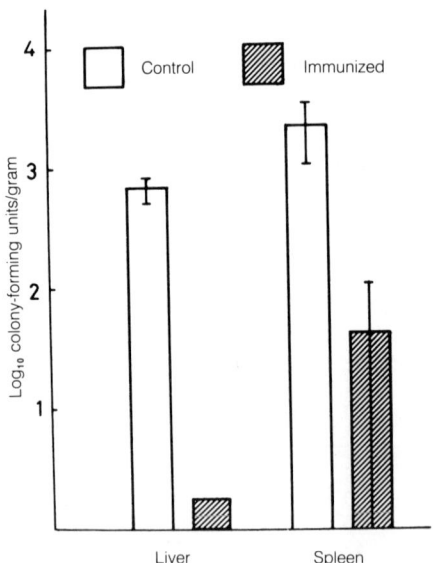

Figure 3-8. Effect of immunization on the growth of S. schenckii in the liver and spleen of BALB/c mice. Mice were immunized with 5×10^5 yeast cells 18 days before intravenous infection with 1×10^6 yeast cells.

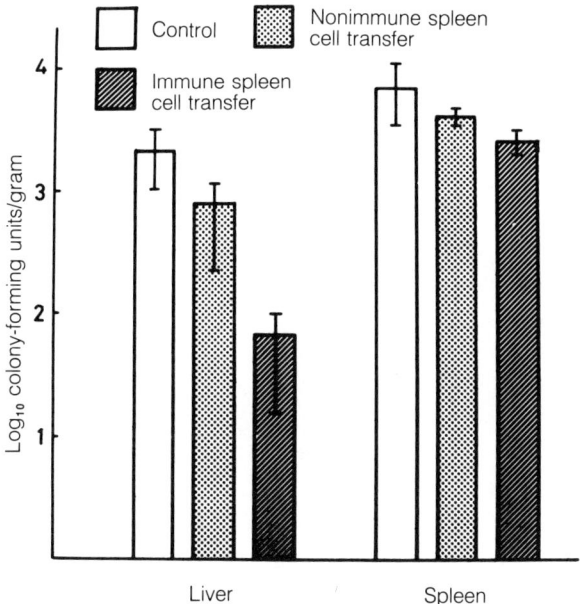

Figure 3-9. Effect of immune and nonimmune spleen cells on the growth of *S. schenckii* in the liver and spleen of *nu/nu* mice. Spleen cells (5×10^7) were transferred 1 day before intravenous infection (1×10^6 yeast cells).

Immune or nonimmune spleen cells (5×10^7) from BALB/c mice were then transferred to *nu/nu* mice which were intravenously challenged on the following day with 1×10^6 cells of *S. schenckii*. Whereas both enhanced the resistance to the infection, the effect of nonimmune cells was weaker than that of immune cells (Figure 3-9).

BALB/c mice that had received a daily intraperitoneal dose of 0.5 KE OK-432 for 4 consecutive days were intravenously challenged with 1×10^6 cells of *S. schenckii*. OK-432 treatment enhanced the resistance to the infection (Figure 3-10). Another in vitro experiment revealed that whereas macrophages from normal mice appeared to be unable to kill this fungus during 4 h of incubation, those from immunized and OK-432-treated mice showed enhanced intracellular killing (Table 3-7).

Table 3-7. Intracellular killing of *S. schenckii* by peritoneal macrophages from BALB/c mice

Origin of macrophages	Colony-forming units	
	0 time	4 hours
Normal mice	8.8×10^5	7.3×10^5
Immunized mice	7.0×10^5	3.0×10^5
OK-432-treated mice	8.3×10^5	1.9×10^5

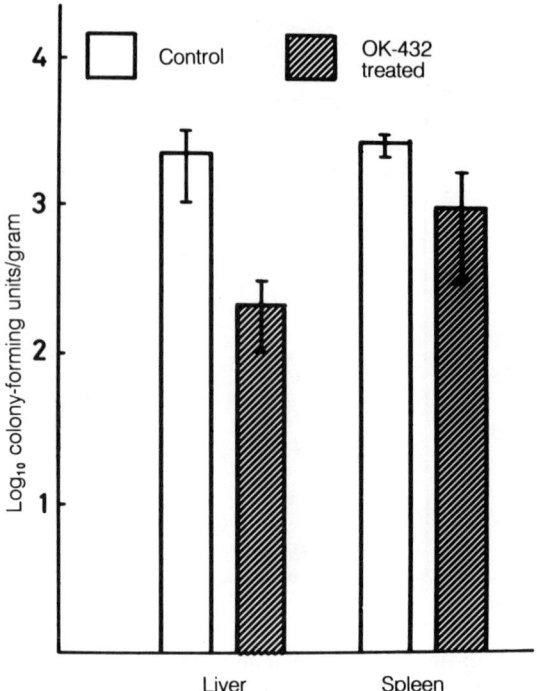

Figure 3-10. Effect of OK-432 treatment on the growth of *S. schenckii* in the liver and spleen of BALB/c mice. OK-432 (0.5 KE) was intraperitoneally administered for 4 consecutive days, and 2 days after the last administration, the mice were intravenously infected with 1×10^6 yeast cells.

Discussion

Candida albicans and *A. fumigatus* are opportunistic fungi; because it is relatively difficult to establish these infections in mice, highly virulent strains must be selected and usually large inocula are needed to infect or kill the mice. Although in vitro *C. albicans* grows in yeast form, in tissue it grows mostly in mycelial form. This hampers the enumeration of its growth in tissue. Therefore, determination of fungal growth and the resistance to this fungus is usually based on survival rate and period.

Our experiments with *C. albicans* infection suggest that at moderate inoculum sizes, T cells play a role in the resistance to this infection. On the other hand, when large doses of the inoculum were introduced intravenously, the animals succumbed, apparently because of candidial septicemia, a short time after infection. As these animals had not had sufficient time to develop any agent-specific form of immunity, the thymus seemed not to participate in the host defense mechanism.

In experimental murine candidiasis, nonspecific defense mechanisms, such as phagocytosis, as manifested by our OK-432 experiments, and fungicidal serum factors, such as transferrin (13), may play a role in the host defense against this infection. Strain virulence, infection route, inoculum size, and parameters used in determining resistance greatly affect the establishment, progression, and evaluation of experimental murine candidiasis.

Regarding the resistance to *A. fumigatus* infection, the experiments reported here detected no difference between nu/nu and $nu/+$ mice. In human chronic mucocutaneous candidiasis, cellular immunity is known to be impaired greatly and in the lesions of chronic lung aspergillosis, cell infiltration occurs. Further work is required, therefore, before human immune response against these two major opportunistic fungi can be related to the experimental murine infection.

On the other hand, the results of experimental murine infections with such pathogenic fungi as *Coccidioides immitis*, *Histoplasma capsulatum*, *Cryptococcus neoformans*, and *Sporothrix schenckii*, appear unequivocal in indicating T-cell function as one of the essential mechanisms in the most defense against these infections. Earlier studies in our laboratory had revealed no remarkable intracellular killing of *C. albicans* and *A. fumigatus* even in immune macrophages. On the other hand, the killing of *S. schenckii* was marked and our findings agreed with those of Yupin (19), who reported hamster immune macrophages to effectively kill *S. schenckii*. This may be attributable to macrophage activation by *S. schenckii* immunization. Further investigations on this point are presently under way in our laboratory.

Acknowledgments

We thank Dr. M. Miyaji for supplying *A. fumigatus* K-2 and *S. schenckii* SP17 strains. This work was supported in part by a Grant-in-Aid for Scientific Research from the Japanese Ministry of Education, Culture and Science.

References

1. Arai, T., and Y. Mikami. Cellular and humoral defense mechanisms of *Candida albicans* infection, pp. 191–200. Yeasts and yeast-like microorganisms in medical science, Proceedings of the second international specialized symposium on yeasts. Tokyo: University of Tokyo Press, 1972.
2. Arai, T., Y. Mikami, and K. Yokoyama. Phagocytosis of *Candida albicans* by rabbit alveolar macrophages and guinea pig neutrophils. Sabouraudia 15:171–177, 1977.
3. Beaman, L., D. Paggagianis, and E. Benjamini. The significance of T-cells in resistance to experimental murine coccidioidomycosis. Infect. Immun. 17:580–585, 1977.
4. Cauley, L.K., and J.W. Murphy. Response of congenitally athymic (nude) and phenotypically normal mice to *Cryptococcus neoformans* infection. Infect. Immun. 23:644–651, 1979.
5. Corbel, M.J., and S.M. Eades. Experimental mucormycosis in congenitally athymic "nude" mice. Mycopathologia 62:117–120, 1977.
6. Cutler, J.E. Acute systemic candidiasis in normal and congenitally thymic-deficient (nude) mice. J. Reticuloendothel. Soc. 20:121–124, 1976.
7. Graybill, J.R., P.C. Craven, L.F. Mitchell, and D.J. Drutz. Interaction of chemotherapy and immune defenses in experimental murine cryptococcosis. Antimicrob. Agents Chemother. 14:659–667, 1978.
8. Mikami, Y., K. Yokoyama, and T. Arai. Fungal infection and macrophage. Jpn. J. Med. Mycol. 19:203–207, 1978.

9. Okamoto, H., Y. Minami, S. Shoin, S. Koshimura, and R. Shimizu. Experimental anticancer studies. XXXI. On the streptococcal preparation having potent anticancer activity. Jpn. J. Exptl. Med. 36:175–186, 1966.
10. Rogers, T.J., E. Balish, and D.D. Manning. The role of thymus-dependent cell mediated immunity in resistance to experimental disseminated candidiasis. J. Reticuloendothel. Soc. 20:291–298, 1976.
11. Sakai, S., K. Rhyoyama, S. Koshimura, and S. Migeta. Studies on the properties of a streptococcal preparation OK-432 (NSC-B116209) as an immunopotentiator. I. Activation of serum complement components and peritoneal exudate cells by group A streptocous. Jpn. J. Exptl. Med. 46:123–133, 1976.
12. Shiraishi, A. Studies on the host defense mechanisms against *Aspergillus* infection. Chiba Med. J. 54:297–304, 1978.
13. Shiraishi, A., Y. Mikami, and T. Arai. Anti-candida activity on tranferrin. Jpn. J. Med. Mycol. 18:108–111, 1977.
14. Shiraishi, A., Y. Mikami, and T. Arai. Protective effect of OK-432 (a streptococcal preparation) on experimental candidiasis. Microbiol. Immunol. 23:549–554, 1979.
15. Shiraishi, A., K. Nakagaki, and T. Arai. Experimental sporotrichosis in congenitally athymic (nude) mice. J. Reticulendothel. Soc. 26:333–336, 1979.
16. Shiraishi, A., K. Nakagaki, and T. Arai. Role of cell mediated immunity in the resistance to experimental sporothrichosis in mice. (In preparation.)
17. Utsumi, T. Studies on antifungal principles in normal human serum. A new assay technique and the distribution of anti-candida activity among human sera. Chiba Med. J. 49:105–111, 1973.
18. Williams, D.M., J.R. Graybill, and D.J. Drutz. *Histoplasma capsulatum* infection in nude mice. Infect. Immun. 21:973–977, 1978.
19. Yupin, C. The role of humoral and cellular immune responses in experimental sporotrichosis in the hamster. Dissertation Absr. Intl. B. 37:1553–1554, 1976.

General Discussion

CUTLER: I have two questions. (1) What is the approximate LD_{100} dose of your strain of *C. albicans*? (2) What is the background mouse strain of your nude mice?

ARAI: Our *C. albicans* killed all the mice by intravenous inoculation of 5×10^6 cells within 14 days. Nude mice used were with BALB/c background.

4

T-independent Antibody Production in Nude Mice Immunized with a Rodent Malaria Parasite (Plasmodium berghei)

Seiji Waki* and Mamoru Suzuki

Department of Parasitology, School of Medicine, Gunma University, Maebashi 371, Gunma, Japan.

Abstract

The development of specific antibody production in nude mice subjected to *Plasmodium berghei* (NK65) infection or immunization was studied by indirect fluorescent antibody test to clarify the relationship between production of the immunoglobulins and participation of T cells. Specific IgM and IgG levels in the immunized nude mice were extremely low compared with those seen in thymus-competent normal mice. Reimmunization, however, elicited a striking secondary antibody response in nude mice as well as in normal mice. In addition, sublethally x-irradiated mice given spleen cells from primary immunized nude mice showed high antibody level following immunization. Although IgG1 production is generally considered to be particularly thymus dependent, the antiplasmodial IgG1 subclass was shown in the immunized nude mice. These results suggest that: (a) At least some portion of malarial antigen stimulates B lymphocytes to produce antibody without participation of T lymphocytes; (b) precursor cells were dominantly generated in nude mice by the primary immunization and such latent precursor cells differentiated to antibody-producing cells following secondary immunization; (c) malarial parasite can be assumed to contain a peculiar antigen that stimulates IgG1 production without helper action by T cells.

Introduction

The indirect fluorescent antibody test (IFAT) has been established as the most reliable standard serologic test for malaria since the test was introduced by Tobie and Coatney (23) and Voller and Bray (24). It has been used practically in detailed investigations on the antibody responses of individuals and in epidemiological studies (2,8,22). Recently, Manawadu and Voller (14) have stand-

---

* To whom correspondence should be addressed.
© 1982 Gustav Fischer New York, Inc.
Proceedings of the Third International Workshop on Nude Mice.

ardized the test to evaluate results objectively with the use of a physical system against a fluorescent standard and preparation of biological standards of malarial antisera and fluorescein-labeled conjugates. Hall et al. (11) examined antigen obtained from long-term *in vitro* cultures of *Plasmodium falciparum* and showed that such cultures are a source of stable IFAT antigen and enable widespread use of the test.

Irrespective of such improvement of the technique, little is known about the mechanism of antibody production, and the biological importance of the antibodies has not yet been satisfactorily clarified. The present study was done with the aim of clarifying the mechanism involved in the antimalarial antibody production at cellular level.

Nude mice are useful experimental animals in studies of the participation of T cells in immunologic responses. In this study, we have compared the development of antibodies in nude and normal mice subjected to *Plasmodium berghei* immunization. Antibody production by a T-independent mechanism is described.

Materials and Methods

Mice. Nude mice and phenotypically normal littermates of both sexes, with a BALB/c background, were raised by mating female normal littermates to nude males in our laboratory. The mice were 6–8 weeks old at the start of each experiment. All the animals were maintained in a closed and air-conditioned animal room in a special cabinet which was supplied with filtered air. They were fed a sterilized diet and water *ad libitum* throughout the experiments.

Parasite. The NK65 strain of *Plasmodium berghei* was maintained by serial intraperitoneal (IP) inoculation of infected blood in mice with occasional freezing of infected blood at $-70°C$. Mice were infected by IP inoculation of 10^7 parasite-infected erythrocytes. Percentage of parasitemia was determined by light microscopic examination of Giemsa-stained thin smears of peripheral blood.

Indirect fluorescent antibody test. Serum levels of IgM and IgG were measured by an indirect fluorescent antibody (IFA) technique described previously (27). An incident light illuminating system fluorescence microscope (Olympus, Model MH-RFL-LB-1) was used in every reading in the present study. Measurement of antiplasmodial IgG subclasses was done using a modified IFA technique. Serial dilutions of test sera from mice were placed on the slide antigen spots and incubated. After washing with phosphate buffered saline (PBS), the slide antigens were reacted with rabbit antimouse total IgG, IgG1, IgG2a, or IgG2b. The slide antigens were stained with a goat antirabbit IgG conjugated with fluorescein isothiocyanate (FITC), washed with PBS, and examined using a fluorescence microscope. Rabbit antimouse IgG subclasses sera and goat antirabbit IgG immunoglobulin conjugated with FITC were obtained from Miles Laboratories Inc., Elkhart, Indiana. Neither the controls of test sera, with or without rabbit antimouse IgG subclasses sera, nor test sera, without rabbit antimouse IgG subclasses sera, gave fluorescence on the antigen spots after being subjected to conjugated serum reactions.

Immunization. Test animals were immunized by infection followed by chemotherapy. Mice were inoculated with 10^7 parasitized erythrocytes. Drug treatment (25 mg/kg/day of pyrimethamine suspended in 2% CMC medium) was given on 4 consecutive days starting on day 3 after infection. Control mice were given 10^7 nonparasitized erythrocytes and pyrimethamine.

X-Ray irradiation. Mice were exposed to a single dose of 800 rads of total-body irradiation by a 180 kV Toshiba deep-therapy apparatus under an operation at 25 mA, filtered through 0.5 mm Al + 0.5 mm Cu, which is equivalent to a half-value layer of 1.11 mm Cu, setting a dose rate of 736 R/min (800 rads/min) at a 20 cm distance.

Spleen cell transfer. Spleens were removed from donor mice, cut into appropriate pieces, teased between two frosted glass plates, and filtered through a 200-gage stainless steel mesh. The resulting cell suspension was washed with Eagle's minimal essential medium (MEM) three times, and diluted to 2.5×10^8 cells/ml. x-Irradiated recipient mice were injected intravenously with 0.2 ml of medium containing 5×10^7 viable spleen cells from immunized nude mice (see groups 1 and 2 in Table 4-1), nonimmunized nude mise (group 3), or nonimmunized normal littermates (group 4), or with 0.2 ml of medium containing no cell (group 5). Primed spleen cells were obtained on day 14 after immunization of donor mice.

Results

The Courses of Antibody Production in Mice Inoculated with the Parasite

Two groups of mice consisting of six nude mice and six phenotypically normal littermates were inoculated with 10^7 parasitized erythrocytes. Figure 4-1 shows

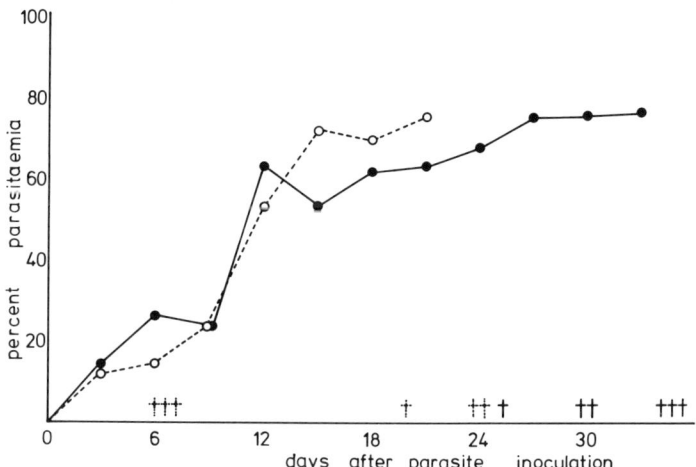

Figure 4-1. Course of parasitaemia in six nude (●) and six phenotypically normal littermates (○) inoculated with 1×10^7 parasitized erythrocytes (*Plasmodium berghei* NK65). Death of nude mice, †; of normal littermates, †.

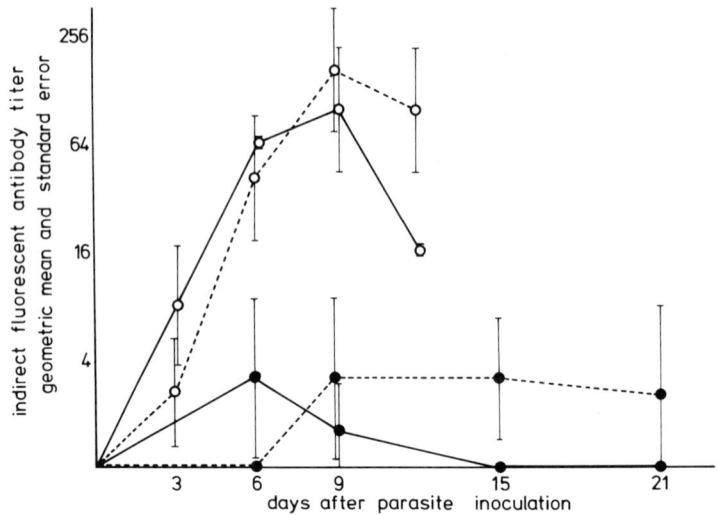

Figure 4-2. Time course of antibody production of six nude (●) and six phenotypically normal littermates (○) inoculated with 1×10^7 parasitized erythrocytes (*Plasmodium berghei* NK65). Solid line, IgM; dashed line, IgG.

the development of parasitemia and survival time of the mice. The time courses of both IgM and IgG classes of fluorescent antibodies in the mice are shown in Figure 4-2. Both IgM and IgG classes of the specific immunoglobulin in thymus-competent normal littermates infected with the parasite reached to maximum titers by the ninth day after inoculation. In nude mice, on the other hand, antiplasmodial antibody production was remarkably low. Both IgG and IgM titers persisted under 1:4 throughout the observation period.

Primary and Secondary Antibody Responses of Nude Mice and Phenotypically Normal Littermates Immunized with the Parasite

Five nude mice and five normal littermates were immunized with 10^7 parasites and specific IgM and IgG levels were examined once a week for 10 weeks. At the seventh week after primary immunization, mice were reimmunized by the same method. As shown in Figure 4-3, specific antiplasmodial antibodies were present in athymic nude mice following primary immunization with the parasite, although titers in nude mice were much lower and more sluggish in development than that seen in thymus-competent normal littermates. However, striking secondary responses were observed in nude mice which were reimmunized with 10^7 parasites 7 weeks after primary immunization. The levels of both IgM and IgG were elevated rapidly as early as the first week after reimmunization. The maximum IgG titer of the secondary response in nude mice was about the same as in normal littermates.

A high level of IgG production in the secondary response without participation of thymus cells was also demonstrated in x-irradiated mice that were transferred with spleen cells from primed nude mice. Spleen cells from immunized nude

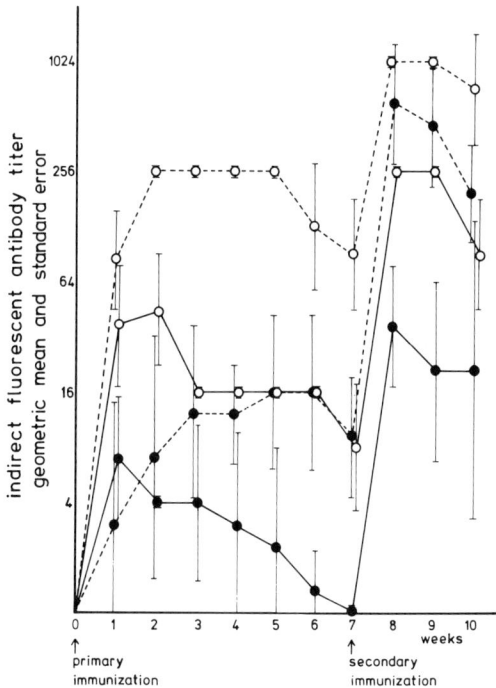

Figure 4-3. Primary and secondary antibody production in five nude mice (○) and five phenotypically normal littermates (●). Note the boosted IgM and IgG response shown by nude mice. Solid line, IgM; dashed line, IgG.

mice (groups 1 and 2), nonimmunized nude mice (group 3), and nonimmunized normal littermates (group 4), or Eagle's culture medium without cells (group 5) were inoculated into sublethally x-irradiated mice. Each group of the recipient mice, except for group 2, was subjected to the immunization (see Materials and Methods) on the same day, and on day 7 after immunization the mice were exsanguinated and antiplasmodial IgG titer in the obtained sera was determined by the indirect fluorescent antibody technique. The mice given spleen cells from primed nude mice demonstrated a high level of specific antibody titers in comparison with the titers shown by the other groups of mice (Table 4-1).

Development of Antiplasmodial IgG Subclasses in Immunized Mice

Titers of antiplasmodial IgG subclasses of the sera from mice on the twenty-eighth day after primary immunization and on the fourteenth day after secondary immunization (Figure 4-3) were examined by the indirect fluorescent antibody technique using specific antimouse IgG subclass sera. It has been generally considered that particularly IgG1 immunoglobulin is produced with the participation of helper action by thymus-derived cells. However, as is shown in Table 4-2, development of antiplasmodial IgG1 not only was demonstrated in immunized normal littermates but also was detected in immunized nude mice.

Table 4-1. Secondary IgG responses following parasite immunization in cell transferred animals[a]

Group of mice	Transferred cell from	Immunization after cell transfer	Mean IFA log$_2$ titer
1	Immunized nude mice	+	6.0
2	Immunized nude mice	−	1.7
3	Non-immunized nude mice	+	—
4	Non-immunized normal littermates	+	3.4
5	—	+	—

[a] All groups of recipient mice were subjected to sublethal x-irradiation. The mice in groups 1–4 were inoculated with 5×10^7 spleen cells from immunized or nonimmunized mice, respectively. Mice of group 5 were inoculated with a medium without suspending cells.

The Relationship between Antiplasmodial IFA Levels and Protective Immunity in Nude Mice and Normal Littermates

Waki (26) reported that mice which were immunized by repeated infection mostly survived the challenge inoculation with the same parasite strain. In the present study, the protective immunity shown by nude mice and by normal littermates, both of which were subjected to the same immunization, were examined by challenge inoculation with 10^6 parasitized erythrocytes 2 weeks after the final immunization. All the immunized nude mice did not resist challenge

Table 4-2. Antiplasmodial IgG subclasses produced in nude mice and phenotypically normal littermates detected by indirect fluorescent antibody technique

Mouse genotype	Sera	IgG subclass	log$_2$ titer
nu/nu	Primary[a]	Total IgG	4
		IgG1	3
		IgG2a	3
		IgG2b	3
	Secondary[b]	Total IgG	9
		IgG1	8
		IgG2a	8
		IgG2b	8
nu/+	Primary[a]	Total IgG	9
		IgG1	8
		IgG2a	7
		IgG2b	7
	Secondary[b]	Total IgG	12
		IgG1	11
		IgG2a	11
		IgG2b	10

[a] 4 weeks after primary immunization.
[b] 2 weeks after secondary immunization.

Table 4-3. Antiplasmodial antibody levels and protective immunity in nude (nu/nu) and phenotypically normal littermates ($nu/+$) vaccinated with viable *Plasmodium berghei* (NK65) parasites

Group	Mouse genotype	No. of mice	Treatment	IFA log 4 IgG titer	Protective immunity[a]
1	nu/nu	6	Vaccinated	4.17 ± 0.40 (N.S.)[b]	6/6
2	$nu/+$	6	Vaccinated	4.83 ± 0.40	0/6
3	nu/nu	4	—	—	4/4
4	$nu/+$	4	—	—	4/4

[a] Protective immunity; Mortality of the mice following 10^6 parasites challenge inoculation done 2 weeks after immunization.

[b] N.S., not significantly different from vaccinated thymus competent mice, $P < 0.05$ (Student's t test).

inoculation, irrespective of a high level of antibody development in the sera (Table 4-3). On the other hand, all the normal mice thus immunized survived. The antibody levels shown by both groups of mice were not significantly different ($P < 0.05$, Student's t test).

Discussion

In the present study, we investigated the mechanism involved in antibody production in athymic nude mice and thymus-competent normal littermates immunized with *P. berghei* parasite. In the primary immunization, specific IgM and IgG titrated in the nude mice persisted at extremely low levels compared with those produced in normal littermates. Hence, antibody production following primary immunization was considered to be apparently thymus dependent. Jayawardena et al. (12) observed that T-cell-deprived mice (thymectomized, x-irradiated, and reconstituted with bone marrow cells) produced low levels of IgM compared with the normal mice during the course of *P. yoelli* infection. Both results suggest that antibodies against the parasite were produced by the participation of T cells.

However, other results also were seen in the present experiment. The antibody levels shown in the athymic nude mice and in thymus-competent normal littermates were similar when the mice were subjected to secondary immunization by the parasite (Figure 4-3). In addition, spleen cells from primed nude donor mice did elicit antibody production after transfer to irradiated mice which were immunized after cell transfer (Table 4-1). The findings suggest that at least some malaria antigen directly stimulated B lymphocytes and induced antibody production without participation of T lymphocytes. It has been documented that some bacterial or viral antigens can induce antibody production without helper action of T cells. Manning et al. (15) and Burns et al. (5) demonstrated that athymic nude mice inoculated with such antigens produced specific antibody at equivalent level to those shown by the normal littermates. Von Eschen and Rudbach (25) showed that *Escherichia coli* lipopolysaccharide (LPS), which has been generally admitted as a T-independent antigen, induced not only primary antibody response but also secondary antibody production in mice. Thus, some thymus-independent antigens work directly on specific clones of B lymphocytes, which, as a consequence, multiply and differentiate into antibody-secreting cells.

It would be reasonable to speculate about two steps of the antibody-producing procedure in the present or similar T-independent antibody-forming system. First, in primary immunization, the antigen directly stimulates the clone of B lymphocyte and drives the cells to multiply. In the second stage, the increased clone of B lymphocytes is switched to differentiate into antibody-secreting cells by the direct exposure to the same antigen without helper action of T lymphocytes.

The features of thymus-independent antigens are not universal. With thymus-independent antigen found in *Brucella abortus*, antigenic stimulus to multiplication stage was more effective than that to differentiation stage; consequently, a large population of memory cells was prepared which was shown by increased secondary antibody response. Another T-independent antigen, levan, works as a stimulator to the differentiation stage, so that a very high proportion of the progeny of the clone of cells undergoes differentiation (1).

Nonspecific mitogenic action of several thymus-independent antigens (LPS: lipopolysaccharide; SIII: type III pneumococcal polysaccharide; POL: polymerized flagellin; levan; PVP: polyvinlypyrrolidone) has been shown by Coutinho and Möller (6), and Jacobs and Morrison (13). Greenwood and Vick (10) demonstrated that a B-cell mitogen was associated with plasmodial parasite, and culture supernatants of *P. falciparum*-infected blood were found to stimulate DNA synthesis in human peripheral blood lymphocytes and mouse spleen cells in vitro. Freeman and Parish (9) showed that the treatment of mice with the mitogenic substance extracted from the parasitized erythrocytes resulted in polyclonal B-cell activation and nonspecific IgM synthesis. These findings support the existence of T-independent antigen involved in *P. berghei* as suggested in the present study.

It has been reported that T-independent antigens predominantly induce production of IgM class antibodies although a few exceptions were also reported (3,17). IgG responses, on the other hand, are considered to be particularly thymus dependent (7). However, one can find several experimental results alternative to the theories. Barthold et al. (4) reported that significant numbers of plaque-forming cells secreting IgG1, IgG2, and IgA were detected in mice immunized with a T-independent antigen which was a type III pneumococcal polysaccharide. In addition to this finding, they also demonstrated that spleen cells of athymic nude mice immunized with the same antigen manifested respective antibody secretion. Stimulation of nude mice with some other antigens, such as DNP-Lys-Ficoll or DNP-AE-dextran, induced not only IgM but also IgG responses (20,21). These findings are comparable to the IgG response in immunized nude mice shown in the present study. It seems, therefore, that IgG production does not always require participation of helper T lymphocytes.

However, one should keep in mind that it has been demonstrated that the spleen of nude mice contains 3–20% of theta-positive cells, which may be switched to differentiate into mature T lymphocytes not only by the action of thymus factor but also by the stimulus of some substances such as polyadenylic–polyuridilyc acid, single-standard DNA, and lipopolysaccharide (16). Other examples of substances inducing thymic activities are the sulfur derivatives levamisole and sodium diethyldithiocarbamate (19). The possibility should be kept in mind that in the present study the malaria parasite contained some similar component, which exerted differentiation of thymus precursor in nude mice and, subsequently, the differentiated T cells helped B cells to secrete IgG. However, at this moment, substantiation of this theory has not been presented in the field of malaria immunology.

In conclusion, antigens involved in some plasmodia share certain common characteristics with many T-independent antigens, and we have demonstrated some results suggesting that some portion of parasite substance may function as an immunogen that evokes antibody production by T-independent mechanism. Burns et al. (5) suggested that the carbohydrate portion of glycoproteins of many viruses might work as thymus-independent immunogens. Such biochemical characterization of the malarial antigen will allow a clear correlation between parasite substances and host reactions at a cellular level.

Many malaria immunologists dream now of developing a safe and effective vaccine method. Purification and characterization of the essential antigen which confers protective immunity in the host should be given top priority in this study. Although we have not yet clarified the function of the proposed T-independent antigen in the parasite, future studies will be focused on investigating the biochemical features of the antigen and the corresponding host responses, especially in relation to protective immunity.

In a previous publication the present authors documented that a high indirect fluorescent antibody titer in the immunized mice did not always reflect protective immunity produced in the mice (27). The indispensable role of thymus or T cells in the establishment of stable immunity in mice was demonstrated in the subsequent experiments employing nude and phenotypically normal littermates (28). In the present study, it was shown that specific IgG titer measured by indirect fluorescent antibody technique did not manifest the protective function provided by T cells (Table 4-3). At this moment, therefore in principle, no immunologic method is available to measure protective in vitro, although Nussenzweig (18) proposed a hopeful method of applying sporozoite antigen in indirect fluorescent antibody tests. In the future, if some parasitic antigen component should be isolated which were to work specifically on T-cell-dependent protection, *in vitro* measurement of protective immunity might be accomplished, and, of course, such antigen would be raised as a hopeful purified vaccine candidate.

Acknowledgments

We are grateful to Miss Ayako Kuroda and Miss Masae Kojima for their technical assistance. The authors also wish to thank Dr. I. Yonome for operating the x-ray instrument. This study received financial support from the World Health Organization (M2/181/145) and from a Scientific Research Grant of the Japan Ministry of Education (057515, 137024).

References

1. Allison, A.C. Interactions of T and B lymphocytes in self-tolerance and autoimmunity, pp. 25–180. *In* D.H. Katz and B. Benacerraf (eds.), Immunological tolerance: Mechanisms and potential therapeutic applications. New York: Academic Press, 1974.
2. Ambroise-Thomas, P. The immunofluorescence reaction in the sero-immunology study of malaria. Bull. WHO 50:267–276, 1974.
3. Baker, P.J., and P.W. Stashak. Quantitative and qualitative studies on the primary antibody response to preumococcal polysaccharides at the cellular level. J. Immunol. 103: 1342–1348, 1969.
4. Barthold, D.R., B. Prescott, P.W. Stashak, D.F. Amsbaugh, and P.J. Baker. Regulation of the antibody response to type III premucoccal polysaccharide III. Role of regulatory T cells in the development of an IgG and IgA antibody response. J. Immunol. 112: 1042–1050, 1974.
5. Burns, W.H., L.C. Billups, and A.L. Notkins. Thymus dependence of viral antigens. Nature (London) 256:654–656, 1975.
6. Coutinho, A., and G. Möller. B cell mitogenic properties of thymus-independent antigens. Nature (London) 245:12–14, 1973.
7. Coutinho, A., and G. Möller. Thymus-independent B-cell induction and paralysis. Adv. Immunol. 21:113–236, 1975.
8. Draper, C.C., A. Voller, and R.G. Carpenter. The epidemiological interpretation of serologic data in malaria. Am. J. Trop. Med. Hyg. 21:696–703, 1972.
9. Freeman, R. R., and C.R. Parish. Polyclonal B-cell activation during rodent malaria infections. Clin. Exptl. Immunol. 32:41–45, 1978.
10. Greenwood, B. M., and R.M. Vick. Evidence for a malaria mitogen in human malaria. Nature (London) 257:592–594, 1975.
11. Hall, C.L., J.D. Haynes, J.D. Chulay, and C.L. Diggs. Cultured *Plasmodium falciparum* used as antigen in a malaria indirect fluorescent antibody test. Am. J. Trop. Med. Hyg. 27:849–852, 1978.
12. Jayawardena, A.N., G.A.T. Targett, R.L. Carter, E. Leuchars, and A.J.S. Davies. The immunological response of CBA mice to *P. yoelii* I. General characteristics, the effects of T-cell deprivation and reconstitution with thymus grafts. Immunology 32:849–859, 1977.
13. Jacobs, D.M., and D.C. Morrison. Dissociation between mitogenicity and immunogenicity of TNP-lipopolysaccharide, a T-independent antigen. J. Exptl. Med. 141:1453–1458, 1975.
14. Manawadu, B.R., and A. Voller. Standardization of the indirect fluorescent antibody test for malaria. Trans. R. Soc. Trop. Med. Hyg. 72:456–462, 1978.
15. Manning, J.K., N.D. Reed, and W. Jutila. Antibody response to *Escherichia coli* lipopolysaccharide and type III pneumococcal polysaccharide by congenitally thymusless (nude) mice. J. Immunol. 108:1470–1472, 1972.
16. Milich, D.R., and M.E. Gershwin. T-cell differentiation and the congenitally athymic (nude) mouse. Dev. Comp. Immunol. 1:289–298, 1977.
17. Mitchell, G.F., F.C. Grumet, and H.O. McDevitt. The effect of thymectomy on the primary and secondary antibody response of mice to poly-L (Tyr, Glu)-poly-D, L-Ala-poly-L-Lys. J. Exptl. Med. 135:126–135, 1972.
18. Nussenzweig, R. S. Paper presented in the advisor group meeting on nuclear methodology and techniques in the study of parasitic diseases of humans, London, 1979.
19. Renoux, G., and M. Renoux. Thymus-like activities of sulphur derivatives on T-cell differentiation. J. Exptl. Med. 145:466–471, 1977.
20. Rüde, E., J. Wrede, and M.L. Gundelach. Production of IgG antibodies and enhanced response of nude mice to DNP-AE-dextran. J. Immunol. 116:527–533, 1976.
21. Sharon, R., P.R.B. McMaster, A.M. Kask, J.D. Owens, and W.E. Paul. DNP-Lys-Ficoll: A T-independent antigen which elicits both IgM and IgG anti-DNP antibody-secreting cells. J. Immunol. 114:1585–1589, 1975.
22. Tobie, J.E., C.C. Abele, G.J. Hill, II, P.G. Contacos, and C.B. Evans. Fluorescent antibody studies on the immune response in sporozoite-induced and blood-induced *vivax*

malaria and the relationship of antibody production to parasitemia. Am. J. Trop. Med. Hyg. 15:676–683, 1966.
23. Tobie, J.E., and G.R. Coatney. Fluorescent antibody staining of human malaria parasites. Exptl. Parasitol. 11: 128–132, 1961.
24. Voller, A., and R.S. Bray. Fluorescent antibody staining as a measure of malarial antibody. Proc. Soc. Exptl. Biol. Med. 110:907–910, 1962.
25. Von Eschen, K.B., and J.A. Rudbach. Immunological responses of mice to native protoplasmic polysaccharide and lipopolysaccharide. Functional separation of the two signals required to stimulate a secondary antibody response. J. Exptl. Med. 140:1604–1614, 1974.
26. Waki, S. Protective immunity of *Plasmodium berghei* in mice induced by repeated infection and chemotherapy. Jpn. J. Parasitol. 25: 441–446, 1976.
27. Waki, S., and M. Suzuki. Development and decline of antiplasmodial indirect fluorescent antibodies in mice infected with *Plasmodium berghei* (NK65) and treated with drugs. Bull. WHO 50:521–526, 1974.
28. Waki, S., and M. Suzuki. A study of malaria immunobiology using nude mice, pp. 37–44. *In* T. Nomura, N. Ohsawa, N. Tamaoki, and K. Fujiwara (eds.), Proceedings of the second international workshop on nude mice. Tokyo: University of Tokyo Press, 1977.

General Discussion

FINERTY: You found antibodies to *P. berghei* in infected nude mice. Have you transferred serum from these nude mice to other mice to determine whether the antibodies made by nudes are protective in other mice?

WAKI: We have not done passive transfer experiments because, in the present study, vaccinated and challenged nude mice did not show any protection, i.e., decreased mortality or decreased level of parasitemia, when compared to unvaccinated, challenged nude mice.

FINERTY: What drug was used to treat the infected mice?

WAKI: We used pyrimethamine as a drug for the radical cure of mice infected with living parasites. In principle, no parasite survived after administration of the drug, and in other studies we have shown that such drug administration had no effect on antibody production in the animals.

5

The Use of CBA/N and Nude Mice to Study the Regulation of B-Cell Activation Following Infection with Trypanosoma rhodesiense

John F. Finerty,* Yvonne J. Rosenberg, Louise Kendrick, Russell P. McKelvin, and Carl T. Hansen†

Laboratory of Microbial Immunity, National Institute of Allergy and Infectious Diseases, National Institutes of Health, Bethesda, Maryland 20205, and Veterinary Resources Branch, DRS,† National Institutes of Health, Bethesda, Maryland 20205.

Abstract

Euthymic and athymic CBA/N and CBA/CaJ mice were infected with *Trypanosoma rhodesiense*; total IgM- and IgG2-secreting cells and total serum IgM and IgG2 levels were determined. *Trypanosoma rhodesiense* infections induced polyclonal B-cell activation in nude and euthymic mice. Type II nudes were different from the other groups because they had the lowest number of IgM-secreting cells and a delay in increase in serum IgM levels after infection. Compared to nude mice, euthymic mice had a larger number of B cells secreting IgG2, indicating that this response was under T-cell regulation. Surprisingly, CBA/N mice were induced to give higher IgM responses than nondeficient CBA/CaJ mice.

Introduction

Naturally occurring immunodeficient animals are extremely useful in studying the immunology of infectious diseases. A prime example is the congenitally athymic (nude) mouse, which does not have to be "pretreated," i.e., thymectomized, prior to infection. Nude mice are therefore useful in delineating T-dependent humoral responses plus T-dependent pathological changes in response

---

* To whom correspondence should be addressed.
Proceedings of the Third International Workshop on Nude Mice.

to infection (8). Whereas nude mice are quite susceptible to infectious agents, evidence exists that they can be reconstituted with thymic grafts, restoring resistance to infection (7).

African trypanosomes, e.g., *Trypanosoma rhodesiense*, are extracellular blood, protozoan parasites. These parasites are quite adaptable to mammalian hosts and readily infect laboratory animals. Trypanosomes initiate an immune response that is nonprotective for the host, assuring parasite survival for transmission to new hosts. Among the changes of the immune system usually associated with these nonprotective responses are lack of cell-mediated immunity (2), polyclonal B-cell activation, high IgM responses, antigenic variation of the parasite, immunosuppression, and clonal B-cell depletion (6).

Our studies on African trypanosomiasis has focused on two aspects of the disease: (a) the interaction between host and trypanosomes, and (b) trypanosomes used as probes of the immune system. In the latter context, trypanosomes can be used to stimulate B cells in euthymic and nude mice. Live trypanosomes present a different type of antigenic stimulation than nonliving antigens because the parasites consist of an expanding antigenic stimulus to the host. Therefore, assaying B-cell activity in these animals may reveal responses not seen with nonliving antigens. Further, few if any data exist on B-cell activiation in nude mice caused by African trypanosomes. This report is the first to compare the B-cell activity in euthymic and nude mice infected with the human parasite, *Trypanosoma rhodesiense*.

Materials and Methods

Animals. Male CBA/N, NIH-white Swiss nudes and (CBA/N × nu/nu) type II nude mice, 8 weeks of age, were obtained from the Animal Production Unit, National Institutes of Health, Bethesda, Maryland. CBA/N mice have the X-linked B-cell defect (Xid) that inhibits IgM antibody response to certain T-independent antigens (1). The type II nude male mice possess this B-cell defect. Male CBA/CaJ mice, normal B-cell responders, 8 weeks of age, were obtaned from The Jackson Laboratory, Bar Harbor, Maine.

Parasite. *Trypanosoma rhodesiense*, strain 1886, maintained as a frozen stabilate at $-90°C$, was thawed rapidly and injected into donor nude mice. Three days later, the mice were bled; the trypanosomes were counted and adjusted to 10^5/ml in phosphate–saline–glucose (PSG), pH 8.0 (5). Experimental mice were given 10^4 trypanosomes per 0.1 ml intraperitoneally.

Parasitemia. Infected mice were bled at appropriate times after infection and the blood was diluted in 0.1% formaldehyde–PBS (Phosphate-buffered saline). Counts were made with a hemocytometer at ×400 magnification.

Immunoglobulin levels. Groups of infected mice were exsanguinated at various times after infection. The blood was collected in heparin and centrifuged at 1435 × g for 10 min at 5°C; the plasma was removed and stored at $-70°C$ until used. Ig isotype levels were assayed by a single-radial gel diffusion technique (3).

Plaque-forming cell assay. Spleen cells were assayed for their ability to secrete IgM and IgG-2 by the reverse plaque-forming cell (PFC) assay as previ-

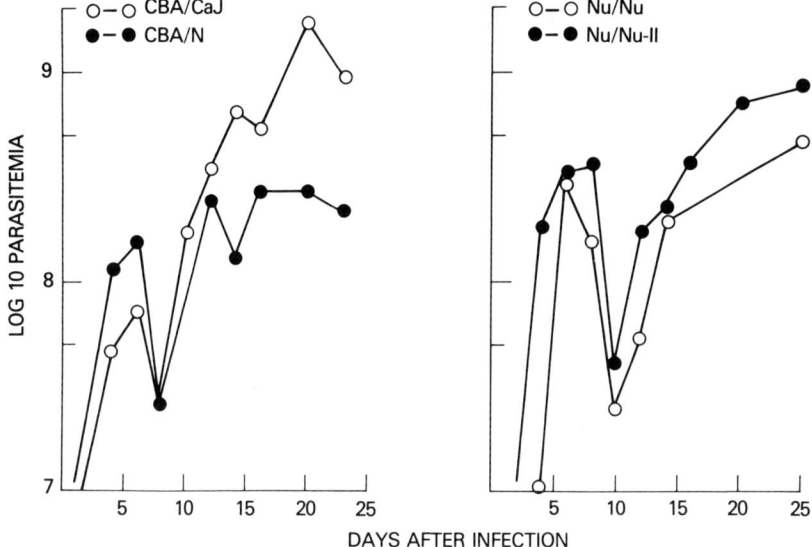

Figure 5-1. *Trypanosoma rhodesiense* 1886 parasitemia in CBA/CaJ, CBA/N, Swiss (*nu/nu*) nude, and type II (*nu/nu*-II) nude mice.

ously described (9). Class-specific rabbit antimouse antisera were used as developing reagents and guinea pig serum as a source of complement. This assay measures total class-specific Ig-secreting cells regardless of antibody specificity.

Results

Course of Infection in Nude and Euthymic Mice

The most striking feature of these infections was the early death of the normal responders, CBA/CaJ (Table 5-1). These in turn, were followed by both strains of nude mice (38–39 days), whereas the CBA/N mice had the longest survival time (61 days). Assessment of the parasitemia (Figure 5-1) showed that both

Table 5-1. Mean survival time of athymic (nude) mice and euthymic control mice infected with *Trypanasoma rhodesiense* 1886

Mouse	Number	Survival time (days)
CBA/CaJ	300	23.2
CBA/N	300	61.5
nu/nu[a]	25	38.8
nu/nu[b]	25	39.9

[a] *nu/nu*, Swiss nudes.
[b] *nu/nu*, II, type II CBA/N nudes.

CBA/CaJ and CBA/N had similar parasitemias during the first 8 days after infection. This was followed by a second wave of trypanosomes, which increased more rapidly in CBA/CaJ mice compared to CBA/N mice, in which the parasites remained controlled. This biphasic pattern in parasitemia was observed in both strains of nude mice, although the initial wave of parasitemia was higher in these mice.

Effect of Infection on Total IgM Production

Spleen cells from infected mice were assayed on days 8 and 11 for total IgM- and IgG2-secreting cells. Table 5-2 shows that on day 8 more than 10-fold increases in the number of IgM PFC were observed in both euthymic and nude mice, although CBA/N showed the most dramatic effects (400-fold increase).

Serum IgM was also assayed at various times after infection to determine whether serum levels expressed cellular activity. Although the serum IgM levels rose rapidly in both CBA/CaJ and CBA/N until day 8 (Figure 5-2), marked strain differences were observed as the infection progressed. Then, while IgM levels rapidly declined in CBA/CaJ mice, the levels in CBA/N mice continued to increase in a biphasic manner. These patterns of increased serum IgM production in CBA/N mice closely coincided with those seen at the cellular level (Table 5-1).

Nude mice showed different kinetics in terms of their serum IgM levels (Figure 5-2). In Swiss nudes, a steady increase occurred until day 10, after which the levels essentially stabilized. However, type II nudes demonstrated a 2-week delay before an increase in their IgM levels was detected. This apparent lack of coincidence between secreting cells and serum levels in nude mice is under investigation.

Table 5-2. Comparison of the total IgM-secreting cell response in spleens of athymic and euthymic mice infected with *Trypanosoma rhodesiense* 1886

Mouse strain	Day 8	Day 11
Infected		
CBA/N	6.213 ± .92 (1,630,000)	6.406 ± .07 (2,520,000)
nu/nu-II	4.996 ± .08 (99,100)	4.871 ± .35 (74,400)
CBA/CaJ	6.286 ± .08 (1,930,000)	5.527 ± .08 (337,000)
nu/nu	5.977 ± .02 (949,000)	5.803 ± .16 (636,000)
Uninfected		
CBA/N	3.602 ± .01 (3,990)	
nu/nu-II	3.628 ± .32 (3,990)	
CBA/CaJ	5.167 ± .08 (147,000)	
nu/nu	4.832 ± .05 (67,900)	

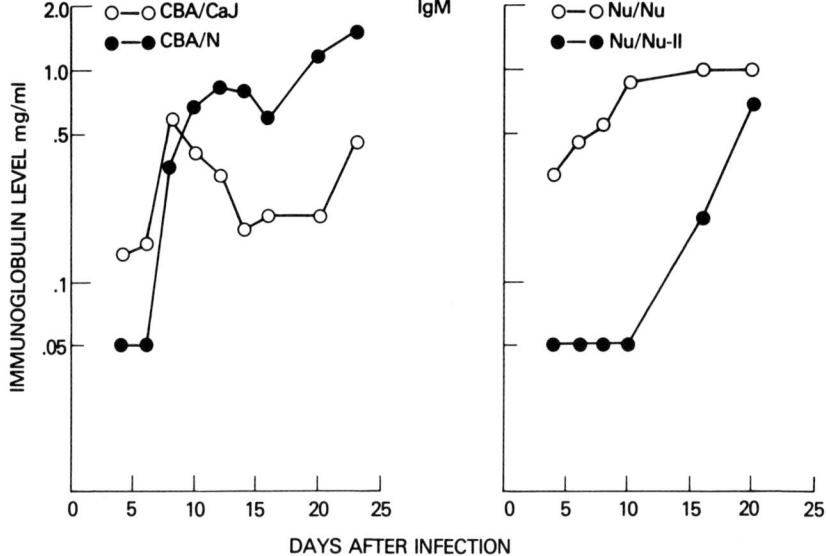

Figure 5-2. IgM serum levels in CBA/CaJ, CBA/N, Swiss (*nu/nu*) nude, and type II (*nu/nu*-II) nude mice infected with *Trypanosoma rhodesiense* 1886.

Effects of Infection on Total IgG2 Levels

In contrast to the increases in IgM levels observed during infection of nude mice, marked differences in the ability of athymic and euthymic mice to produce IgG were predictably observed (Table 5-3). Therefore, although the effects of

Table 5-3. Comparison of the total IgG2 secreting cell response in spleens of athymic and euthymic mice infected with *Trypanosoma rhodesiense* 1886

Mouse strain	Day 8	Day 11
Infected		
CBA/N	6.208 ± .05	6.187 ± .02
	(1,620,000)	(1,530,000)
nu/nu-II	4.460 ± .08	3.742 ± .13
	(28,900)	(5,520)
CBA/CaJ	6.106 ± .02	5.550 ± .13
	(1,170,000)	(354,000)
nu/nu	4.469 ± .09	4.185 ± .09
	(44,300)	(15,300)
Uninfected		
CBA/N	3.903 ± .03	
	(8,200)	
nu/nu-II	3.301 ± .08	
	(12,300)	
CBA/CaJ	4.079 ± .07	
	(12,300)	
nu/nu	3.301 ± .02	
	(2,000)	

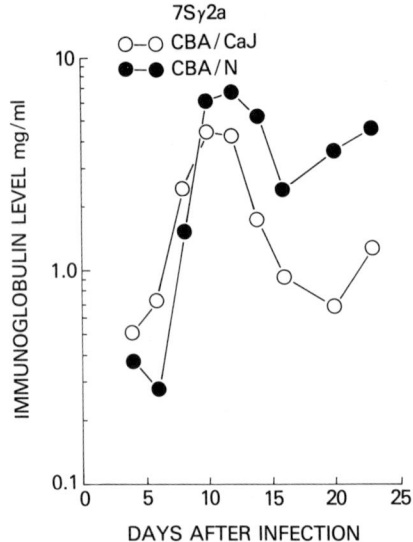

Figure 5-3. IgG2a (7Sγ2a) serum levels in CBA/CaJ, CBN/N, Swiss (*nu/nu*) nude, and type II (*nu/nu*-II) nude mice infected with *Trypanosoma rhodesiense* 1886.

infection were less on spleen IgG2 PFC of nude mice, large numbers of B cells were obtained to IgG2 in CBA/CaJ and CBA/N mice, resulting in more than 100-fold increases in IgG2 PFC. These increases were also reflected by high serum IgG2 levels in CBA/CaJ and CBA/N mice, both of which showed peak levels on day 10 (Figure 5-3). In a situation similar to that seen with host IgM reduction, IgG2 at the cellular and serum levels decreased more rapidly in CBA/CaJ mice than in the defective CBA/N mice, indicating a reduction in host B-cell function in the more susceptible CBA/CaJ mice.

Discussion

These results indicate that *T. rhodesiense* infections in nude and euthymic mice induce a marked polyclonal B-cell activation. All four strains of mice demonstrated increased IgM serum levels after infection, indicating that part of the response is independent of T cells. The kinetics in type II nudes were different from the other groups in that they had the lowest number of IgM-secreting cells and exhibited no increase in serum IgM levels for 2 weeks after infection. This may indicate that type II nudes had a slower developing clone of IgM-secreting B cells. The larger number of B cells secreting IgG2 plus the increase in serum IgG2 observed in euthymic mice, compared to nude mice, indicates that this response was under T-cell regulation. Further, these studies show that, despite the low background levels of IgM in the defective CBA/N, and other reports (1) indicating a deficiency in these mice to give an IgM response to certain T-dependent antigens, CBA/N mice can be induced to give higher IgM response than nondeficient CBA/CaJ. In addition, these studies have shown a reduced ability of CBA/N mice to be induced to nonspecific IgG1 activation (data not shown), suggesting a strain-dependent and class-restricted regulation of polyclonal B-cell activation.

The rapid rise in parasitemia plus the short survival time of CBA/CaJ mice concomitant with a rapid reduction in Ig-secreting cells suggest a decline in the immune response to these mice. A possible explanation for this reduction is the role of T-suppressor cells or autoantibody in inhibiting T-cell function (4). The characteristic response observed in CBA/N mice, combined with the longer survival time, indicates a positive role for B-cell activity in mice infected with *T. rhodesiense*. The data suggest that if T-suppressor cells inhibit B-cells in CBA/CaJ mice, then CBA/N may lack a subset of T-suppressor cells. Alternatively, the inability of defective mice to give certain immune responses, e.g., immune complexes, may enhance their survival.

One of the more interesting observations was the parasitemia curve of the type II nudes. The parasite kinetics plus very low B-cell activity indicate that the initial wave of parasitemia was not under immune control. It was proposed that antibody induced cycling and changes in the surface antigenic determinants of African trypanosomes (6). The low B-cell activity suggest little or no antibody to trypanosomes in these mice, which will be a useful model by which to study the effects of antibody on trypanosome populations.

In conclusion, it is obvious that *T. rhodesiense* induces polyclonal B-cell activation, the kinetics of which differ in euthymic mice CBA/CaJ versus CBA/N, plus nude mice, Swiss versus type II nudes. The PFC data combined with the serum IgM and IgG2 levels indicate that B cells in normally responding CBA/CaJ mice decreased as the infection progressed and this contributed to the shortest survival time. The longer survival time of nude mice indicates that IgM antibody may play an important role in host survival. However, T-cell-mediated pathological effects (4,8), lacking in nude mice, may contribute to their longevity. These and other possibilities are under investigation.

References

1. Amsbaugh, D.F., C.T. Hansen, B. Prescott, P.W. Stashak, D.R. Barthold, and P.J. Baker. Genetic control of the antibody response to type III pneumococcal polysaccharide in mice. I. Evidence that an X-linked gene plays decisive role in determining responsiveness. J. Exptl. Med. 136:913–949, 1972.
2. Finerty, J.F., E.P. Krehl, and R.L. McKelvin. Delayed-type hypersensitivity in mice immunized with *Trypanosoma rhodesiense* antigens. Infect. Immun. 20:464–467, 1978.
3. Finerty, J.F., J.E. Tobie, and C.B. Evans. Antibody and immunoglobulin synthesis in germfree and conventional mice infected with *Plasmodium berghei*. Am. J. Trop. Med. Hyg. 21:499–505, 1972.
4. Kobayakawa, T., J. Louis, S. Izui, and P.H. Lambert. Autoimmune response to DNA, red blood cells, and thymocyte antigens in association with polyclonal antibody synthesis during experimental African trypanosomiasis. J. Immunol. 122:296–301, 1979.
5. Lanham, S.M., and D.G. Godfrey. Isolation of salivarian trypanosomes from man and other mammals using DEAE-cellulose. Exptl. Parasitol. 28:521–534, 1970.
6. Mansfield, J.M. Immunobiology of African trypanosomiasis. Cell. Immunol. 39:204–210, 1978.

7. Roberts, D.W., R.G. Rank, W.P. Weidanz, and J.F. Finerty. Prevention of recrudescent malaria in nude mice by thymic grafting or by treatment with hyperimmune serum. Infect. Immun. 16:821–826, 1977.
8. Roberts, D.W., and W.P. Weidanz. Splenomegaly, enhanced phagocytosis and anemia are thymus dependent responses to malaria. Am. J. Trop. Med. Hyg. 28:1–3, 1978.
9. Rosenberg, Y.J. Autoimmune and polyclonal B cells responses during murine malaria. Nature (London) 274:170–172. 1978.

General Discussion

HONG: I recall that trypanosomiasis induces a 7S IgM response. Do you have information on the size of your IgM antibody? Also, if you have 7S IgM you might not detect such an immunoglobulin as a PFC, thus missing a polyclonal response. Further, a 7S IgM response would give false high values in a diffusion assay.

FINERTY: We have not detected any 7S IgM during the 20 days after infection.

MITCHELL: Do you have information on the ratio of antitrypanosome antibody to nonspecific Ig in the IgM and IgG3 isotype fractions in the serum of any of the trypanosome-infected mice?

FINERTY: Antitryp IgM antibody was detected in both CBA/CaJ and CBA/N mice; some of the Swiss nude mice were positive for IgM antibody, whereas in type II nudes, IgM antibody was not detected. IgG3 antibody is under investigation. While exact ratios have not been established, approximately 10% of the IgM is antibody.

6

Trypanosoma musculi *Infection of Nude Mice*

Bradford O. Brooks* and Norman D. Reed†

Department of Microbiology, Montana State University, Bozeman, Montana 59717.

Abstract

Trypanosoma musculi produces in mice a self-limiting parasitemia which confers long-lasting immunity. Studies done in normal mice suggest that parasitemia is initially controlled through the combination of a trypanocidal antibody and an ablastic (parasite reproduction-inhibiting) serum factor. Final elimination of parasitemia appears to be mediated through a cellular mechanism. We studied *T. musculi* infections in nude mice (nu/nu), their thymus-bearing littermates (NLM: nu/+ or +/+), and thymus gland-grafted nude mice (TG-Nu). NLM and TG-Nu mice limited both the level of parasitemia and parasite reproductive activity and cleared *T. musculi* from the blood by day 24 post-inoculation. In contrast, nude mice had higher levels of parasitemia and parasite reproductive activity and maintained this high level of infection. Adoptive transfer of spleen cells or thymus cells from immune or normal mice restored *T. musculi* elimination potential to nude mice. Passive transfer of immune serum severely limited both parasite reproductive activity and levels of parasitemia but did not restore *T. musculi* elimination potential to nude mice. Although the ablastic serum factor is thymus dependent and has the physicochemical properties of IgG1, its successful absorption with homologous live trypanosomes or trypanosome antigens has not been reported. Using nude mice, known to be deficient for IgG1 and ablastin production, as an *in vivo* assay of ablastin activity, we found that ablastin activity was absorbed from immune serum by a dividing form-enriched parasite population (prepared in irradiated mice) but not by a nondividing population.

---

† To whom correspondence should be addressed.
© 1982 Gustav Fischer New York, Inc.
Proceedings of the Third International Workshop on Nude Mice.
* NIH Postdoctoral Fellowship awardee AI 05976-01. Present Address: Department of Preventive Medicine, New York State College of Veterinary Medicine, Cornell University, P.O. Box 786, Ithaca, New York 14850.

Introduction

In recognition of the global economic and medical importance of tropical disease, the World Health Organization (WHO) has selected six diseases as initial targets of intensive research efforts (6). Trypanosomiasis, one of six diseases included in the WHO Special Programme, affects some 10 million people, 8 million in the western hemisphere (6). Despite its significance, current knowledge of mechanisms of immunity to trypanosome infections remains fragmentary and poorly understood. In those models which have been studied most extensively, conclusions have frequently been contradictory (4,12). The number and complexity of antigens of individual trypanosomes and their ability to undergo cyclic antigenic variation (3) have contributed to the relative dearth of knowledge concerning host immunologic responses to trypanosomes.

The availability of the *Trypanosoma musculi*–mouse model and recent advances in immunology have provided an approach to understanding the complex mechanisms involved in host immunity to trypanosomes. It should be emphasized that *T. musculi* is a natural parasite of the mouse (11); passage of trypanosomes through an abnormal host frequently results in a disproportional increase in virulence (10), which can significantly affect host immune responses. Previous studies on acquired immunity to *T. musculi* have been conducted in experimentally immunosuppressed mice (9,19,20); however, models involving irradiation, thymectomy, and drug or antilymphocyte administration often suffer from nonspecific or incomplete effects from such treatments. The availability of thymus-deficient (nude) mice allows a more critical evaluation of the importance of thymus competence in the immune response to *T. musculi*. It is our long-range objective to initiate a systematic examination of the host immune response to *T. musculi* by using nude mice as an aid in determining those factors requisite for trypanosome elimination.

Materials and Methods

Mice. Congenitally thymus deficient nude mice (nu/nu) and their normal thymus-bearing littermates (NLM) were from our colony, in which cross–intercross mating is in progress to derive a line of nude and NLM mice congenic with BALB/c mice.

Parasites. Mice were infected by intraperitoneal (IP) injection of 3×10^4 *T. musculi* (Partinico II strain obtained from the American Type Culture Collection, Rockville, Maryland) in 0.1 ml of buffered saline. Parasitemia and percentage dividing forms were monitored by counting parasites on Giemsa-stained thin blood smears; data are expressed as organisms per high field (O/HPF) and percentage dividing forms (%DF) (8).

To produce a population of parasites with a high percentage of dividing forms (dividing trypomastigotes and epimastigotes), NLM mice were sublethally irradiated (550 rads, ^{60}Co) 24 h before infection with *T. musculi*. The techniques used for obtaining a dividing population (DP = 35–50%DF) and a nondividing population (NDP = <5%DF) of trypanosomes have been previously described (1).

Thymus implants. Nude mice, anesthetized with sodium pentabarbitol, were implanted with one neonatal BALB/c thymus gland under each renal capsule using a previously described technique (7). To confirm thymus function, thymus gland-grafted nude mice (TG-Nu) were immunized with sheep erythrocytes. The immune response of TG-Nu mice approached the responses of NLM mice, whereas nude mice produced significantly lower response to sheep erythrocytes.

Adoptive transfer. Spleens or thymus glands were removed from BALB/c donor mice 2–3 weeks old and teased over 80-mesh stainless steel screens in chilled saline containing 2% normal mouse serum (NMS). Cells were enumerated and assayed for viability; subsequently, 1×10^8 viable thymus or spleen cells were injected intravenously into each recipient nude mouse on the day of *T. musculi* infection.

Immune serum. Serum with ablastic activity was obtained from NLM mice 18 days postinfection (PI) (1,17); this serum will hereafter be designated ablastin (Abl). Serum containing only trypanocidal activity was obtained on day 28 PI (1,17); this serum will hereafter be designated immune recovered serum (IRS). Serum obtained from uninfected NLM mice (NMS) was used as a control. Blood was allowed to clot overnight at 4°C. Serum was heated at 56°C for 30 min before storage at −70°C.

Absorption and administration of serum. Pooled Abl was split into three portions: One portion was absorbed with a dividing population (DP) and designated DP-Abs-Abl; a second portion was absorbed with a nondividing population (NDP) and designated NDP-Abs-Abl; the remaining portion was not absorbed and served as a control for ablastic activity. Serum absorption was carried out at 4°C for 30 min using 1×10^9 trypanosomes per milliliter of serum and was repeated three times.

Mice received intravenously (IV) 0.25 ml of appropriate serum on the day of *T. musculi* infection; subsequent to infection, mice received a regimen of 0.25 ml of appropriate serum every 5 days for the duration of the experiment.

In Figures 6-1 through 6-8, each value is the arithmetic mean of three experiments, with no fewer than six mice in each experimental group.

Results

Thymus Dependency of T. musculi *Elimination from Mice*

Trypanosoma musculi produces a self-limiting parasitemia which confers long-lasting immunity in normal mice. Parasitemia is thought to be initially controlled through the combination of a trypanocidal antibody and an ablastic (parasite reproduction-inhibiting) serum factor (20). Final elimination of parasitemia appears to be mediated through a cellular mechanism (13,19). Early investigation of this system showed that functional immunity to *T. musculi* was clearly thymus dependent (2,14); NLM and TG-Nu mice limited the level of parasitemia and cleared *T. musculi* by day 24 PI, whereas nude mice maintained chronically high levels of parasitemia throughout experimental observation (2). Because it

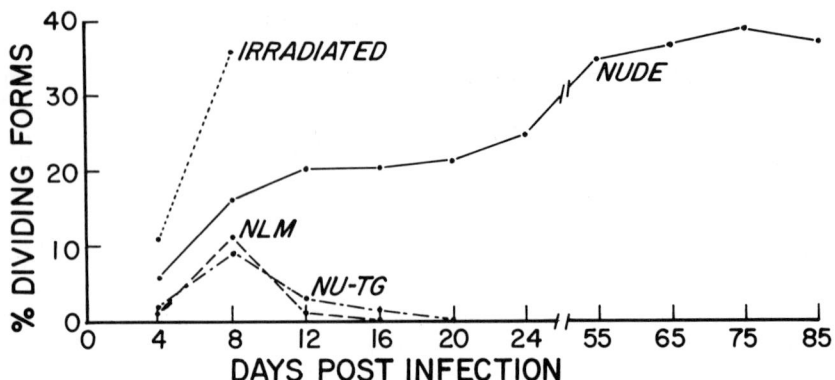

Figure 6-1. Percentage of dividing forms in *T. musculi* parasitemia in the peripheral blood of normal mice (NLM), nude mice (nude), thymus gland-grafted nude mice (TG-NU), and irradiated mice (550 rads, ^{60}Co).

has been hypothesized that ablastic factors play a role in the initial control of *T. musculi* parasitemia, it was important to monitor parasite reproductive activity in NLM, nude, and TG-Nu mice. The results (Figure 6-1) show that nude mice maintained levels of dividing trypanosomes throughout infection. In marked contrast, both NLM and TG-Nu mice severely limited parasite reproductive activity by day 12 PI. These data clearly demonstrate the thymus dependency of the ablastic phenomena (5,16).

Adoptive Transfer of Immune or Normal Thymus or Spleen Cells

Adoptive transfer experiments were designed to identify those cell populations capable of generating *T. musculi* elimination potential in nude mice. Either immune or normal spleen or thymus cells were adoptively transferred IV into nude mice at the time of *T. musculi* infection. The results (Figure 6-2) show that all cell populations investigated were able to generate *T. musculi* elimination potential. Nude mice that did not receive cells developed increasing parasitemia that culminated in their death by day 13 PI. Immune spleen cells were more effective in generating elimination potential than any other cell population investigated.

Passive Transfer of Immune Serum

It has been suggested that the initial control of *T. musculi* parasitemia is antibody mediated (20). This event, referred to as the first crisis, is thought to be the result of the cooperative action of a thymus-independent trypanocidal antibody and thymus-dependent ablastic serum factor (13,16–20). Previous investigations have suggested that both ablastin and trypanocidal antibodies are present in the serum of NLM on day 18 PI (Abl) (1,17). Serum obtained from NLM mice on day 28 PI has been reported to contain only trypanocidal antibody (IRS) (1,17).

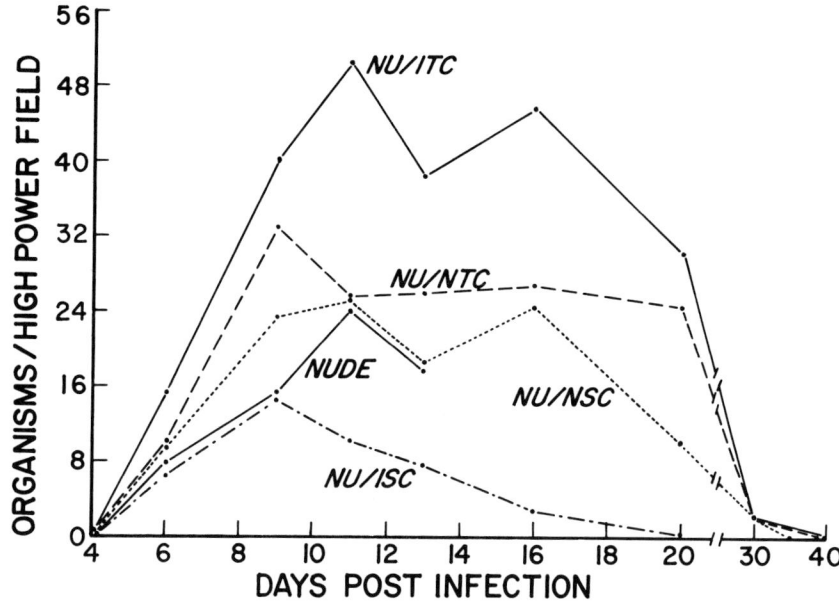

Figure 6-2. *Trypanosoma musculi* parasitemia (organisms per high-power field) in nude mice that received either immune thymus cells (Nu/ITC), normal thymus cells (Nu/NTC), immune spleen cells (Nu/ISC), normal spleen cells (Nu/NSC), or no cells (nude).

In an attempt to clarify those mechanisms responsible for the first crisis, passive transfer experiments were designed to examine the effects of immune sera on *T. musculi* parasitemia in nude and NLM mice. Groups of nude and NLM mice received 0.25 ml of Abl, IRS, or NMS IV at the time of infection. Mice were maintained on a regimen of 0.25 ml of appropriate serum IV every 5 days for the extent of experimental observation. The results (Figure 6-3) show that as early as day 8 PI there was a difference between nude mice that received Abl (nude/Abl) and nude mice that did not receive Abl (nude, nude/NMS). The effect of Abl was especially apparent in *T. musculi* infections of nude mice; both O/HPF (Figure 6-3) and %DF (Figure 6-4) were markedly reduced throughout the infection. In spite of this conspicuous reduction of O/HPF and %DF, the passive transfer of Abl did not provide *T. musculi* elimination potention to nude mice. Abl had a dramatic effect on %DF in NLM mice as well; normal reproductive activity was never detected (Figure 6-4).

Nude mice that received an IRS regimen (nude/IRS) also showed a restricted parasitemia as early as day 8 PI (Figure 6-5); in spite of this reduction in O/HPF, *T. musculi* elimination potential was not evident. In marked contrast to nude/Abl mice and NLM/Abl mice (Figure 6-4), nude/IRS and NLM/IRS mice did not exhibit an obvious diminution of parasite reproductive activity (Figure 6-6).

Collectively, these data confirm the usefulness of the nude mouse as an in vivo model for the assay of ablastin activity; unlike recipient normal mice, recipient

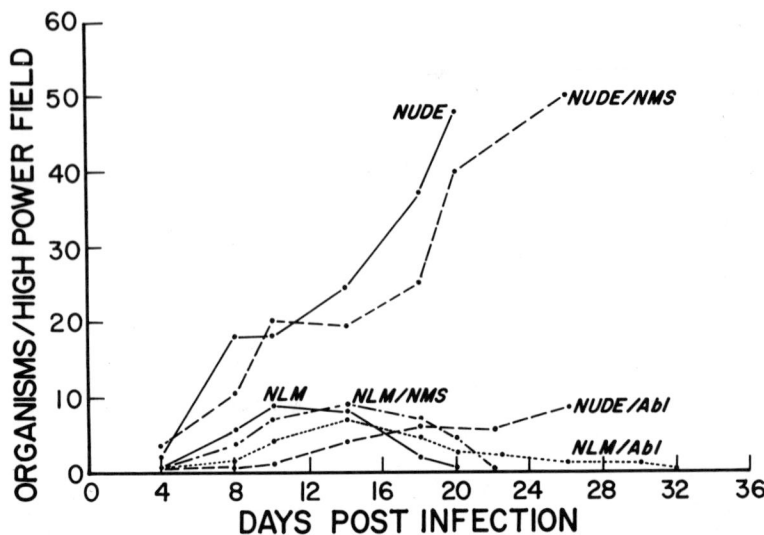

Figure 6-3. *Trypanosoma musculi* parasitemia (organisms per high-power field) in nude mice and normal mice that received either ablastic serum (nude/Abl and NLM/Abl), normal mouse serum (nude/NMS and NLM/NMS), or no serum (nude and NLM).

nude mice do not actively make ablastin to complicate this assay. The lack of ablastic activity in IRS (Figure 6-6) confirms predictions made in previous studies (17). It is noteworthy that Abl or IRS regimes were able to profoundly

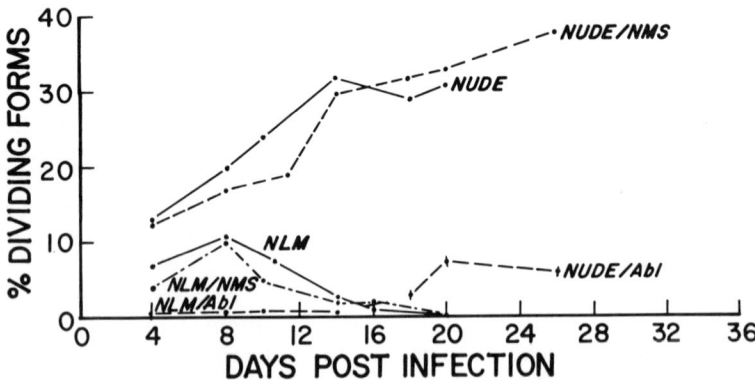

Figure 6-4. Percentage of dividing forms in *T. musculi*-infected nude and normal mice that received either ablastic serum (nude/Abl and NLM/Abl), normal mouse serum (nude/NMS and NLM/NMS), or no serum (nude and NLM).

Trypanosoma musculi Infection 117

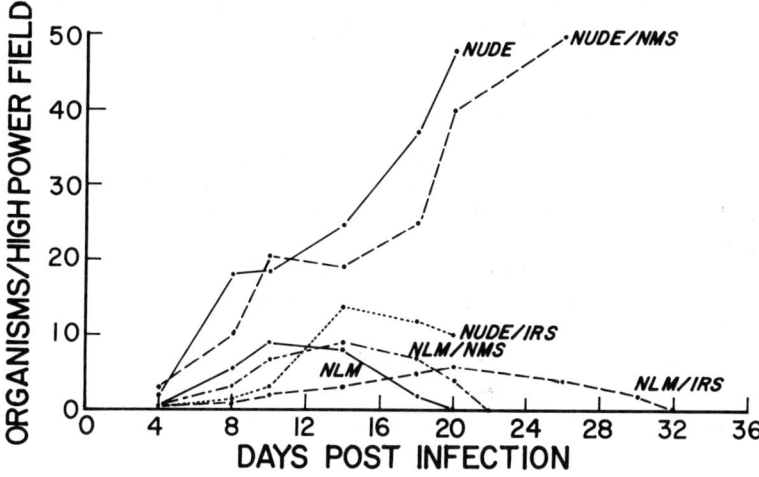

Figure 6-5. *Trypanosoma musculi* parasitemia (organisms per high-power field) in nude mice and normal mice that received either immune recovered serum (nude/IRS and NLM/IRS), normal mouse serum (nude/NMS and NLM/NMS), or no serum (nude and NLM).

modify *T. musculi* parasitemia in mice when injected IV but were not able to provide *T. musculi* elimination potential to nude mice or prevent their death. This observation lends credence to the proposal that final elimination of *T. mus-*

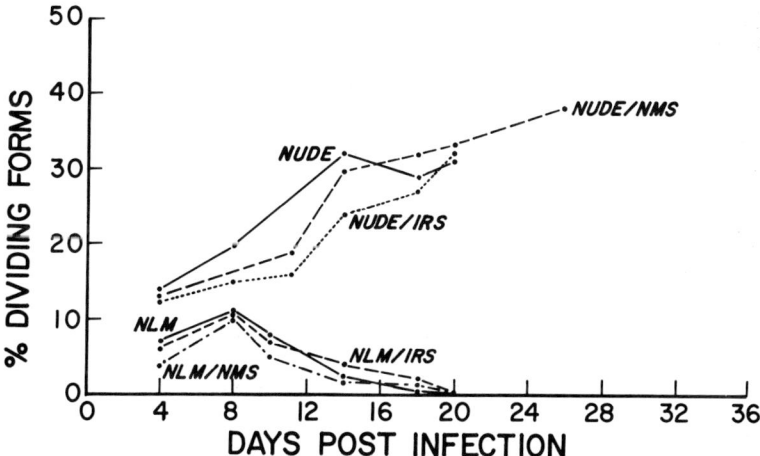

Figure 6-6. Percentage of dividing forms in *T. musculi*-infected nude and normal mice that received either immune recovered serum (nude/IRS and NLM/IRS), normal mouse serum (nude/NMS and NLM/NMS), or no serum (nude and NLM).

culi from peripheral blood of mice is dependent on a cellular mechanism and that this mechanism is nonfunctional in nude mice (20).

Absorption of Ablastic Activity from Immune Serum

The ablastic factor produced by *T. musculi*-infected NLM mice has been a historically perplexing compound. Ablastin has been reported to be physicochemically similar to immunoglobulin, specifically IgG1 (9). However, its nonabsorbability with homologous trypanosomes (9) and the lack of characterization of its eliciting antigens have cast serious doubt on the antibody nature of ablastin. Because ablastin is proposed to inhibit reproduction without destroying trypanosomes (5), the possibility that dividing trypanosomes are the source of the eliciting antigen has seemed plausible. If this is the case, a population of *T. musculi* enriched for dividing forms should present an efficient antigen preparation for absorption of ablastic serum. To test this hypothesis two different trypanosome populations were prepared for absorption of ablastic serum: (a) a trypanosome preparation with 35–50%DF (DP) and (b) a trypanosome preparation with less than 5%DF (NDP). Serum containing ablastic activity (Abl) was split into portions and absorbed with either DP or NDP or was not absorbed. These variously absorbed Abl sera were then employed in passive transfer experiments using nude mice to see whether absorption had altered ablastic activity. The results (Figures 6-7 and 6-8) show that as early as day 8 PI there was a difference between groups of nude mice that received Abl or NDP-Abs-Abl and

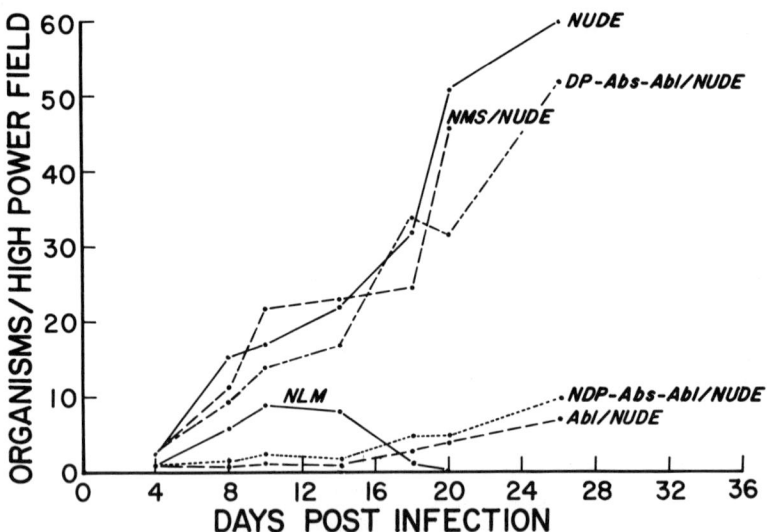

Figure 6-7. *Trypanosoma musculi* parasitemia (organisms per high-power field) in nude mice that received either ablastic serum absorbed with a dividing parasite population (DP-Abs-Abl/nude), ablastic serum absorbed with a nondividing parasite population (NDP-Abs-Abl/nude), ablastic serum (Abl/nude), or no serum (nude). O/HPF was also monitored in *T. musculi*-infected normal mice (NLM).

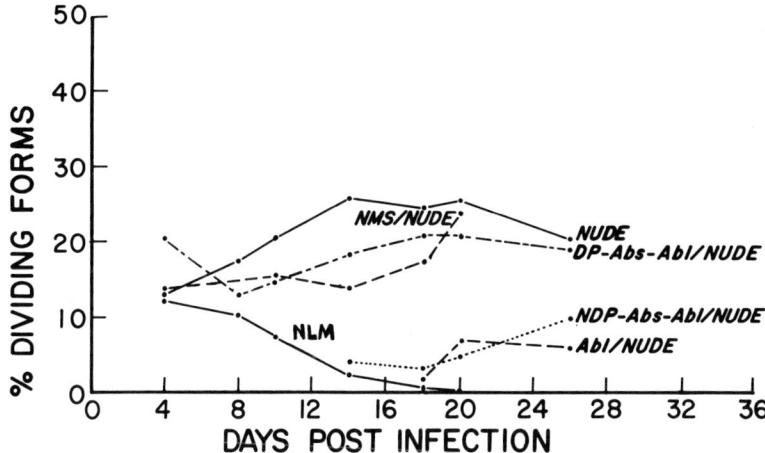

Figure 6-8. Percentage of dividing forms in *T. musculi*-infected nude mice that received either ablastic serum absorbed with a dividing parasite population (DP-Abs-Abl/nude), ablastic serum absorbed with a nondividing parasite population (NDP-Abs-Abl/nude), ablastic serum (Abl/nude), or no serum (nude). %DF was also monitored in *T. musculi*-infected normal mice (NLM)

groups of nude mice that received DP-Abs-Abl, NMS, or no serum. Abl and NDP-Abs-Abl mice exhibited low parasitemia (Figure 6-7) and very low levels of parasite reproductive activity (Figure 6-8). In marked contrast, nude mice that received DP-Abs-Abl, NMS, or no serum maintained high levels of both O/HPF and %DF throughout experimental observation. NLM mice exhibited a typical *T. musculi* parasitemia with marked reduction in %DF by day 14 PI and elimination of parasitemia by day 20 PI. Absorption of unrelated antiserum (mouse antisheep erythrocyte) with either DP or NDP did not alter hemagglutination titers obtained against sheep erythrocytes (data not shown). This control indicated that neither DP nor NDP had any effect on unrelated antiserum.

These data show that ablastin activity can be absorbed from serum using a trypanosome preperation containing 35–50% dividing forms; absorption with a trypanosome preparation containing 5% dividing forms did not appreciably alter ablastic activity. These data suggest that ablastin is indeed an antibody that can be absorbed onto homologous trypanosomes.

Discussion

The thymus dependency of functional immunity to *T. musculi* (2,14) does not distinguish between the role of cell-mediated or humoral factors in mechanisms of immunity to this parasite. Previous studies have suggested that initial acquired immune response to *T. musculi* is effected by the combined action of a thymus-independent trypanocidal antibody and a thymus-dependent ablastic serum factor (20). Final clearance of *T. musculi* from the peripheral blood of mice has been hypothesized to be dependent on a cell-mediated mechanism rather than a direct antibody effect (13). Our data (Figures 6-1 through 6-8) confirm and extend these theories. Nude mice, which have been previously shown to be incapable of *T. musculi* elimination (1,14), were also found to be deficient in ablastin production potential (Figures 6-1 and 6-4). Nude mice that received by adoptive transfer immune or normal populations of spleen or thymus cells prior to *T. musculi* infection were able to generate parasite elimination potential (Figure 6-2). Conversely, nude mice that received passive transfer regimens of either Abl or IRS serum demonstrated a marked reduction in parasitemia but were not able to generate *T. musculi* elimination potential (Figures 6-3 through 6-6). It is interesting to note that passive transfer of Abl to nudes did generate a remarkable reduction in %DF and O/HPF (Figures 6-3 and 6-4), whereas passive transfer of IRS had little effect on %DF yet reduced O/HPF dramatically (Figures 6-5 and 6-6). These data are in agreement with the hypothesis (19,20) that day 18 PI serum contains both ablastic and trypanocidal antibodies, whereas day 28 PI serum contains only trypanocidal antibodies (19,20).

Since the original description of the ablastin phenomena (15), the concept of a serum factor that controls reproduction without actually killing parasites has intrigued many investigators. Experimental evidence indicates that ablastin is an immunoglobulin, possibly IgG1 (9). However, the nonabsorbability of ablastin activity from serum with homologous trypanosomes and the inability to isolate its eliciting antigen(s) (9) has presented a historically perplexing problem in understanding the nature of ablastin and the mechanisms of its action. One of the problems in the study of ablastin has been the lack of dependable in vivo assays for ablastic activity. Our data show that nude mice are not capable of producing ablastin (Figure 6-1), but that passively transferred Abl has a profound effect on *T. musculi* parasitemia in nude mice (Figures 6-3 and 6-4). These observations suggest the utility of the nude mouse as an in vivo assay of ablastin activity. Data presented in Figures 6-4 and 6-6 confirm this proposal and demonstrate the reliability of this assay.

Because ablastin is supposedly interacting with dividing forms of *T. musculi* (5,9), it seemed that division forms might be a primary source of antigen(s) that elicits the production of ablastin. If this supposition was true, a dividing form-enriched population of *T. musculi* should result in an appropriate antigen preparation for the absorption of ablastin from serum. To test this hypothesis, a dividing population (DP = 35–50%DF) of trypanosomes was used to absorb Abl serum. The data (Figures 6-7 and 6-8) from passive transfer experiments in nude mice indicate that a dividing population can absorb ablastic activity for Abl serum, whereas a nondividing population (NDP = <5%DF) cannot. These data provide the crucial evidence needed to establish the antibody nature of ablastin. Furthermore, these data suggest that ablastin is a "normal" antibody that can be absorbed from serum with homologous trypanosomes. The results of these experiments also support the hypothesis that antigens that elicit ablastin production are present in appreciable amounts in a dividing parasite population.

Use of column chromatography techniques to selectively deplete ablastic serum of immunoglobulin subclasses and subsequent analysis of Abl serum in the nude mouse *in vivo* ablastin assay could be a useful approach to identifying those specific immunoglobulin subclasses involved in the ablastin phenomenon. Identification of these specific immunoglobulin subclasses is desirable because they, in turn, could be used to identify those relevant antigens responsible for ablastin production.

Acknowledgments

This work was supported by U.S. Public Health Service Research Grants CA 24443, AI 12854, and RR 09135.

References

1. Brooks, B.O. Humoral and cellular aspects of immunity to *Trypanosoma musculi* in mice. Dis. Abst 79-25, 063, 1979. Ph.D. Thesis, Montana State University, Bozeman, Montana.
2. Brooks, B.O., and N.D. Reed. Thymus dependency of *Trypanosoma musculi* elimination from mice. J. Reticuloendothel. Soc. 22:605–608, 1977.
3. Brown, K.N. Nature and variation of parasite antigen, pp. 21–25. *In* Pan American Health Organization Scientific Publication No. 150, 1967.
4. Cox, F.E.G. Mechanisms of parasite immunity. Nature (London) 270:387–390, 1977.
5. D'Alesandro, P.A. Ablastin: The phenomenon. Exptl. Parasitol. 38:303–308, 1975.
6. Dorozynski, A. The attack on tropical disease. Nature (London) 262:85–86, 1976.
7. Dukor, P., J.F.A.P. Miller, W. House, and V. Allman. Regeneration of thymus grafts: II. Histological and cytological aspects. Transplantation 3:639–643, 1965.
8. Dusanic, D.G. *Trypanosoma musculi* infections in complement-deficient mice. Exptl. Parasitol. 37:205–210, 1975.
9. Dusanic, D.G. Immunosuppression and ablastin. Exptl. Parasitol. 38:322–337, 1975.
10. Hoare, C.A. Evolutionary trend in mammalian trypanosomes. Adv. Parasitol. 5:47–51, 1967.
11. Kendall, A.I. A new species of trypanosomes occurring in the mouse *Mus musculus*. J. Infect. Dis. 3:228–234, 1906.
12. Playfair, J.H.L. Why doesn't the immune system protect us against parasites? Lab Lore 8:497–500, 1978.
13. Pouliot, P., P. Viens, and G.A.T. Targett. T. lymphocytes and the transfer of immunity to *Trypanosoma musculi* in mice. Clin. Exptl. Immunol. 27:507–511, 1977.
14. Rank, R., D. Roberts, and W. Weidanz. Chronic infection with *Trypanosoma musculi* in congenitally athymic nude mice. Infect. Immun. 16:715–718, 1977.
15. Taliaferro, W.H. Ablastic and trypanocidal antibodies against *Trypanosoma duttoni*. 35:303–309, 1938.
16. Targett, G.A.T., and P. Viens. Ablastin: Control of *Trypanosoma musculi* infections in mice. Exptl. Parasitol. 38:309–316, 1975.
17. Viens, P. Immunological responses of mice to *Trypanosoma musculi* infection. Ph.D. Thesis, University of London, 1972.
18. Viens, P., P. Pouliot, and G.A.T. Targett. Cell-mediated immunity during the infection of CBA mice with *Trypanosoma musculi*. Can. J. Microbiol. 20:105–106, 1974.
19. Viens, P., and G.A.T. Targett. Trypanosoma musculi infections in intact and thymectomized CBA mice. Trans. R. Soc. Trop. Med. Hyg. 65:424–429, 1971.
20. Viens, P., G.A.T. Targett, E. Leuchars, and A.J.S. Davies. The immunological response of CBA mice to *Trypanosoma musculi*: Initial control of infection and the effect of T-cell deprivation. Clin. Exptl. Immunol. 16:279–285, 1974.

General Discussion

MITCHELL: This seemed to be such a clear-cut demonstration of the fact that "ablastin" is an antibody directed against dividing forms of the parasite that I was surprised that your conclusions were somewhat guarded. Do you have any reservations about the data?

BROOKS: I'm just a youngster and it seemed appropriate that I should act timid— No, I have no reservations.

MITCHELL: Have you yet treated dividing form *T. musculi*—derived from an infected nude and therefore theoretically deficient for any immunoglobulin on the parasite surface—with ablastic sera plus a tagged antiimmunoglobulin to determine the site of binding of ablastic antibodies?

BROOKS: No—but this would be interesting.

7

Experimental Pneumocystis carinii *Infection in Nude and Steroid-Treated Normal Mice*

Peter D. Walzer* and Ralph D. Powell, Jr.†

VA Medical Center and Division of Infectious Diseases,† Departments of Medicine and Pathology, University of Kentucky, College of Medicine, Lexington, Kentucky 40507.*

Abstract

Our early studies demonstrated that rat *Pneumocystis carinii* infection could be transmitted to outbred Swiss *nu/nu* mice by intrapulmonary injection of fresh infected lung homogenates and via the environment. In later studies performed at another institution with a different strain of Swiss *nu/nu* mice obtained from a different source, rat *P. carinii* infection could not be established in the mice by any mode of transmission. However, these studies suggested that *P. carinii* infection could be transmitted from corticosteroid-treated normal mice to *nu/nu* and *nu/+* mice by close contact. Steroid-induced pneumocystis pneumonia was studied in eight different strains of normal mice, and considerable variation in intensity of the infection among the mouse strains was found. The data suggest that both hereditary and environmental factors influence host susceptibility to infection with *P. carinii*. The studies also illustrate the problems in developing the nude mouse as an experimental model for *P. carinii* and other organisms of low virulence. Future studies with nude mice should use well-defined inbred nude mouse strains obtained from a single source with well-characterized endogenous microbial flora.

Introduction

Pneumocystis carinii is a well-known cause of pneumonia in the compromised host (27), but the epidemiology and pathogenesis of this infection are poorly understood. Pneumocystis pneumonia can be produced in rats by the adminis-

---

* To whom correspondence should be addressed.
© 1982 Gustav Fischer New York, Inc.
Proceedings of the Third International Workshop on Nude Mice.

tration of corticosteroids and apparently represents reactivation of latent infection (8). Similar studies have been performed in rabbits (20), but other animal species have not been developed as experimental models of this infection.

We have been interested in studying experimental *P. carinii* infection in mice. In our initial studies at Memorial Sloan Kettering Cancer Center (MSKCC), we were able to transmit rat and human *P. carinii* infection to outbred Swiss athymic nude (nu/nu) mice (31). At the Lexington Veterans Administration Medical Center and University of Kentucky, we produced pneumocystis pneumonia in different stains of normal mice by steroid administration (28).

The present report describes our studies of *P. carinii* infection in outbred Swiss nude mice of different genetic background than the mice used at MSKCC. The results of these studies have been compared with our previous studies in nude and steroid-treated normal mice.

Materials and Methods

Source of nude mice. Outbred female Swiss nude nude (nu/nu) and haired ($nu/+$) littermates at least 5 weeks old were used in these studies. Male mice were also sometimes used. The mice were developed at the National Institutes of Health (NIH) and were obtained through an interagency agreement between the National Cancer Institute (NCI) of NIH and Veterans Administration (VA). The NCI contracts with commercial breeding laboratories to develop certain strains of mice. The mice used here were raised in a closed conventional colony at Harlan Industries (Indianapolis, Indiana) as part of this interagency agreement and shipped to Lexington at regular intervals. At our institution, the mice were housed in standard cages without tops in Bioclean Units (Fieldstone Corp., Cincinnati, Ohio) in a room apart from other animals. The mice ate conventional food and drank tap water ad libitum. Cages and bedding were changed weekly. A successful program of breeding was established by mating nu/nu males with $nu/+$ females. The offspring were healthy and were used in inoculation experiments when they reached maturity.

Earlier studies at MSKCC used an ICR strain of outbred Swiss nude mice raised and maintained at MSKCC (31).

Source of normal mice. Eight different strains of normal mice obtained through the interagency agreement with NCI were used in these studies, as described previously (28).

Experimental protocols for nude mice. Several basic protocols were used. Pneumocystis pneumonia was produced in adult male Sprague-Dawley rats by the administration of steroids, a low-protein (8%) diet, and tetracycline in the drinking water (29). A heavily infected rat was sacrificed and the lungs were aseptically removed, cut into small pieces, homogenized with a teflon pestle and fine wire mesh screen, and washed with a balanced salt solution (with or without antibiotics). The inoculum size was calculated by placing 1-μl dosages of the homogenate on a glass slide, staining with cresyl echt violet, and counting the number of *P. carinii* cysts, as described previously (30).

In one type of experiment, 0.05–0.1 ml of the homogenate was injected into groups of mice intranasally, intratracheally, or directly into one or both lungs.

In some instances pentobarbital anesthesia was used. In selected studies mice were also administered steroids and/or a low-protein diet.

Another type of experiment involved the use of aerosols. The mice were placed in a modified Henderson exposure chamber and the lung homogenate was aerosolized for varying periods of time with a DeVilbiss nebulizer.

The inoculated mice were then placed in a separate Bioclean Unit or in germ-free isolators, whereas control mice remained in the original Bioclean Unit. Both inoculated and control mice ate regular diet and drank tap water with tetracycline 1 mg/ml to suppress bacterial infection.

Environmental transmission studies were performed in germ-free isolators. Groups of mice and steroid-treated rats were placed in the same cage and separated by a metal screen to study the transmission of *P. carinii* by close contact. The mice were sacrificed at sequential time intervals to determine the incubation period of *P. carinii*. In airborne transmission studies, the mice and steroid-treated rats were placed in separate germ-free isolators; a unidirectional air flow system was established from the rat to mouse isolator.

Environmental transmission studies were also performed with steroid-treated normal mice. Pneumocystis pneumonia was produced in C3H/HeN mice by the administration of cortisone acetate injections or dexamethasone in the drinking water, low-protein diet, and tetracycline in the drinking water. These mice were placed in the same cage with *nu/nu* and *nu/+* mice but separated by a metal screen. The *nu/nu* and *nu/+* mice received no steroids, ate regular diet, and drank water with tetracycline.

Histopathology. At death or sacrifice, mouse lung sections were prepared by standard techniques and stained with methenamine silver and hematoxylin and eosin (12). A standardized procedure, which had been developed for steroid-treated mice and rats (14,28,29), was used to grade the intensity of *P. carinii* infection in nude mice. At least three blocks of lung (one each from the upper and lower portions of the right lung, and one from the midportion of the left lung) were stained with methenamine silver, coded, and read blindly. The following scoring system was used: 0, no *P. carinii* found; 0.5+, minimal infection, <1% alveoli involved; 1+, light, 1–25% alveoli involved; 2+, moderate, 25–50% alveoli involved; 3+, heavy, 50–75% alveoli involved; 4+, very heavy, >75% alveoli involved. For comparison, the mean score of a particular mouse group was calculated only from those mice in which *P. carinii* was demonstrated (i.e., with a score $\geq 0.5+$).

Results

Transmission of Rat P. carinii by Inoculation

These studies are outlined in Table 7-1. Since the mice obtained from Harlan Industries and the mice born in our colony exhibited similar susceptibility to *P. carinii*, these animals have been pooled in the analysis of all studies. Intranasal and intratracheal injection of rat *P. carinii* failed to produce *P. carinii* infection in *nu/nu* and *nu/+* mice despite a rather wide variation in inoculum size. Intrapulmonary injection resulted in infection in a few *nu/nu* and *nu/+*

Table 7-1. Transmission of rat *Pneumocystis carinii* (PC) infection to nude mice by different methods of inoculation

Method of inoculation	Dose range (no. cysts)	No. infected/ total	Mean PC score
Intranasal	5.4×10^3–7.8×10^5		
nu/nu		0/9	0.0
nu/+		—	—
Intratracheal	5.4×10^3–7.8×10^5		
nu/nu		0/13	0.0
nu/+		0/5	0.0
Intrapulmonary	5.4×10^3–1.5×10^6		
nu/nu		5/80	0.5
nu/nu[a]		2/24	0.5
nu/+		4/28	0.5
Aerosol	9.5×10^5–7.9×10^{8b}		
nu/nu		1/55	0.5
nu/+		1/16	0.5
Controls	—		
nu/nu		3/78	0.5
nu/nu[a]		0/18	0.0
nu/+		0/28	0.0

[a] Received steroids.
[b] Per milliliter of homogenate.

mice, but the success of this techique could not be related to inoculum size. Since no differences were noted between unilateral and bilateral intrapulmonary injections, these results have been pooled. Aerosolization of infected rat lung homogenates was also largely unsuccessful. The presence of *P. carinii* in a few control *nu/nu* made it impossible to conclude that successful transmission of the organism took place by any of the modes of inoculation.

A variety of maneuvers was carried out in an attempt to increase the success of transmission. The administration of steroids (cortisone acetate, 1 mg, injected subcutaneously twice weekly or dexamethasone, 0.5—1.0 mg per 1000 ml drinking water) did not enhance susceptibility to *P. carinii* in inoculated or control *nu/nu* mice. The substitution of a low-protein diet for the regular diet shortened the life span of the *nu/nu* mice and led to cannibalism. The use of different methods of lung homogenization, salt solutions, substances (e.g., Ca^{2+}, and Mg^{2+}, antibiotics) in the salt solution or different injection techniques all failed to enhance transmission.

Environmental Transmission of Rat P. carinii

Transmission of *P. carinii* from steroid-treated rats to nude mice by close contact or the airborne route was unsuccessful (Table 7-2).

Environmental Transmission of Mouse P. carinii

Pneumocystis carinii was found in 4 (17%) of 23 *nu/nu* and 6 (53%) of 14 *nu/+* mice exposed to steroid-treated C3H/HeN mice, suggesting that trans-

Table 7-2. Environmental transmission of *Pneumocystis carinii* (PC) infection from rats to nude mice

Mode of transmission	No. infected/ total	Mean PC score
Close contact		
nu/nu	0/17	0.0
nu/+	0/11	0.0
Airborne		
nu/nu	2/48	0.5
nu/+	0/19	0.0
Controls		
nu/nu	1/65	0.5
nu/+	1/16	0.5

mission of the infection might have occurred (Table 7-3). The mean ± standard error survival in the isolators was 40 ± 3 days for *nu/nu* mice, compared with 59 ± 4 days for *nu/+* mice.

Original Studies at Memorial Sloan-Kettering

Investigations involving rat *P. carinii* have been summarized in Table 7-4 for comparison. Studies involving human *P. carinii* are not included. Intrapulmonary injection (either unilateral or bilateral) of fresh injected rat lung homogenates resulted in *P. carinii* in 11 (65%) of 17 *nu/nu* mice, compared with only 1 (7%) of 14 *nu/+* mice; intranasal inoculation was unsuccessful. Environmental transmission studies were performed in germ-free isolators with steroid-treated rats in a manner similar to the present studies with the following exceptions: (a) The mice were not sacrificed at sequential intervals; (b) for airborne transmission studies the mice and rats were placed in separate cages within the same isolator. The results showed a high frequency *P. carinii* infection in exposed *nu/nu* mice. No *P. carinii* infection was found in control *nu/nu* or *nu/+* mice raised apart from other animals, but *P. carinii* was found in 6 (15%) of 39 *nu/nu* mice housed in the rat colony room.

Table 7-3. Environmental transmission of *Pneumocystis carinii* (PC) infection from steroid-treated C3H/HeN mice to nude mice

Mode of transmission	No. infected/ total	Mean PC score
Close contact		
nu/nu	4/23	0.5
nu/+	6/14	0.6
Controls		
nu/nu	1/18	0.5
nu/+	0/13	0.0

Table 7-4. Initial studies of transmission of rat *Pneumocystis carinii* (PC) infection to nude mice

Mode of transmission	No. infected/ total	Mean PC score
Intranasal		
nu/nu	0/8	0.0
Intrapulmonary		
nu/nu	11/17	0.8
nu/+	1/14	0.5
Close contact		
nu/nu	9/10	1.1
Airborne		
nu/nu	6/10	0.8
Controls		
nu/nu	0/84	0.0
nu/nu[a]	6/39	0.8
nu/+	0/10	0.0

[a] In same room with rats.

Steroid-Treated Normal Mice

In earlier studies a variety of steroid dosage regimens was evaluated in attempts to find the optimal dose to produce pneumocystis pneumonia. These regimens have been combined here in the summary of the results (Table 7-5). *Pneumocystis carinii* infection was found in all strains of steroid-treated mice. The intensity of *P. carinii* infection varied considerably among the different strains of mice. C3H/HeN mice had the highest mean *P. carinii* score (1.9);

Table 7-5. *Pneumocystis carinii* infection in different strains of normal mice

Mouse strain	No. infected/ total	Mean PC score
Steroid treated		
C3H/HeN	27/30	1.9
BALB/cAnN	31/38	1.5
AKR/J	17/17	1.3
C57BL/6N	15/20	1.2
B10.A(2R)	10/14	1.1
Swiss Webster	13/16	1.1
DBA/2N	5/14	0.8
DBA/1J	4/17	0.8
Controls		
C3H/HeN	2/12	0.5
BALB/cAnH	4/19	0.5
AKR/J	2/6	0.5
C57BL/6N	1/6	0.5
B10.A(2R)	1/6	0.5
Swiss Webster	1/9	0.5
DBA/2N	0/6	0.0
DBA/1J	0/3	0.0

DBA/2N and DBA/1J mice had the lowest score (0.8) and also the lowest frequency of infection; other strains of mice had intermediate scores. A few control mice of each strain (except DBA/2N and DBA/1J) had scattered foci of infection.

Histopathology

The histological features of *P. carinii* infection on methenamine silver-stained lung sections of *nu/nu*, *nu/+*, and steroid-treated normal mice were similar; the major difference was that the intensity of the infection was consistently greater in the steroid-treated mice. *Pneumocystis* organisms in all the mice were morphologically indistinguishable from rat or human *P. carinii*. In light infections, the organisms lined up singly or in small groups along walls of alveoli; in heavier infection, there was more alveolar filling. The host inflammatory responses to *P. carinii* on hematoxylin and eosin-stained lung sections of nude and steroid-treated normal mice consisted primarily of a mild, nonspecific interstitial mononuclear cell response.

Discussion

Our early studies demonstrated that rat and human *P. carinii* infection could be transmitted to *nu/nu* mice by intrapulmonary injection and via the environment (31). These studies suggested that the nude mouse might be a useful experimental model for *P. carinii* because steroids were not needed to produce the infection. Since most of the nude mice did not appear ill and had light infection, this animal appeared to be valuable mainly in studies of epidemiological features of *P. carinii*.

By contrast, the nude mice derived from the NCI appeared considerably more resistant to rat *P. carinii* (Tables 7-1 and 7-2). Despite a variety of modes of transmission and other maneuvers, the frequency of *P. carinii* in *nu/nu* mice was no greater than that achieved in *nu/+* littermates or in uninoculated controls.

The reasons for these disparate results are unclear, but several hypotheses can be offered. One is that there are strain differences in rat *P. carinii*. There is little data on this subject, but no morphologic or immunologic differences have been reported in *P. carinii* isolated from rats (24,25,32). The rats used in the early and current studies were primarily adult Sprague-Dawley rats but were obtained from different breeders. Most of the rats in the current studies were obtained from a single breeder; in a few instances rats were obtained from other sources but there were no differences in the success of transmission of *P. carinii* to *nu/nu* mice.

Another possibility is that the results represent different experimental conditions. The current studies were designed primarily to extend the results of the earlier studies. Greater attention was devoted to such factors as establishing the incubation period of *P. carinii*, calculating the minimal infecting dose, increasing the intensity of the infection, and developing new modes of transmission (e.g., aerosols). However, some of the current intrapulmonary injection studies virtually duplicated the earlier studies in terms of inoculum size and technique, and transmission of *P. carinii* was still unsuccessful.

A third possibility is that the results represent strain differences among the mice in susceptibility to *P. carinii*. Although the nude mice used in these studies were outbred, they were of different genetic backgrounds. Differences among mouse strains have been found in susceptibility to *Toxoplasma gondii* (1), an organism that shares some similarities with *P. carinii*. Differences in *P. carinii* infection among the strain of steroid-treated mice also support this hypothesis; however, these results must be interpreted with some caution because the mechanism is reactivation of latent infection rather than exogenous challenge.

A fourth possibility is that the results represent different environmental conditions. Nude mice raised in different environments have exhibited differences in studies of immune function (19). *Nu/nu* mice can be more susceptible or more resistant than *nu/+* littermates to infection depending on such factors as the organism involved, the route of inoculation, the age of the mice, and preexisting microbial flora (3,4,6,15). A variety of factors contribute to the concept of "nonspecific resistance" in nude mice, which is expressed at least in part by the activity of macrophages (5,16); this resistance can be modified by such factors as the administration of antibiotics, which alters the mouse microbial flora (9,16). The nude mice in our early and current studies were raised in different environments. On the other hand, no differences in susceptibility to *P. carinii* were noted in nude mice obtained from Harlan Industries and nude mice born at our institution.

Studies of exposure to steroid-treated normal mice (Table 7-3) were the only instance suggesting that transmission of *P. carinii* infection occurred. It is unclear whether the higher frequency of *P. carinii* in *nu/+* mice represents longer survival or perhaps greater sensitivity to the organism. This higher frequency of infection again raises the question of species or strain differences in *P. carinii*. No morphological differences have been found in *P. carinii* from a variety of animal hosts, and previous attempts to transmit the organisms to immunologically intact animals of several species have been unsuccessful (26).

The only other studies of *P. carinii* in *nu/nu* mice have come from Ueda et al., who found naturally occurring pneumocystis pneumonia in a BALB/c nude mouse colony (23). Mice more than 6 months old apparently died from *P. carinii* infection, whereas mice 2–3 months old had light infection and exhibited no clinical signs of illness. These authors also produced *P. carinii* infection in *nu/+* BALB/c mice by administration of steroids of cyclophosphamide and transmitted the infection by close contact to *nu/nu* mice in a manner similar to our studies.

We have attempted to study *P. carinii* infection in BALB/c AnN nude mice obtained in the NCI–VA sharing agreement. Our efforts have been frustrated by problems in obtaining adequate numbers of mice, by the variable state of health of the mice which are received, and by our inability to breed the mice at our institution.

The nude mouse is a unique animal whose potential as an experimental model for infection agents is just beginning to be realized (2). Until recently, studies of the *nu/nu* tract in relation to other genetic factors relating to host susceptibility in infection have been limited because nude mice were only available in a few inbred mouse strains. The NIH breeding program of establishing the *nu/nu* gene in a variety of well-characterized mouse strains should overcome this problem. There is also an emerging consensus that environmental conditions vitally influence the value of the nude mouse in biomedical research (Nomura and Kagiyama, Chapter II, this volume). Thus, nude mice should be obtained from a single source with well-defined endogenous microbial flora to ensure the validity and reproducibility of the results obtained. These factors can best be controlled by the individual investigator.

In conclusion, our studies illustrate some of the problems in working with the nude mouse as an experimental model for infection with *P. carinii* and other organisms of low virulence. Future studies with this system should be conducted according to the guidelines outlined above. Other experimental techniques may improve the development of animal models for *P. carinii* infection. These include the use of other types of immunodeficient mice and nude rats (7,13,21,22); the use of substances (e.g., mucin) that enhance the infectivity of organisms (17; Lamborn et al., Chapter 2, this volume); and the use of substances or procedures (e.g., silica, cyclophosphamide, radiation) that impair alveolar macrophages or other host defenses in nude mice and other experimental animals (10,11,15,18).

Acknowledgments

This work was supported in part by a grant from the Medical Research Service, U.S. Veterans Administration; by an American Cancer Society Institutional Research Grant; and by U.S. Public Health Service Biomedical Research Support Grant RR 05374 from the Division of Research Facilities and Resources, National Institutes of Health.

We gratefully acknowledge the assistance of Mary Ellen Rutledge and Darrell Winsett.

References

1. Araujo, F.G., D.M. Williams, F.C. Grumet, and J.S. Remington. Strain-dependent differences in murine susceptibility to toxoplasma. Infect. Immun. 13:1528–1530, 1976.
2. Armstrong, D. and P.D. Walzer. Experimental infections in the nude mouse, pp. 477–489. In J. Fough, and B. Giovanella (eds.), The nude mouse in experimental and clinical medicine. New York: Academic Press, 1978.
3. Baeman, B.L., M.E. Gershwin, and S. Maslan. Infectious agents in immunodeficient murine models: pathogenicity of *Noscardia asteroides* in congenitally athymic (nude) and hereditary asplemic (Dh/+) mice. Infect. Immun. 20:381–387, 1978.
4. Baeman, B.L., E. Goldstein, M.E. Gershwin, S. Maslan, and W. Lippert. Lung response of congenitally athymic (nude), heterozygous, and Swiss Webster mice to aerogenic and intranasal infection by *Nocardia asteroides*. Infect. Immun. 22:867–877, 1978.
5. Cheers. C., and R. Waller. Activated macrophages in congenitally athymic mice to bacterial pathogens. Infect. Immun. 18:636–645, 1977.
6. Emmerling, P., H. Finger, and J. Bockemuhl. *Listeria monocytogenes* infection in mice. Infect. Immun. 12:437–439, 1975.
7. Festing, M.F.W., D. May, T.A. Connors, D. Lovell, and S. Sparrow. An athymic nude mutation in the rat. Nature (London) 274:365–366, 1978.
8. Frenkel, J.K., J.T. Good, and J.A. Schultz. Latent *Pneumocystis* infection of rats, relapse, and chemotherapy. Lab. Invest. 15:1559–1577, 1966.

9. Hellstrom, P.B., and E. Balish. Effect of tetracycline, the microbial flora, and the athymic state on gastrointestinal colonization and infection of BABL/c mice with *Candida albicans*. Infect. Immun. 23:764–774, 1979.
10. Hunninghake, G.W., and A.S. Fauci. Immunological reactivity of the lung IV: effect of cyclophosphamide on alveolar macrophage cytoxic effector function. Clin. Exptl. Immunol. 27:555–559, 1977.
11. Irvin, A.D. Tumor induction in nude mice by bovine lymphoid cells transformed by a protozoan parasite, pp. 45–52. *In* T. Nomura, N. Ohsawa, N. Tamaoki, and K. Fujiwara (eds.), Proceedings of the second international workshop on nude mice, Tokyo: University of Tokyo Press, 1977.
12. Lim, J.K., W.T. Hughes, and S. Feldman. Studies of morphology and immunofluorescence of *Pneumocystis carinii*. Proc. Soc. Exptl. Biol. Med. 141:304–309, 1972.
13. Lozzio, B.B. The lasat mouse: A new model for transplantation of human tissues. Biomedicine 24:144–147, 1976.
14. Milder, J.E., P.D. Walzer, J.D. Coonrod, and M.E. Rutledge. Comparison of histologic and immunologic techniques for the detection of *Pneumocystis carinii* in bronchial lavage fluid of rats. J. Clin. Microbiol. 11:409–417, 1980.
15. Mogensen, S.C., and H.K. Anderson. Role of activated macrophages in resistance of congenitally athymic nude mice to hepatitis induced by Herpes Simplex virus type 2. Infect. Immun. 19:792–798, 1978.
16. Nikol, A.D., and P.F. Bonventure. Anomolous high native resistance of athymic mice to bacterial pathogens. Infect. Immun. 18:636–645, 1977.
17. Nungester, W.J., L.F. Jourdonais, and A.A. Wolfe. The effect of mucin on infection by bacteria. J. Infect. Dis. 59:11–21, 1936.
18. Reynolds, H.W., J.A. Kazmierowski, and D.C. Dale. Changes in the composition of canine respiratory cells obtained by bronchial lavage following irradiation or drug immunosuppression. Proc. Soc. Exptl. Biol. Med. 151:756–761, 1976.
19. Salmon, J.C., and N. Lunch. Discrepancies in nude mice. Biomedicine 26:77–78, 1977.
20. Sheldon, W.H. Experimental pulmonary *Pneumocystis carinii* infection in rabbits. J. Exptl. Med. 110:147–160, 1959.
21. Sher, I., A.D. Steinberg, A.K. Berning, and W.B. Paul. X-linked B-lymphocyte defect in CBA/N mice. J. Exptl. Med. 142:637–650, 1975.
22. Shultz, L., and M.C. Green. Motheaten, an immunodeficient mutant of the mouse. J. Immunol. 116:936–1943, 1976.
23. Ueda, K., U. Goto, S. Yamazaki, and K. Eugiwara. Chronic fatal pneumocystosis in nude mice. Jpn. J. Exptl. Med. 47:475–482, 1977.
24. Vavra, J., and K. Kucera. *Pneumocystis carinii* in Delanoe, its ultrastructure and ultrastructural affinities. J. Protozool. 17:463–483, 1970.
25. Vossen, M.E.M.H., P.J.A. Beckers, J.H.E.T.H. Menwissen, and A.M. Stradboulders. Developmental biology of *Pneumocystis carinii*, and alternative view on the life cycle of the parasite. Z. Parasitenkd. 55:101–118, 1978.
26. Walzer, P.D. *Pneumocystis carinii* pneumonia. South. Med. J. 70:1330–1337, 1977.
27. Walzer, P.D., D.P. Perl, D.J. Krogstad, and M.G. Schultz. *Pneumocystis carinii* pneumonia in the United States. Ann. Intern. Med. 80:83–93, 1974.
28. Walzer, P.D., R.D. Powell, and K. Yoneda. Experimental *Pneumocystis carinii* pneumonia in different strains of cortisonized mice. Infect. Immun. 24:939–947, 1979.
29. Walzer, P.D., R.D. Powell, K. Yoneda, M.E. Rutledge, and J.E. Milder. Growth characteristics and pathogenesis of experimental *Pneumocystis carinii* pneumonia. Infect. Immun. 27:928–937, 1980.
30. Walzer, P.D., M.E. Rutledge, K. Yoneda, and B.J. Stahr. *Pneumocystis carinii*: New separation method from lung tissue. Exptl. Parasitol. 47:356–368, 1979.
31. Walzer, P.D., V. Schnelle, D. Armstrong, and P.P. Rosen. Nude mouse: A new experimental model for *Pneumocystis carinii* infection. Science 197:177–179, 1977.
32. Yoneda, K., and P.D. Walzer. Interaction of *Pneumocystis carinii* with host lung: an ultrastructural study. Infect. Immun. 29:692–703, 1980.

8

Characteristic Responses of Nude Mice in Angiostrongyliasis and Echinococcosis

Masao Kamiya,* Yuzaburo Oku, Haruo Kamiya, and Tatsuji Nomura†

Department of Parasitology, Faculty of Veterinary Medicine, Hokkaido University, Sapporo, Japan, and Central Institute for Experimental Animals,† 1430 Nogawa, Takatsu, Kawasaki 213, Japan.

Abstract

Characteristic host responses of nude mice to infections with three zoonotic helminths, *Angiostrongylus cantonensis*, *A. costaricensis*, and *Echinococcus multilocularis*, were studied. (a) *Angiostrongylus cantonensis*: Nine days after infection with the larvae of *A. cantonensis*, a peak of peripheral eosinophilia was observed in male and female *nu/+* mice and female *nu/nu* mice, but not in male *nu/nu* mice. A relative impairment in the development of tissue eosinophilia in the subarachnoidal space was observed in *nu/nu* mice. Twenty-eight days postinfection, body length and reproductive organs of the worm in the brain of *nu/nu* mice were larger than those in *nu/+* mice. (b) *Angiostrongylus costaricensis*: The egg granuloma developed in the intestinal wall of *nu/+* and *nu/nu* mice infected with *A. costaricensis*. Infiltration of mast cells and eosinophils in the intestinal muscle of *nu/nu* and *nu/+* mice was not much different, although the extent of infiltration was less in *nu/nu* mice. However, a marked difference was observed in globule leukocytes, e.g., an increase of the globule leukocytes in the intestinal epithelium of *nu/+* but not *nu/nu* mice. As early as 6 weeks postinfection, precipitation antibody was detected in the sera of *nu/+* mice, but not in serum of *nu/nu* mice, whereas circulating antigens were found in the sera of *nu/nu* mice by the double-diffusion method. (c) *Echinococcus multilocularis*: In *nu/nu* mice, the formation of protoscolices was initiated as early as 4 weeks after the intraperitoneal administration of brood capsules, 0.05–0.2 mm in diameter, whereas it took more than 8 weeks in *nu/+* mice. In the two helminth infections (i.e., *A. cantonensis* and *E. multilocularis*) rapid growth of the helminths with little tissue reaction was observed in nude mice. However, nude mice could not become definitive hosts.

---

* To whom correspondence should be addressed.
© 1982 Gustav Fischer New York, Inc.
Proceedings of the Third International Workshop on Nude Mice.

Introduction

In general, a remarkable peripheral blood and local eosinophilia appears in many helminthic infections. Recent studies have shown that thymic lymphocytes play an important role in eosinophilia. Since several functions of eosinophils in helminthic infections, such as antibody-dependent damage of schistosomules and resistance to *Trichinella spiralis*, have been noted (1–3,8,10,18), the nude mouse is considered useful as a host model in studies on the response and function of eosinophils and related cells.

Angiostrongylus cantonensis, a metastrongylid nematode parasitizing the pulmonary artery of rats, is an etiological agent of eosinophilic meningoencephalitis (EM) of man in Southeast Asia and the Pacific Islands (5,21,24,26). The characteristic features of human EM caused by the worm have also been observed in experimental infection in the mouse (11,12). Growth of *A. cantonensis* and eosinophil kinetics in infected nude mice were investigated in the present studies.

Angiostrongylus costaricensis was first found in man in Costa Rica (19), causing abdominal granuloma especially in children. It was the second species of the genus *Angiostrongylus* known to be capable of producing an important human parasitosis; subsequently, its normal definitive host, the cotton rat, which is naturally parasitized in the mesenteric artery, was obtained in the field. Unlike *A. cantonensis*, *A. costaricensis* has a wide range of definitive hosts, including man and the cotton rat as natural hosts, and laboratory mice and rats as animal models. Kinetics of eosinophils, mast cells, and globule leukocytes in the infected nude mice were studied in the present work.

Echinococcus multilocularis larvae cause one of the most hazardous tapeworm diseases in man. This metacestode, like a malignant tumor, infiltrates the organs of the host with features known as alveolar hydatid disease. Among various experimental animals used as disease models (13,14,25), inbred mice in particular were studied concerning susceptibility to the worm from different aspects such as host age, sex, and strain. Recently, it was shown that complement of host animals induce lytic and cestocidal reaction in vitro in the protoscolex (17) and preadult (15), and the reaction is also correlated with the extent of host resistance to infection with *E. multilocularis*. In the present studies, we attempted to clarify the characteristic growth of *E. multilocularis* in nude mice from the aspect of host resistance.

Materials and Methods

Mice. Athymic *nu/nu* mice and heterozygous littermates *nu/+* of BALB/cA genetic background 8 weeks of age, from the Central Institute for Experimental Animals, Kawasaki, Japan, were used except for the experiments on worm recovery and host mortality, in which 7-month-old mice were employed. They were kept under a barrier system and supplied with sterilized diet and water throughout the experiments.

Parasites. *Angiostrongylus cantonensis*, supplied initially by Dr. Leon Rosen of Hawaii, was maintained via the snail, *Biomphalaria glabrata*. *Angiostrongylus costaricensis*, donated by Professor Pedro Morera, University of Costa Rica, San Jose, Costa Rica, was maintained via the same snail as *A. cantonensis*. *Echinococcus multilocularis*, given by Dr. R. L. Rausch of Alaska, was maintained for 22 years in our laboratory mainly by secondary echinococcosis using the Mongolian gerbil.

Inoculation of the parasites. Third-stage larvae of *A. cantonensis* and *A. costaricensis* were used for oral inoculation (16). *Echinococcus multilocularis* larvae were administered to animals intraperitoneally as a protoscoleces suspension (25). Oral administration of the larvae was also applied to *nu/nu* and *nu/+* mice.

Ouchterlony double diffusion (DD) was used for detection of circulating antigens in serum from mice infected with *A. costaricensis*. Also, for detection of circulating antibodies from *E. multilocularis*-infected nude mice reconstituted by thymus gland implantation, the DD and immunofluorescent antibody methods were used. For the classification of eosinophils, mast cells, and globule leukocytes, Dominici staining was employed in addition to H&E staining methods.

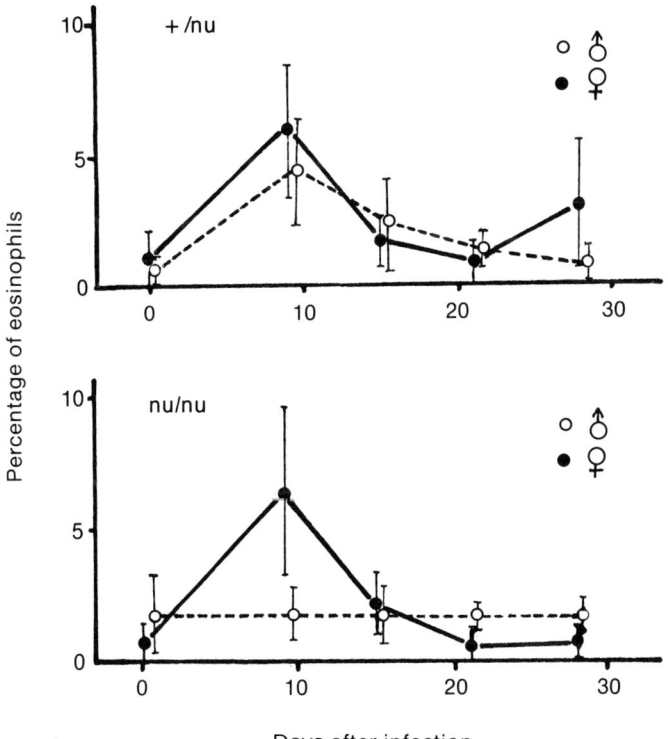

Figure 8-1. Average percentage of peripheral blood eosinophils of BALB/cA-*nu/nu* and -*nu/+* mice infected with *Angiostrongylus cantonensis*.

Results

Angiostrongylus cantonensis

Migration of the worm from the brain to the pulmonary artery was observed in one out of nine *nu/nu* mice and in one out of seven *nu/+* mice. In cases 9 days after infection with 50 infective-stage larvae, a peripheral eosinophilia was observed in approximately 5% of male and female *nu/+* and female *nu/nu* mice, but not in male *nu/nu* mice (Figure 8-1). An impairment in the development of tissue eosinophilia in the subaracnoidal space was observed in *nu/nu* mice (Figure 8-2). Twenty-two days postinfection, body length and the distances of reproductive organs from the vulva to the anterior extremity of the ovary in female and from the cloaca to the anterior extremity of testes in male worms in the brain of *nu/nu* and *nu/+* mice were more or less the same, but 28 days postinfection, those from *nu/nu* were bigger than those from *nu/+* mice (Figure 8-3). Numbers of worms recovered from the nine infected *nu/nu* and *nu/+* mice were 256 (infectivity: 56%) and 116 (infectivity: 37%), respectively. The mortality rate of *nu/nu* and *nu/+* mice up to 35 days postinfection was 33% (3/9) and 67% (6/9), respectively. More intact worms were recovered from *nu/nu* mice than from *nu/+* mice.

Angiostrongylus costaricensis

A peak of tissue eosinophilia was observed in the cecum walls and mesenteries in *nu/+* mice 3 weeks after infection, but little was found in the nude mice (Table 8-1). Extramedullary eosinophilopoiesis was observed in *nu/nu* and *nu/+* mice 6–15 weeks after infection (Table 8-1). Marked increases in the mast cells in the intestinal wall of both *nu/nu* and *nu/+* mice were observed as

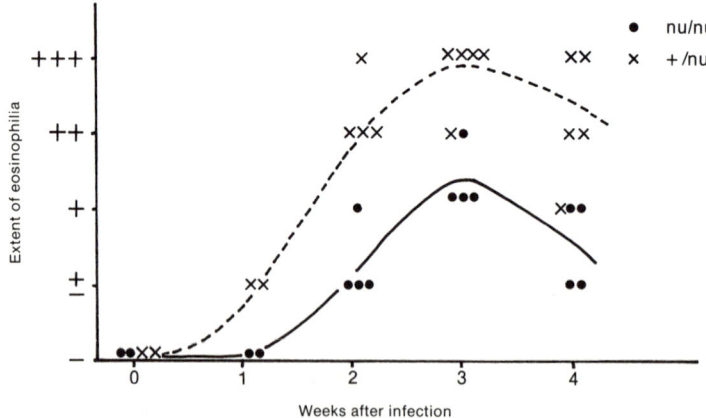

Figure 8-2. Kinetics of eosinophils in the brain of BALB/cA-*nu/nu* (●) and -*nu/+* (×) mice infected with *Angiostrongylus cantonensis*.

Table 8-1. Increase of eosinophils in nude mice (BALB/cA-*nu/nu*) and heterozygous mice (BALB/cA-*nu/+*) infected with *A. costaricensis*

Organ	Genotype	Control	\multicolumn{8}{c}{Weeks after infection}

Organ	Genotype	Control	1	2	3	6	9	12	15
Brain, kidneys, pancreas, stomach, and muscles of the femur	*nu/+*	−[a]	−	−	−	−	−	−	N.E.[b]
	nu/nu	−	−	−	−	−	−	−	−
Upper small intestine	*nu/+*	±	−	+	++	±	+	+	N.E.
	nu/nu	±	±	±	±	±	±	±	±
Lower small intestine	*nu/+*	±	−	++	+++	++	+	+	N.E.
	nu/nu	±	±	±	++	++	±	±	±
Cecum	*nu/+*	±	±	±	+++	+++	±	±	N.E.
	nu/nu	±	±	±	++	++	±	±	±
Upper colon	*nu/+*	−	−	−	±	+	+	+	+
	nu/nu	−	−	−	±	±	±	±	N.E.
Lungs	*nu/+*	−	−	−	−	−	−	−	−
	nu/nu	−	−	−	−	−	−	−	N.E.
Liver	*nu/+*	−	−	++	++	++	++[c]	++[c]	N.E.
	nu/nu	−	−	+	+	+	±	±	±
Mesentery	*nu/+*	±	+	++	++	++[c]	++[c]	++[c]	N.E.
	nu/nu	±	±	±	±	±	±	±	±
Spleen	*nu/+*	−	−	−	±	++[c]	++[c]	++[c]	N.E.[c]
	nu/nu	−	−	−	±	±	±	±	+

[a] Relative number of cells: (−) none to (+++) many.
[b] Not examined.
[c] Extramedullary islands of eosinophil precursors were observed.

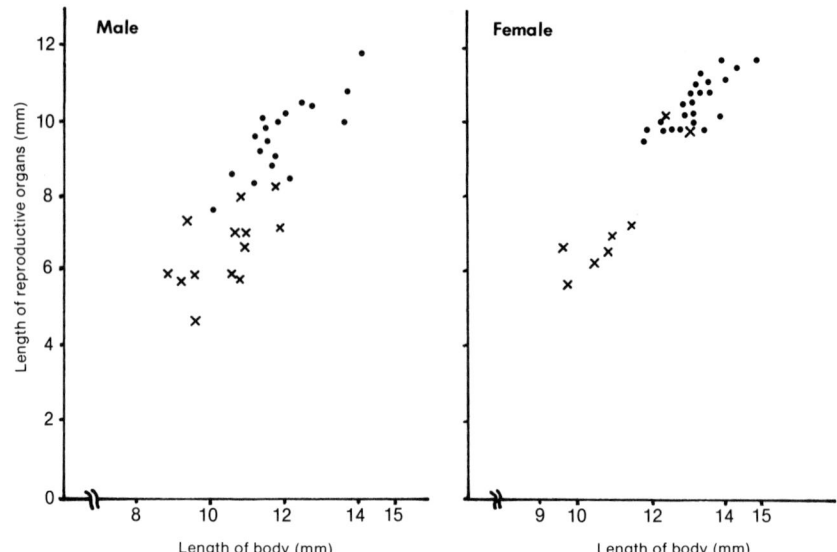

Figure 8-3. Length of body and reproductive organs of worms recovered from BALB/cA-*nu/nu* (●) and -*nu/+* (×) mice 28 days after infection with *Angiostrongylus cantonensis*.

early as 6 weeks thereafter (Table 8-2). An increase of globule leukocytes was noted in the intestinal epithelium of *nu/+* mice as early as 2 weeks thereafter, but it was not found in *nu/nu* mice (Table 8-3, Figure 8-4). None of three types of cells showed significant increases in the brain, lungs, spleen, kidney, femoral muscle, and stomach (Table 8-1). By 3 weeks after infection, *nu/nu* mice showed

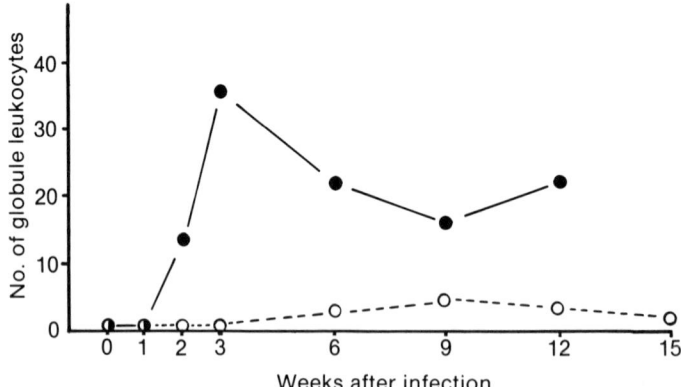

Figure 8-4. Kinetics of globule leukocytes (number per field of microscope vision, ocular ×10, objective ×40) in the cecal mucosa of nude (BALB/cA-*nu/nu*, ●) and heterozygous (BALB/cA-*nu/+*, ○) mice infected with *Angiostrongylus costaricensis* (10 L$_8$).

Table 8-2. Increase of mast cells in nude mice (BALB/cA-*nu/nu*) and heterozygous mice (BALB/cA-*nu/+*) infected with *A. costaricensis* (10 L₃)

Organ	Genotype	Control	Weeks after infection[a]						
			1	2	3	6	9	12	15
Lower small intestine	*nu/+*	±	±	±	±	++	+++	+++	N.E.[b]
	nu/nu	±	±	±	±	±	+++	+++	+++
Cecum	*nu/+*	±	±	±	+	±	++++	+++	N.E.
	nu/nu	±	±	±	±	±	++	++	++
Upper colon	*nu/+*	±	±	±	±	++	++	++	N.E.
	nu/nu	±	±		±	±	++	++	++

[a] Symbols as in Table 8-1.
[b] Symbols as in Table 8-1.

Table 8-3. Increase of globule leukocytes in nude mice (BALB/cA-*nu/nu*) and heterozygous mice (BALB/cA-*nu/+*) infected with *A. costaricensis* (10 L3)

Organ	Genotype	Control	Weeks after infection[a]						
			1	2	3	6	9	12	15
Lower small intestine	*nu/+*	−[a]	−	+	+	±	+	+	N.E.[b]
	nu/nu	−	−	−	−	−	±	−	−
Cecum	*nu/+*	−	−	++	+++	++	++	++	N.E.
	nu/nu	−	−	−	−	±	±	±	±
Upper colon	*nu/+*	−	−	+	+	+	++	++	N.E.
	nu/nu	−	−	−	−	−	±	−	−

[a] Symbols as in Table 8-1.
[b] Symbols as in Table 8-1.

a higher mortality than $nu/+$ mice. With DD, the antibodies were detected in $nu/+$ mice after 6 weeks, whereas in nu/nu mice no antibodies were detected but there were circulating antigens in the sera.

Echinococcus multilocularis

Parasites 200–230 μm in length were found in the intestinal villi of nu/nu and $nu/+$ mice 1 day after inoculation. Subsequently, no parasites could be detected in either nu/nu or $nu/+$ mice 2–54 days after inoculation. Focuses of secondary echinococcosis were observed in the abdominal cavity of two out of three female nu/nu mice 24 and 48 days after inoculation, respectively.

The development of the parasite in the abdominal cavity of nude mice was very rapid and host tissue reaction was very slight when compared with those of $nu/+$ mice (Table 8-4, Figure 8-5). A few macrophages, eosinophils, and fibroblasts appeared in adventitious tissue in nude mice 4 weeks after inoculation (Figure 8-5, 3). On the other hand, numerous macrophages, eosinophils, and lymphocytes were observed around the parasites in $nu/+$ mice 4 weeks after inoculation (Figure 8-5, 4). Brood capsules and mature protoscoleces were formed in nude mice as early as 2 and 4 weeks after infection, respectively, but no brood capsule formation was observed in free cysts in the abdominal cavity of $nu/+$ mice 8 weeks after infection.

Specific fluorescence was detected on the tegument of worms incubated with infected sera of $nu/+$ and thymus-implanted nude mice. Also, specific circulating antibody was detected by gel diffusion in the sera of $nu/+$ and thymus-implanted mice.

Discussion

Peripheral blood and tissue hypereosinophilia are characteristics of invasive helminthic infections in animals and humans (7). The absence of eosinophils in nude mice infected with *Trichinella spiralis* (23) and *Ascaris suum* (20) has been reported. However in the *Schistosoma mansoni* and *Toxocara canis* infection, thymus-bearing mice showed two peaks of peripheral blood hypereosinophilia but only the first peak was observed in nude mice (6). The present results of peripheral blood hypereosinophilia observed in female nude mice infected with *A. cantonensis* are similar to those for *S. mansoni* and *T. canis*. Various eosinophil chemotactic factors have been reported and classified (7). Among them, chemotactic factors derived from mast cells and T lymphocytes can be neglected in the present cases because of the absence of mast cells in the brain tissue and the lack of mature T lymphocytes in nude mice. From our observations that the death of the worms and the decrease of eosinophils in the brain started at almost the same time, 4 weeks after infection, we suggest that the chemotactic factor responsible for hypereosinophilia in the brain originates directly from living *A. cantonensis*. It was suggested by *in vitro* studies that mast cells, the origins of which are still uncertain, originate from the thymus and thymus-derived lymphocytes (9). Since nu/nu mice possess pre-T cells, it is understandable that mast cells infiltrate both nu/nu and $nu/+$ mice infected with *A. costari-*

Table 8-4. Development and number of echinococcal cysts free in the abdominal cavity of nude mice and total weight of echinococcal tissue

Host	Weeks after inoculation	No. examined	No. of free cysts in abdominal cavity (mean)	Brood capsule formation	Protoscolex formation	Total weight of echinococci (mean) (g)
Nude	1	1	93	−	−	Not weighed
	2	2	407	+	−	0.63
	3	2	435	+	−	1.08
	4	2	637	+	+	2.43
	5	2	165	+	+	4.0
	6	2	1	+	+	5.45
	8	4	13	+	+	7.0
	15[a]	2	0	+[b]	+[b]	18.5
	18[a]	1	0	+[b]	+[b]	18.0
	19[a]	1	0	+[b]	+[b]	18.0
Control (heterozygote)	2	1	27	−	−	0.25
	3	1	17	−	−	0.1
	4	1	10	−	−	0.4
	8	2	18	+[b]	+[b]	1.25

[a] Many echinococci spread by metastasis to abdominal organs. It was difficult to weigh cysts precisely.
[b] Examination of cysts attached to abdominal organs.

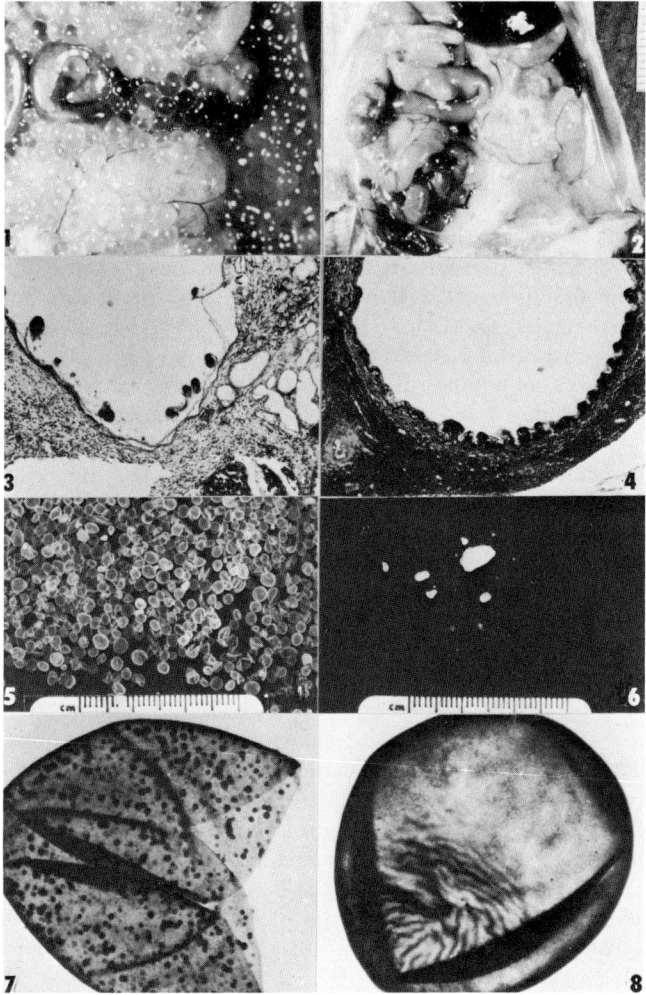

Figure 8-5. (1) Female nude mouse, 4 weeks after inoculation of suspended echinococcal tissue (ISET), showing numerous spherical and transparent cysts free in the abdominal cavity. (2) Female $nu/+$ mouse, 4 weeks after ISET, showing a few echinococcal clusters attached to abdominal organs. (3) Male nude mouse, 4 weeks after ISET, showing hepatic lesion accompanied by protoscolex formation and slight host tissue reaction. ×33. (4) Female $nu/+$ mouse, 4 weeks after ISET, showing no brood capsule formation and severe host tissue reaction with accumulation of many macrophages and eosinophils. ×33. (5) Female nude mouse, 4 weeks after ISET, showing numerous cysts with brood capsule formation, removed from the abdominal cavity. (6) Female $nu/+$ mouse, 4 weeks after ISET, showing a few small and opaque cysts in the abdominal cavity. (7) Female mouse, 4 weeks after ISET, showing a cyst with numerous brood capsules. ×11. (8) Female $nu/+$ mouse, 4 weeks after ISET, showing a cyst with thick cuticle and without brood capsule formation. ×11.

censis although the extents differ slightly. An increase in intestinal mast cells was reported in $nu/+$ mice infected with *T. spiralis* and *Nippostrongylus brasiliensis*, but not in nu/nu mice (22). On the other hand, in this study, mast cells were observed in the intestinal muscles of nu/nu and $nu/+$ mice infected with *A. costaricensis*. It has been reported that nonintestinal mast cells are abundant in nu/nu and $nu/+$ mice (22). Therefore, we would like to suggest that the infiltration process of intestinal mast cells might differ from that of nonintestinal mast cells.

Globule leukocytes, which were not found in the present nu/nu mice infected with *A. costaricensis*, are known to be abundant during parasitic infections. Although their functions and origin are also subjects for further investigations, several researchers (4) have suggested that intestinal globule leukocytes may be derived from subepithelial mast cells and have proposed a function related to an immune process containing IgE. Among the histocytes mentioned in this study, the infiltrating tendency of globule leukocytes was unique, but characteristic responses obtained in the present study seemed mutually correlated. Further investigations on both angiostrongyliasis and echinococcosis are expected using W/W^v mice, which are almost completely deficient in mast cells. Nude mice cannot act as the definitive host for the present helminths, *A. cantonensis* and *E. multilocularis*, and it can be concluded that the nude mouse has a protective mechanism against helminths which does not involve mature T lymphocytes.

Acknowledgments

We wish to thank Professor Y. Tajima, Central Institute for Experimental Animals, Dr. N. Ohsawa, Faculty of Medicine, University of Tokyo, and Professor M. Ohbayashi, Faculty of Veterinary Medicine, Hokkaido University, who helped in many aspects of the work. Thanks are also due to Mrs. Y. Sakata for bibliographic assistance.

This study was supported in part by Grant-in-Aid No. 487072 from the Ministry of Education, Science and Culture of Japan and a grant from the Science and Technology Agency of Japan.

References

1. Butterworth, A.E. The eosinophil and its role in immunity to helminth infection. Curr. Top. Microbiol. Immunol. 77:127–168, 1977.
2. Butterworth, A.E., R.F. Sturrock, V. Houba, A.A.F. Marmoud, A. Sher, and P.H. Rees. Eosinophils as mediators of antibody-dependent damage to schistosomula. Nature (London) 256:727–729, 1975.
3. Butterworth, A.E., R.F. Sturrock, V. Houba, and P.H. Rees. Antibody-dependent cell-mediated damage to schistomula in vitro. Nature (London) 252:503–505, 1974.
4. Carrire, R., and M. Buschke. The influence of thyroid and testiculer hormones on globule leucocytes in the rat duodenal crypt epithelium. Anat. Rec. 192:407–422, 1978.
5. Chen, S.N. Blood and cerebrospinal fluid findings in eosinophilic meningoencephalitis: Immunoglobulins and antibody to *Angiostrongylus cantonensis*. Am. J. Trop. Med. Hyg. 21:415–420, 1975.

6. Colley, D.G., S.P. Katz, and S.K. Wikel. Schistosomiasis: An experimental model for the study of immunopathological mechanisms which involve eosinophils. Adv. Biosci. 12:653–666, 1973.
7. Goetzl, E.J. and K.F. Austen. Cellular characteristics of the eosinophil compatible with a dual role in host defense in parasitic infections. Am. J. Trop. Med. Hyg. 26:142–150, 1977.
8. Grove, D.I., A.A.F. Mahmoud, and K.S. Warren. Eosinophils and resistance to *Trichinella spiralis*. J. Exptl. Med. 145:755–759, 1977.
9. Ishizaka, T., H. Okudaira, L.E. Mauser, and K. Ishizaka. Development of rat mast cells in vitro. 1. Differentiation of mast cells from thymus cells. J. Immunol. 116:747–754, 1976.
10. James, S.L. and D.G. Colley. Eosinophil-mediated destruction of *Schistosoma mansoni* eggs. J. Reticuloendothel. Soc. 20:359–374, 1976.
11. John, D.T. The biology of *Angiostrongylus cantonensis* in the white mouse. Ph.D. Dissertation. Abst 31B page 6431–6432 microfilm 71-11, 710. University of North Carolina, Chapel Hill, North Carolina, 1970.
12. John, D.T. and A.J. Martinez. Animal model of human disease. Central nervous infection with the nematode *Angiostrongylus cantonensis*. Am. J. Pathol. 80:345–348, 1975.
13. Kamiya, H. Studies on echinococcosis. XXIV. Age difference in resistance to infection with *Echinococcus multilocularis* in AKR strain of mouse. Jpn. J. Vet. Res. 20:69–76, 1972.
14. Kamiya, H. Observation on differences of susceptibility to larval *Echinococcus multilocularis* among uniform strains of the mouse. Jpn. J. Parasitol. 22:294–299, 1973.
15. Kamiya, H., M. Kamiya, and M. Ohbayashi. Studies on the host resistance to infection with *Echinococcus multilocularis*. II. Lytic effect of complement and its mechanism. Jpn. J. Parasitol. 29:167–179, 1980.
16. Kamiya, M., and H. Tanaka. Hemagglutination test in rats infected with *Angiostrongylus cantonensis*. Jpn. J. Exptl. Med. 39:593–599, 1969.
17. Kassis, A.I., and C.E. Tanner. The role of complement in hydatid disease: in vitro studies. Intl. J. Parasitol. 6:25–35, 1976.
18. Mahmoud, A.A.F., K.S. Warren, and P.A. Peters. A role for the eosinophil in acquired resistance to *Schistosoma mansoni* infection as determined by anti-eosinophil serum. J. Exptl. Med. 142:805–813, 1975.
19. Morera, P., and R. Cespedes. *Angiostrongylus costaricensis* n. sp. (Nematoda: Metastrongyloidea), a new lungworm occurring in man in Costa Rica. Rev. Biol. Trop. 18: 173–185, 1971.
20. Nielsen, K., L. Fogh, and S. Andersen. Eosinophil response to migrating *Ascaris suum* larvae in normal and congenitally thymusless mice. Acta Pathol. Microbiol. Scand. (B) 82:919–920, 1974.
21. Rosen, L., R. Chappell, G.L. Lagueur, G.D. Wallace, and P.P. Weinstein. Eosinophilic meningoencephalitis caused by a metastrongylid lung-worm of rats. J. Am. Med. Assoc. 179:620–624, 1962.
22. Ruitenberg, E.J., and A. Elgersma. Absence of intestinal mast cell response in congenitally athymic mice during *Trichinella spiralis* infection. Nature (London) 264:258–260, 1976.
23. Ruitenberg, E.J., A. Elgersma, W. Kruizinga, and F. Leentra. *Trichinella spiralis* infection in congenitally athymic (nude) mice: Parasitological, serological and haematological studies with observations on intestinal pathology. Immunology 33:581–587, 1977.
24. Tungkanak, R., S. Sirisinha, and S. Punyagupta. Serum and cerebrospinal fluid in eosinophilic meningoencephalitis: Immunoglobulins and antibody to *Angiostrongylus cantonensis*. Am. J. Trop. Med. Hyg. 21:415–420, 1972.
25. Yamashita, J., M. Ohbayashi, and S. Konno. Studies on echinococcosis. VI. Secondary echinococcosis multilocularis in mice. Jpn. J. Vet. Res. 5:197–202, 1957.
26. Yii, C.Y. Clinical observations on eosinophilic meningitis and meningoencephalitis caused by *Angiostrongylus cantonensis* in Taiwan. Am. J. Trop. Med. Hyg. 25:233–249, 1976.

9

Thymus Dependence of Specific Homocytotropic Antibody Production and Intestinal Mast Cell Accumulation

Norman D. Reed,* Patricia K. Crowle,† Rheta S. Booth, and John J. Munoz‡

Departments of Microbiology and Biology,† Montana State University, Bozeman, Montana 59717, and Laboratory of Microbial Structure and Function,‡ National Institute of Allergy and Infectious Diseases, Rocky Mountain Laboratory, Hamilton, Montana 59840.

Abstract

Thymus dependence of specific homocytotropic antibody production and intestinal mucosal mast cell accumulation was studied by using congenitally athymic (nude-nu/nu) mice and their euthymic (-$nu/+$ or -$+/+$) littermates. When immunized with the thymus-dependent antigens hen egg albumen, crude extract of *Ascaris suum*, keyhole limpet hemocyanin, saline extract of *Bordetella pertussis*, or living *Nippostrongylus brasiliensis*, euthymic mice made antigen-specific IgE (72-h passive cutaneous anaphylaxis, PCA) and, in every case tested, antigen-specific IgG1 (2-h PCA); athymic mice made neither IgE nor IgG1 when immunized with these antigens. Neither euthymic mice nor athymic mice made homocytotropic antibody when immunized with the thymus-independent antigens type III pneumococcal polysaccharide, *Escherichia coli* 0113 lipopolysaccharide, or Vi antigen. Similarly, in Schultz-Dale reactions uteri of immunized euthymic but not athymic mice contracted in the presence of specific thymus-dependent antigen. Although normal or elevated numbers of mast cells (MC) were found in the skin of athymic mice and athymic mice were as effective as euthymic mice as recipients in MC-dependent PCA reactions, nude mice were deficient in intestinal mucosal MC. Furthermore, the dramatic increase in intestinal mucosal MC seen in *N. brasiliensis*-infected euthymic mice did not occur in infected athymic mice. The defects in homocytotropic antibody production and mucosal mast cell response were both repaired by implantation of thymus glands or injection of thymus cells into athymic mice. These results indicate that production of antigen-specific IgE and IgG1 homocytotropic antibody and the mucosal mast cell response to infection with the intestinal parasite *N. brasiliensis* are rigorously thymus dependent.

---

* To whom correspondence should be addressed.
© 1982 Gustav Fischer New York, Inc.
Proceedings of the Third International Workshop on Nude Mice.

Introduction

Because nude mice have become popular tools in immunoparasitology (18; Mitchell and Holmes, Chapter I, this volume), we have evaluated the ability of nude mice to make specific homocytotropic antibody and to accumulate intestinal mast cells—two responses that have been proposed to play a role in the regulation of certain host–parasite relationships.

The ability of athymic nude mice to produce homocytotropic antibody is not clear. Although in some studies nude mice failed to produce specific IgE when sensitized with ovalbumin, bovine serum albumin, or human gammaglobulin with *Bordetella pertussis* vaccine as adjuvant (16) and failed to produce specific IgE or IgG when infected with *Shistosoma mansoni* (8), nude mice immunized with ovalbumin adsorbed to $Al(OH)_3$ gel were reported to produce specific IgE (2). In a recent study (Roberts et al., Chapter 20, this volume), nude mice immunized with tetanus toxoid plus $Al(OH)_3$ failed to produce specific IgE; however, when total IgE was measured nude mice had significant levels.

Similarly, mast cell production of nude mice has not been well defined. Athymic nude mice have normal numbers of mast cells in lip and tongue (21) and normal or increased numbers in skin (12,25). In contrast, in the intestinal mocusa of nude mice the number of mast cells is markedly reduced compared to euthymic mice (21), and the impressive increase in numbers of intestinal mucosal mast cells seen in nematode-infected euthymic mice does not occur in infected athymic mice (21,23).

In this study, we used various immunization schedules to immunize athymic and euthymic mice with combinations of eight different thymus-dependent or thymus-independent antigens with four different adjuvants. No group of athymic mice made any specific homocytotropic antibody detectable by passive cutaneous anaphylaxis or Schultz-Dale assays, whereas euthymic mice made specific homocytotropic antibody to thymus-dependent but not to thymus-independent antigens. We also infected mice with the nematode *Nippostrongylus brasiliensis* and evaluated the intestinal mucosal mast cell response at different times after infection; the mucosal mast cell response occurred in euthymic but not nude mice. The defects in homocytotropic antibody production and mucosal mast cell response were both repaired by implantation of thymus glands or injection of thymus cells into nude mice.

Materials and Methods

Mice. Congenitally athymic (nude-*nu/nu*) mice and their normal thymus-bearing (euthymic-$+/+$ or -*nu/+*) littermates were from our colony in which cross–intercross mating is in progress to derive a line of athymic mice congenic with BALB/c mice. The techniques used for thymus gland implantation and thymus cell injection of nude mice have been described (9).

Antigens and adjuvants. Bovine gammaglobulin (BGG: Armour, Phoenix-

Arizona), hen egg albumen (OVA: Nutritional Biochemical Corp., Cleveland, Ohio), concanavalin A (ConA: Miles-Research Products Division, Cleveland, Ohio), and aluminum amonium sulfate (Al: Humco, Texarkana, Texas) were used as purchased from the commercial sources indicated. Keyhole limpet hemocyanin (KLH) was purchased from Pacific Bio-marine and dissociated using the method of Weigle (26). Type III pneumococcal polysaccharide (SIII) prepared using a modification of Felton's procedure (7) was a gift from Dr. P.J. Baker, U.S. National Institutes of Health, Bethesda, Maryland. The Vi antigen was prepared by continuous-flow paper curtain electrophoresis (11) and donated by Dr. F.G. Jarvis, Idaho State University, Pocatella, Idaho. *Escherichia coli* 0113 lipopolysaccharide (LPS), extracted by the phenol–water method (27), was given to us by Dr. Jon Rudbach, University of Montana, Missoula, Montana. Dr. Herbert Morse III, U.S. National Institutes of Health, made and donated the crude *Ascaris* extract. Saline extract of *Bordetella pertussis* was prepared as described by Munoz and Hestekin (19).

Details of immunization are given in the footnotes to tables.

Infection of mice. Third-stage larvae (L_3) of a mouse-adapted strain of *Nippostrongylus brasiliensis* were prepared and used to infect mice as described elsewhere (10).

Assay of homocytotropic antibodies. Worm-specific IgE responses of *N. brasiliensis*-infected mice were determined by passive cutaneous anaphylaxis (PCA) tests in rats (22); we have published elsewhere the details of this test (15). All other PCA responses were done in mice using procedures described by Clausen et al. (5) in which the 2-h reaction results from homocytotropic IgG1 and the 72-h reaction results from IgE.

Schultz-Dale reactions (20) were done using immune mouse uteri incubated at 37°C in Tyrode's solution; for challenge, specific antigen or serotonin was added in the amounts indicated in Table 9-2.

Enumeration of mast cells. Mice were killed by cervical dislocation and the entire small intestine was removed. The proximal and distal portions of the intestine were cut away; the middle 20-cm portion was cut along the mesenteric attachment, flattened on a strip of paper, folded into ten 2-cm folds, and immersed in Carnoy's fixative (6). Tissue was fixed 3–8 h prior to dehydration and infiltration with paraffin. Tissues were sectioned at 10-μm thickness and stained with eosin and the Astra blue method described by Bloom and Kelly (3). A villus–crypt unit was defined according to Miller and Jarrett (17). Counts of intraepithelial and lamina propria mast cells were combined and treated as a single population of cells.

Results and Discussion

Thymus Dependence of Homocytotropic Antibody Production

The results of many experiments collectively involving eight antigens, four adjuvants, and eleven different immunization schedules are summarized in Table 9-1. Although euthymic mice regularly produced specific IgE in response to

Table 9-1. Specific homocytotropic antibody responses of immunized euthymic and athymic nude mice[a]

Group	Antigen[b]	Adjuvant[b]	Day serum collected for PCA test[c]	PCA, 2 h[c] Athymic	PCA, 2 h[c] Euthymic	PCA, 72 h[c] Athymic	PCA, 72 h[c] Euthymic
Thymus dependent							
A.	OVA, 125 µg day 0	PE, 50 µg day 0	21	0	5	0	5
B.	OVA, 10 µg days 0 and 20	Al, 10 mg day 0	10	ND	ND	0	32
			20	ND	ND	0	32
			30	ND	ND	0	32
C.	OVA, 100 µg day 0	ConA, 400 µg day 0	10	0	32	0	64
			33	0	4	0	16
D.	OVA, 10 µg day 0	Al, 1 mg day 0 and LPS 20 µg day 0	10	0	16	0	32
E.	CAE, 10 µg days 0 and 20	Al, 10 mg day 0	10	ND	ND	0	128
			20	ND	ND	0	32
			30	ND	ND	0	16
F.	KLH, 50 µg day 0 and 5 µg days 14 and 21	PE, 50 µg day 0 and 5 µg days 14 and 21	21[d]	0	5	0	5
			27[d]	0	125	0	5
			21[e]	0	5	0	5
			27[e]	0	25	0	5
G.	Nippostrongylus brasiliensis[f] infection (300 L₃) day 0		9	ND	ND	0	0
			20	ND	ND	0	64

Thymus independent

	Antigen	Adjuvant	Day						
H.	LPS, 1 μg days 0 and 14	None	10	0	0	0	0	0	0
			20	0	0	0	0	0	0
I.	Vi, 1 μg days 0 and 10	Al, 10 mg day 0	10	0	0	ND	ND	0	0
			20	0	0	ND	ND	0	0
J.	SIII, 0.5 μg days 0 and 21	PE, 25 μg days 0 and 21	21	0	0	ND	ND	0	0
			30	0	0	ND	ND	0	0
K.	SIII, 0.5 μg days 0 and 21	None	21	0	0	0	0	0	0
			30	0	0	0	0	0	0

[a] Abbreviations: Al, aluminum ammonium sulfate; CAE, crude extract of *Ascaris suum*; ConA, concanavalin A; KLH, keyhole limpet hemocyanin; LPS, *Escherichia coli* 0113 lipopolysaccharide; ND, not done; OVA, hen egg albumen; PCA, passive cutaneous anaphylaxis; PE, saline extract of *Bordetella pertussis*; SIII, type III pneumococcal polysaccharide; Vi, Vi antigen.

[b] All antigens and adjuvants were given intraperitoneally.

[c] Serum taken at various times from immunized athymic and euthymic mice was diluted and injected intracutaneously to passively sensitize recipient CFW mice; 2 or 72 h after passive sensitization the recipient mice were injected intravenously with specific antigen and Evans blue dye and PCA titers were determined 30 min later. Data are reported as the reciprocals of the highest serum dilution giving a positive PCA reaction in the recipient mice.

[d] Recipient mice were challenged with KLH and Evans blue dye.

[e] Recipient mice were challenged with PE and Evans blue dye; PE functions both as an antigen and an adjuvant.

[f] PCA titers determined in rats.

Table 9-2. Schultz-Dale reactions with uteri of immunized euthymic and athymic mice[a]

Immunization schedule	Mice	Contraction of uteri in presence of[c]	
		Specific antigen	Serotonin
100 μg BGG and	Athymic	BGG —	+
50 μg PE day 0 (20)[b]	Euthymic	BGG +	+
	Athymic	PE —	+
	Euthymic	PE +	+
50 μg KLH and	Athymic	KLH —	+
50 μg PE day 0;	Euthymic	KLH +	+
5 μg KLH and	Athymic	PE —	+
5 μg PE days	Euthymic	PE +	+
14 and 21 (27)			
1 μg LPS days	Athymic	LPS —	+
0 and 14 (20)	Euthymic	LPS —	+

[a] Abbreviations: see footnote a, Table 1.
[b] Day of sacrifice for Schultz-Dale test shown in parentheses.
[c] For Schultz-Dale reactions, uteri were placed in 20 ml Tyrode's solution at 37°C; for challenge, 0.5 ml of a solution of serotonin (1 mg/ml), BGG (2 mg/ml), KLM (2 mg/ml), PE (2 mg/ml), or LPS (1 mg/ml) was added to the bath.

thymus-dependent antigens (groups A–G), no athymic mouse group immunized with these antigens produced antigen-specific IgE detectable by the 72-h PCA assay. Similarly, in every case examined (groups A, C, D, and F) euthymic mice but not athymic mice produced specific IgG1 (2-h PCA) in response to immunization with thymus-dependent antigens. When mice were immunized with the thymus-independent antigens used in our study, neither athymic nor euthymic mice made antibody detectable by PCA assay (groups H–K). Failure of thymus-independent antigens to induce specific IgE production by euthymic animals has been reported by others (24) also.

Using mice immunized with some antigens we not only examined the serum for homocytotropic antibody by the PCA assay but also attempted to induce contraction of isolated uterine smooth muscle using serotonin or specific antigen. Although serotonin caused contraction of uterine smooth muscle from both immunized athymic and immunized euthymic mice, specific antigen caused contraction of uteri from immunized euthymic mice only (Table 9-2). LPS was as ineffective in causing active sensitization of uterine muscle (Table 9-2) as it was in stimulating production of antibody detectable by PCA assay (Table 9-1).

The defect in the ability of athymic nude mice to produce specific homocytotropic antibody was corrected when nude mice were implanted with syngeneic neonatal thymus glands or injected with syngeneic thymus or spleen cells from young adult donors (Table 9-3).

We conclude that nude mice do not produce specific homocytotropic antibody in response to the immunization procedures commonly used to stimulate production of such antibody.

Table 9-3. Specific homocytotropic antibody responses of immunized euthymic, athymic, and reconstituted athymic mice[a]

Mice[b]	Day 10 serum titer		Day 25 serum titer	
	2-h PCA[b]	72-h PCA	2-h PCA	72-h PCA
Euthymic	32	64	2048	128
Athymic	0	0	0	0
Athymic-TG	0	128	256	64
Athymic-TC	0	0	8	8
Athymic-SC	0	0	512	32

[a] Mice were injected intraperitoneally with 10 μg hen egg albumen (OVA) and 10 mg alum on day 0 and 10 μg OVA and 1 mg alum on day 15; serum samples were taken for PCA tests on days 10 and 25.

[b] Abbreviations: athymic-TG, nude mice implanted with one syngeneic neonatal thymus gland under each renal capsule on day −52; athymic-TC, nude mice injected intravenously on day −1 with 10^8 thymus cells from 8-week-old syngeneic donors; athymic-SC, nude mice injected intravenously on day −1 with 10^8 spleen cells from 8-week-old syngeneic donors; PCA, passive cutaneous anaphylaxis.

Thymus Dependence of Mast Cell Production

Others have reported (12,21) and we have observed that nude mice have normal or elevated numbers of mast cells in skin, lip, and tongue. Mast cells in the skin of nude mice appear to have at least some of the functional capabilities of mast cells from normal mice because nude mice were as effective as normal mice when used as recipients in mast cell-dependent PCA reactions (Table 9-4) (14). These experiments suggest that mast cells in skin of nude mice are present in normal numbers and can bind IgG1 and IgE homocytotropic antibody and release vasoactive mediators. In contrast, athymic nude mice are reported to be deficient in intestinal mucosal mast cells and intestinal histamine (21), and the dramatic increase in intestinal mucosal mast cells and intestinal histamine which occurs in nematode-infected euthymic mice is not seen in infected athymic mice (21,23).

We counted mucosal mast cells in intestines from uninfected and *N. brasiliensis*-

Table 9-4. Passive cutaneous anaphylaxis reactions in euthymic and athymic nude mice

Serum[a] pool	Recipient mice	PCA titer of serum	
		2-h PCA	72-h PCA
A	Athymic	64	64
	Euthymic	64	32
B	Athymic	ND	64
	Euthymic	ND	64

[a] Serum from euthymic mice immunized with hen egg albumen (OVA) in alum was diluted and injected intracutaneously to passively sensitize euthymic or athymic mice; 2-h or 72-h after passive sensitization the recipient mice were injected intravenously with OVA and Evans blue dye and PCA titers were determined 30 min later.

Table 9-5. Intestinal mast cell response to *Nippostrongylus brasiliensis* infection in euthymic, athymic, and reconstituted athymic mice

Mice	Number of mast cells/10 villus crypt units			
	0^a	8^a	10^a	13^a
Experiment 1				
Euthymic	6	168	237	278
Athymic	0	0	0	0
Athymic-TG[b]	0	7	50	124
Experiment 2				
Euthymic	ND[c]	197	301	123
Athymic	ND	ND	0	ND
Athymic-TG[b]	ND	0	233	72
Athymic-TC[d]	ND	105	127	106

[a] Number of days postinfection.
[b] Nude mice were implanted with one syngeneic neonatal thymus gland under each renal capsule 6–8 weeks before infection with 500 *N. brasiliensis* larvae.
[c] ND, not done.
[d] 5×10^7 thymocytes were injected via the lateral tail vein of each athymic mouse 30 days before infection with 500 *N. brasiliensis* larvae.

infected euthymic and athymic mice. Mucosal mast cells were found in intestines of uninfected euthymic but not athymic mice (Table 9-5). By 8–13 days postinfection the number of mast cells in infected euthymic mice had increased markedly but mast cells were not seen in the intestinal mucosa of infected athymic mice (Table 9-5). Intestinal mastocytosis did occur in athymic mice implanted with syngeneic neonatal thymus glands or injected with syngeneic thymus cells and subsequently infected with *N. brasiliensis* larvae (Table 9-5).

Burnet (4) suggested that in mice there are two distinct types of mast cells. The results reported here (Tables 9-4 and 9-5) and other data recently reviewed (1) support the idea of distinct types of mast cells. One type, the connective tissue mast cell, is found throughout the extravascular tissue spaces—especially around blood vessels—and the number of these cells in a given tissue is relatively static (1). Recent studies (13) have shown clearly that connective tissue mast cells are derived from precursors in fetal liver and bone marrow. The data given here and data reported elsewhere (12,21,25) indicate that this type of mast cell is thymus independent; e.g., connective tissue mast cells are abundant in athymic nude mice.

Another type of mast cell is the mucosal mast cell which, as shown here (Table 9-5), rapidly increases in number in response to certain stimuli, such as infection by nematodes. Although it is clear that thymus-derived cells are involved in the genesis of mucosal mast cell accumulations (Table 9-5) (21,23), the origin of the mucosal mast cell is not clear. These cells may be derived from nonthymus precursor cells, which differentiate under thymic influence, or they may be derived directly from thymus-derived precursor cells.

We plan to use nude mice, mast cell-aberrant beige mice (*bg/bg*), and mast cell-deficient W/W^V mice (13) in reconstitution studies to determine the origin of mucosal mast cells and the mechanism of regulation of mucosal mast cell accumulation and function.

Acknowledgments

This work was supported by U.S. Public Health Service Research Grants CA 24443, AI 12854, and RR 09135.
We thank Pat Healow and Beth Wilmot for valuable technical assistance.

References

1. Askenase, P.W. Immunopathology of parasitic diseases: Involvement of basophils and mast cells. Springer Semin. Immunopathol. 2:417–442, 1980.
2. Barnett, J., and C.J. Wust. Production of IgE in congenitally athymic mice. Fed. Proc. 34:1000, 1975.
3. Bloom, G., and J.W. Kelly. The copper phthalocyanin dye "astrablau" and its staining properties, especially the staining of mast cells. Histochemie 2:48–57, 1960.
4. Burnet, F.M. The probable relationship of some or all mast cells to the T-cell system. Cell. Immunol. 30:358–360, 1977.
5. Clausen, C.R., J. Munoz, and R.K. Bergman. Reaginic-type of antibody in mice stimulated by extracts of *Bordetella pertussis*. J. Immunol. 103:768–777, 1969.
6. Enerbach, L. Mast cells in rat gastrointestinal mucosa. I. Effects of fixation. Acta Pathol. Microbiol. Scand. 66:289–302, 1966.
7. Felton, L.D., G. Kauffmann, and H.F. Stahl. The precipitation of bacterial polysaccharides with calcium phosphate. J. Bacteriol. 29:149–161, 1935.
8. Hsu, C.K., and S.H. Hsu. Immunopathology of schistosomiasis in athymic mice. Nature (London) 262:397–399, 1976.
9. Isaak, D.D., R.H. Jacobson, and N.D. Reed. The course of *Hymenolepis nana* infections in thymus-deficient mice. Intl. Arch. Allergy Appl. Immunol. 55:504–513, 1977.
10. Jacobson, R.H., and N.D. Reed. The immune response of congenitally athymic (nude) mice to the intestinal nematode *Nippostrongylus brasiliensis*. Proc. Soc. Exptl. Biol. Med. 147:667–670, 1974.
11. Jarvis, F.G., M.T. Mesenko, and J.E. Kyle. Electrophoretic purification of the Vi antigen. J. Bacteriol. 80:677–682, 1960.
12. Keller, R., M.W. Hess, and J.F. Riley. Mast cells in the skin of normal, hairless and athymic mice. Experientia 32:171–172, 1976.
13. Kitamura, Y., S. Go, M. Shimada, H. Matsuda, K. Hatanaka, and M. Seki. Distribution of mast cell precursors in hematopoietic and lymphopoietic tissues of mice. J. Exptl. Med. 150:482–490, 1979.
14. Lynch, N.R., J. Prin, and J.C. Salomon. Passive cutaneous anaphylaxis and vasoactive amine challenges in nude (nu/nu) mice. Immunology 32:89–94, 1977.
15. Manning, D.D., J.K. Manning, and N.D. Reed. Suppression of reaginic antibody (IgE) formation in mice by treatment with anti-μ antiserum. J. Exptl. Med. 144:288–292, 1976.
16. Michael, J.G., and I.L. Bernstein. Thymus dependence of reaginic antibody formation in mice. J. Immunol. 111:1600–1601, 1973.
17. Miller, H.R.P., and W.F.H. Jarrett. Immune reactions in mucous membranes. I. Intestinal mast cell response during helminth expulsion in the rat. Immunology 20:277–288, 1971.
18. Mitchell, G.F. Metazoan and protozoan parasitic infections in hypothymic nu/nu ("nude") mice. Contemp. Top. Immunobiol. 8:55–67, 1978.
19. Munoz, J., and B.M. Hestekin. Method of preparing histamine sensitizing factor. Nature (London) 196:1192–1193, 1962.
20. Munoz, J., and M. Maung. Anaphylaxis in *Bordetella pertussis*-treated mice. IV. Schultz-Dale reaction. Proc. Soc. Exptl. Biol. Med. 106:70–73, 1961.
21. Olson, C.E., and D.A. Levy. Thymus dependency of mast cell response to *N. brasiliensis* in mice. Fed. Proc. 35:491, 1976.

22. Ovary, Z., S.S. Caiazza, and S. Kojima. PCA reactions with mouse antibodies in mice and rats. Intl. Arch. Allergy Appl. Immunol. 48:16–21, 1975.
23. Ruitenberg, E.J., and A. Elgersma. Absence of intestinal mast cell response in congenitally athymic mice during *Trichinella spiralis* infection. Nature (London) 264: 258–260, 1976.
24. Shinohara, N., and T. Tada. Hapten-specific IgM and IgG antibody response in mice against a thymus-independent antigen (DNP-Salmonella). Intl. Arch. Allergy Appl. Immunol. 47:762–776, 1974.
25. Viklicky, V., P. Sima, and H. Pritchard. On the origin of mast cells in adult life. Folia Biol. (Praha) 19:247–251, 1973.
26. Weigle, W.O. Immunochemical properties of hemocyanin. Immunochemistry 1:295–302, 1964.
27. Westphal, O., O. Lüderitz, and F. Bister. Über die Extraktion von Bakterien mit Phenol/Wasser. Z. Naturforsch. (B) 7:148–155, 1952.

10

Inhibition of Encephalomyocarditis Virus Replication by Macrophages from Athymic (Nude) Mice Sensitized with Nonviable Mycobacterium tuberculosis

Donald L. Lodmell,* Robert R. Cent, Jr.,†
Anne M. Pusateri,† and Larry C. Ewalt

U.S. Department of Health, Education, and Welfare, Public Health Service, National Institutes of Health, National Institute of Allergy and Infectious Diseases, Laboratory of Persistent Viral Diseases, Rocky Mountain Laboratories, Hamilton, Montana 59840, and The Department of Microbiology,† University of Montana, Missoula, Montana 59801.

Abstract

The immunologic defects of congenitally athymic (nude) mice were utilized to determine the importance of T lymphocytes in the nonspecific inhibition of encephalomyocarditis virus (EMCV) replication in mouse embryo fibroblast (MEF) monolayers. It is shown that unstimulated peritoneal cells (PC) from nude mice sensitized with an oil-droplet emulsion of nonviable *Mycobacterium tuberculosis*, when used at a PC/MEF ratio of 20, were (a) >1000-fold more effective in inhibiting viral replication than PC from nude mice that did not receive mycobacteria and (b) were >10-fold more effective than PC from their mycobacteria-sensitized euthymic littermates or C57BL/10ScN mice. At PC/MEF ratios of 2 and 1, PC of mycobacteria-sensitized nude mice inhibited replication >90 and 90%, respectively, whereas similar PC from euthymic mice were ineffective. Cell enrichment techniques consisting of hypaque-density gradients, plastic adherence, nylon wool columns, or removal of carbonyl iron phagocytic cells with magnetic force suggested that a macrophage or a macrophage-like cell was responsible for the inhibition. Furthermore, an antiviral mediator similar to immune (INF-γ) interferon has been detected in supernatant fluids prepared from PC of mycobacteria-sensitized nude mice. These results indicated that functionally mature T lymphocytes were not essential for induction of nonspecific resistance to EMCV by nonviable mycobacteria.

---

* To whom correspondence should be addressed.
Proceedings of the Third International Workshop on Nude Mice.

Introduction

Previous studies in our laboratory (9,10) have indicated that C57BL/10ScN mice sensitized with nonviable *Mycobacterium tuberculosis* in an oil-droplet emulsion were markedly resistant to a lethal challenge of encephalomyocarditis virus (EMCV), and that peritoneal cells (PC) from similarly sensitized mice inhibited EMCV replication in mouse embryo fibroblast monolayers (18). Inhibition of viral replication by the PC appeared to be caused by an interferon that is similar but not identical to classical mouse immune (INF-γ) interferon.

To further ascertain the mechanism(s) of this nonspecific inhibition of viral replication, the immunologically compromised congenitally athymic (nude) mouse (nu/nu) was studied. These mice provide an excellent model for analysis of the immune response because they are deficient in mature T-lymphocyte functions (4,14) and lack T-helper cell activity (1). As a result, nude mice maintain skin xenografts of widely divergent phylogenetic origin (13), as well as xenogenic tumors (19). Furthermore, immunoglobulin responses of nude mice to thymus-dependent antigens are either reduced or absent (17).

We have established that PC of mycobacteria-sensitized nude mice have greater nonspecific activity against EMCV than PC from their similarly sensitized heterozygous ($nu/+$) littermates or C57BL/10ScN mice. Functionally mature T lymphocytes, therefore, were not essential for this nonspecific resistance. A macrophage or a macrophage-like cell appeared to be the effector cell.

Materials and Methods

Tissue culture and media. Mouse L cells were maintained in Eagle's minimum essential medium (MEM) supplemented with 10% heat-inactivated fetal calf serum, 200 U of penicillin G per milliliter, and 1 μg of amphotericin B per milliliter (MEM-10). Mouse embryo fibroblast (MEF) monolayers prepared from Carworth Farms Webster mice reared at the Rocky Mountain Laboratories (RML) (CFW/R) were grown in MEM-10. Dulbecco's phosphate-buffered saline (PBS) with Ca^{2+} and Mg^{2+} was used to wash monolayers and cell suspensions (11).

Viruses. EMCV, obtained from Dr. Michael Ross, National Institutes of Health, Bethesda, Maryland, was titrated in a plaque assay on mouse L cells with a 0.75% methylcellulose overlay prepared in 2× concentrated MEM-10. Vesicular stomatitis virus (VSV) (Indiana strain), obtained from the American Type Culture Collection, Rockville, Maryland, was grown in MEF monolayers.

Mice. Three strains of 6- to 12-week-old mice were used: BALB/c ($nu/+$), athymic (nude) (nu/nu, produced by successive cross–intercrossing onto a BALB/c background), and C57BL/10ScN. All mice were derived from stock maintained at the RML. The nude mice were maintained in a special animal room in conventional autoclaved cages with filter caps using sterilized bedding and water bottles with sterilized food and water supplied ad libitum.

Preparation of nonviable mycobacteria oil-droplet emulsions. *Mycobacterium tuberculosis* strain Jamaica was obtained from Dr. Carl Larson, University of Montana, Missoula, Montana. The culture was isolated from a fatal case of tuberculosis in 1933 by Drs. J. Freund and E. Opie (personal communication, Dr. George Kubica, Center for Disease Control, Atlanta, Georgia). Preparation of the oil-droplet emulsions has been described in detail (9). Mice were sensitized intraperitoneally (IP) with 0.2 ml of emulsion containing 500 μg of mycobacteria. Control mice were untreated or injected IP with the oil emulsion without mycobacteria.

Harvest of peritoneal cells. Mice were killed by spinal dislocation and then injected IP with 4 ml of Dulbecco's PBS containing 5 units of heparin per milliliter. After massage, the peritoneum was pierced and peritoneal cells (PC) collected. The PC were washed twice, treated with 0.83% Tris-NH_4Cl to lyse any remaining erythrocytes, washed again, and resuspended in MEM supplemented with 2% fetal calf serum and antibiotics (MEM-2). Trypan blue determinations indicates that the PC were consistently >90% viable.

Incubation of PC with EMCV-infected MEF monolayers. Secondary CFW/R MEF monolayers grown in trays containing 96 flat-bottomed wells (0.28 cm^2) were washed once and incubated at 37°C in 5% CO_2 in air with two to four PFU of EMCV (0.00004 PFU/cell) suspended in 0.03 ml of MEM. At the end of 2 h, inocula were aspirated, monolayers were washed, and either PC suspended in 0.15 ml of MEM-2 or 0.15 ml of MEM-2 was added to the wells; the PC were not stimulated by the addition of specific mycobacterial antigens to the cultures. In most instances, a ratio of 20 PC to MEF cells was used. After an 18-h incubation, entire cultures were harvested, frozen once, and titrated for infectious EMCV on ML cells. Initial experiments showed no effect on results if entire cultures, media only, or monolayers only were assayed for infectious virus. Because we were interested in all virus (extracellular and intracellular) present in cultures, entire cultures were harvested. Quadruplicate samples were included in all assays and each experiment was repeated a minimum of three times.

Preparation of supernatant fluids from PC. Peritoneal cells suspended MEM-2 at 5×10^6 cells/ml were incubated at 37°C in 5% CO_2 in air for 18 h. Supernatant fluids then were removed, centrifuged at $300 \times g$ for 5 min, and stored at $-70°C$ until assayed for antiviral activity.

Assay for antiviral activity in supernatant fluids from PC cultures. Secondary CFW/R MEF monolayers in trays containing 24 wells (16-mm diameter) were washed, infected with EMCV, and washed again as stated above. One milliliter of supernatant fluid diluted 1:2 in MEM-2 was added to the wells. Eighteen hours postinfection the cultures were harvested, frozen once, and assayed for infectious EMCV on ML cells. Duplicate samples were included in all assays and each experiment was repeated a minimum of three times.

Hypaque-density gradient centrifugation. The procedure of Mahmoud (12) with various modifications was used. Briefly, 3×10^7 PC suspended in 10 ml of heparinized PBS were gently overlaid on 10 ml of a hypaque-density gradient consisting of 1.0 part 50% hypaque/2.1 parts distilled H_2O. The gradient was centrifuged at $400 \times g$ for 40 min at 4°C. After centrifugation, cells at the interface and the pellet were harvested, washed in heparinized PBS, and resuspended in MEM-2. Eleven and 10% of the PC that were placed on the gradient were recovered at the interface and pellet, respectively. Both populations were >95% viable as determined by trypan blue exclusion.

Nylon wool columns. The method of Julius et al. (7) with minor modifications was used. Five cubic centimeters of compressed nylon wool in a 20-ml syringe were saturated with warm MEM-10 and allowed to incubate at 37°C in 5% CO_2 in air for 2 h. Peritoneal cells, 7×10^7 in 1 ml MEM-10, then were layered on the column and allowed to sink in; the column was then overlain with MEM-10. At 15-min intervals 1 ml of MEM-10 was added to the column. After 45 min incubation at 37°C in 5% CO_2 in air the nonadherent cells were collected by slow elution (25 min) with 20 ml of warm MEM-10, washed, and resuspended in MEM-2. Adherent cells were recovered by transferring the nylon wool to a 100-ml beaker containing PBS without Ca^{2+} and Mg^{2+} and teasing the wool apart with sterile forceps. The cells were collected, washed, and resuspended in MEM-2. Fourteen percent of the PC that were added to the column was recovered: 2% was nonadherent and 12% was adherent. Both populations were $\geqq 70\%$ viable as determined by trypan blue exclusion.

Carbonyl iron separation of phagocytic cells. The procedure described by Greenberg et al. (6), as modified by Sanderson et al. (20), was used. Carbonyl iron powder was coated with protein by incubating 25 mg of iron for 1 h at 37°C in 1 ml of MEM containing 50% heat-inactivated FCS. After incubation of 1×10^8 PC with the protein-coated carbonyl iron for 45 min at 37°C, magnets were applied to the tube and the iron-free nonphagocytic cells were collected. The procedure was repeated with the nonphagocytic cells separated from the first treatment. Twenty-two percent of the treated PC was recovered as nonphagocytic cells; >99% was viable.

Cell identifications. Cells were identified by three techniques: (a) the conventional Wright's stain, (b) phagocytosis of polystyrene latex particles (21), with minor modifications, and (c) the nonspecific esterase stain (8).

Results

Inhibition of EMCV Replication by PC from Mycobacteria-Sensitized Athymic (nu/nu), BALB/c (nu/+), and C57BL/10ScN Mice

It recently has been shown in our laboratory that nude mice and neonatally thymectomized C57BL/10ScN mice sensitized with nonviable *M. tuberculosis* suspended in an oil emulsion are resistant to a lethal challenge of EMCV (10). The data in Table 10-1 indicate that this nonspecific resistance also can be demonstrated in vitro with PC from mycobacteria-sensitized mice. At a PC/MEF ratio of 20, PC from all strains of mycobacteria-sensitized mice inhibited viral replication. Interestingly, PC from nude mice inhibited EMCV replication to the greatest degree in that 3.1 $\log_{10}$, 1.6 $\log_{10}$, and 1.2 $\log_{10}$ inhibition was detected with PC from athymic (nude), C57BL/ScN, and BALB/c mice, respectively.

To determine how much more effective the nude mice PC were in inhibiting EMCV replication, PC were used at PC/MEF ratios of 2 and 1. It can be seen

Table 10-1. Inhibition of EMCV replication by PC from mycobacteria-sensitized athymic (nu/nu), BALB/c (nu/+) and C57BL/10ScN mice[a]

Mice	Ratio of PC to MEF cells		
	20	2	1
	(PFU/0.2 ml, $\log_{10}$)		
nu/nu, sensitized	3.5	5.1	5.7
	(3.1)[b]	(1.6)	(0.8)
nu/nu, unsensitized	6.6	6.7	6.5
nu/+, sensitized	5.1	6.8	6.9
	(1.2)	(0)	(0)
nu/+, unsensitized	6.3	6.8	6.8
C57BL/10ScN, sensitized	4.7	6.7	6.9
	(1.6)	(0)	(0)
C57BL/10ScN, unsensitized	6.3	6.8	6.8

[a] EMCV-infected MEF monolayers were incubated at different PC/MEF ratios with PC from unsensitized or mycobacteria-sensitized mice. After an 18-h incubation, entire cultures were harvested and titrated for virus. In the absence of PC, the virus titer was $10^{6.9}$ PFU/0.2 ml. The data are representative of three similar experiments.

[b] Numbers in parentheses indicate the $\log_{10}$ inhibition of EMCV replication as determined by comparing virus titers of cultures incubated with similar ratios of PC from unsensitized and mycobacteria-sensitized mice of the same strain.

in Table 10-1 that PC from nude mice inhibited viral replication 1.6 $\log_{10}$ (>90%) at a ratio of 2, and approximately 1 $\log_{10}$ (90%) with only one PC per MEF. In marked contrast, PC from C57BL/10ScN mice and the heterozygous littermates were ineffective at these lower ratios.

Inhibition of EMCV Replication by Hypaque-Density Gradient Separated PC from Mycobacteria-Sensitized Athymic (Nude) Mice

In our initial inhibition assays we identified high concentrations of neutrophils in the PC. Consequently, we selectively enriched for these cells with a hypaque-density gradient in which neutrophils primarily are found at the bottom of the tube in the pellet. The data in Table 10-2 show that cells at the interface, which consisted of only 11% of the cells that were added to the gradient, inhibited viral replication as well as the unseparated PC (1.5 $\log_{10}$ versus 1.7 $\log_{10}$). In contrast, little protection was associated with cells in the pellet (0.4 $\log_{10}$ inhibition). Differential stains indicated that 41% of the cells in the pellet were neutrophils, whereas no neutrophils were detected at the interface of the gradient.

Inhibition of EMCV Replication by Nylon Wool Adherent PC from Mycobacteria-Sensitized Athymic (Nude) Mice

Our previous in vitro studies with mycobacteria-sensitized C57BL/10ScN mice (unpublished data) suggested that PC which did not adhere to nylon wool (T lymphocytes) were unimportant for mycobacteria-induced nonspecific in-

Table 10-2. Inhibition of EMCV replication by hypaque-density gradient enriched PC from mycobacteria-sensitized athymic (nude) mice[a]

Origin of PC	Virus titer (PFU/0.2 ml, $\log_{10}$)	Inhibition of viral replication ($\log_{10}$)[b]
Mycobacteria-sensitized mice		
Unseparated	5.0	1.7
Interface	5.2	1.5
Pellet	6.3	0.4
Unsensitized mice		
Unseparated	6.7	—
None (media only)	6.7	—

[a] EMCV-infected MEF monolayers were incubated at a PC/MEF ratio of 20 with unseparated PC or PC that had been selectively enriched on a hypaque-density gradient (see Materials and Methods). Eleven and 10% of the PC that were placed on the gradient were recovered at the interface and pellet, respectively. Both populations were >95% viable. After an 18-h incubation, entire cultures were harvested and titrated for virus. The data are representative of three similar experiments.

[b] Inhibition of EMCV replication was determined by comparing virus titers of cultures incubated with PC from unsensitized mice with the other test groups.

hibition of EMCV replication, but that, in contrast, cells which did adhere to the wool were markedly effective. To ascertain whether the effector cell of mycobacteria-sensitized nude mice also adhered to nylon wool, similar columns were used to separate the PC.

The data in Table 10-3 show that the adherent cells, which consisted of only 12% of the cells that were added to the column, were as effective in inhibiting EMCV replication as the unseparated PC (1 $\log_{10}$ inhibition). In contrast, the nonadherent cells were ineffective. Latex phagocytosis studies and an esterase

Table 10-3. Inhibition of EMCV replication by nylon wool adherent PC from mycobacteria-sensitized athymic (nude) mice[a]

Origin of PC	Virus titer (PFU/0.2 ml, $\log_{10}$)	Inhibition of viral replication ($\log_{10}$)[b]
Mycobacteria-sensitized mice		
Unseparated	4.0	1.0
Adherent	4.0	1.0
Nonadherent	5.1	—
Unsensitized mice		
Unseparated	5.0	—
None (media only)	5.3	—

[a] EMCV-infected MEF monolayers were incubated at a PC/MEF ratio of 20 with unseparated PC or PC that had been selectively enriched on a nylon wool column (see Materials and Methods). Fourteen percent of the PC that were added to the column was recovered: 2% was nonadherent and 12% was adherent. Both populations were ≥70% viable. After an 18-h incubation, entire cultures were harvested and titrated for virus. The data are representative of three similar experiments.

[b] Inhibition of EMCV replication was determined by comparing virus titers of cultures incubated with PC from unsensitized mice with the other test groups.

stain, which is specific for monocytes and macrophages, indicated that 42% of the nylon wool adherent cells were macrophages as compared to <1% of those that were nonadherent (data not shown). Additional data also suggested that the effector cell adhered to plastic because cells that did not adhere to plastic were ineffective in the inhibition of EMCV replication (data not shown).

Effect of Nonphagocytic PC from Mycobacteria-Sensitized Athymic Nude Mice on EMCV Replication

To study in more detail the possibility that the effector cell was a macrophage, PC were incubated with carbonyl iron and the phagocytic cells were removed by magnetic force. It was determined that treatment with carbonyl iron removed the effector cell; unseparated PC inhibited replication 2 $\log_{10}$, whereas the nonphagocytic cells, which represented 22% of the PC, were essentially ineffective (0.3 $\log_{10}$ inhibition) (Table 10-4). Evidence that macrophages had been depleted by this technique was shown in the latex phagocytosis studies and esterase stains: phagocytic cells had been reduced from 71% to 6% and <1% of the cells were positive for esterase (data not shown).

Effect of Supernatant Fluids Prepared from Unstimulated PC Cultures of Mycobacteria-Sensitized Athymic (nu/nu), BALB/c (nu/+), and C57BL/10 ScN Mice on EMCV Replication

Recent in vitro studies (18) have shown that supernatant fluids prepared from PC cultures of mycobacteria-sensitized C57BL/10ScN mice inhibited EMCV replication. An immune (INF-γ) interferon similar but not identical to classical INF-γ was identified in these fluids. The data in Table 10-5 show that supernatant fluids harvested from PC of mycobacteria-sensitized nude mice also inhibited EMCV replication, and that this inhibition was as dramatic as the inhibition induced by supernatant fluids prepared from the euthymic C57BL/

Table 10-4. Effect of nonphagocytic PC from mycobacteria-sensitized athymic (nude) mice on EMCV replication[a]

Origin of PC	Virus titer (PFU/0.2 ml, $\log_{10}$)	Inhibition of viral replication ($\log_{10}$)[b]
Mycobacteria-sensitized mice		
Unseparated	2.8	2.0
Nonphagocytic	4.5	0.3
Unsensitized mice		
Unseparated	4.8	—
None (media only)	5.3	—

[a] EMCV-infected MEF monolayers were incubated at a PC/MEF ratio of 20 with unseparated PC or PC that did not phagocytose carbonyl iron (see Materials and Methods). Twenty-two percent of the treated PC was recovered as nonphagocytic cells; >99% was viable. After an 18-h incubation, entire cultures were harvested and titrated for virus. The data are representative of six similar experiments.

[b] Inhibition of EMCV replication was determined by comparing virus titers of cultures incubated with PC from unsensitized mice with the other test groups.

Table 10-5. Effect of supernatant fluids prepared from PC cultures of mycobacteria-sensitized athymic (nu/nu), BALB/c ($nu/+$), and C57BL/10ScN mice on EMCV replication[a]

Mice	Origin of supernatant fluids		Inhibition of viral replication ($\log_{10}$)[b]
	Mycobacteria-sensitized mice	Unsensitized mice	
	(PFU/0.2 ml, $\log_{10}$)		
nu/nu	5.0	6.9	1.9
$nu/+$	7.0	7.0	—
C57BL/10ScN	5.0	7.0	2.0

[a] Supernatant fluids were harvested from 18-h cultures of mycobacteria-sensitized and unsensitized PC, centrifuged, mixed with an equal volume of MEM-2, and subsequently overlaid on EMCV infected MEF monolayers. Eighteen hours later, cultures were harvested and virus titers determined. Cultures incubated with only MEM-2 had a virus of $10^{6.9}$ PFU/.02 ml. The data are representative of three similar experiments.

[b] Inhibition of EMCV replication was determined by comparing virus titers of cultures incubated with supernatant fluids from unsensitized and mycobacteriasensitized mice of the same strain.

10ScN mice. We are uncertain as to why supernatant fluids prepared from PC of BALB/c ($nu/+$) mice did not inhibit EMCV replication. Characterization of the mediator released from the PC of mycobacteria-sensitized nude mice is in progress. Preliminary results indicate that the mediator resembles INF-γ in that it inhibits EMCV and VSV replication in ME and ML cells, is labile after heating for 1 h at 56°C, and is partially labile to pH 2.0 after treatment at 4°C for 24 h. Most importantly, it is not neutralized by antiserum to (INF-β).

Discussion

The results reported here have shown that PC from athymic (nude) mice previously sensitized with nonviable mycobacterium suspended in an oil emulsion inhibited EMCV replication in MEF monolayers. Furthermore, at a PC/MEF ratio of 20, PC from mycobacteria-sensitized nude mice inhibited replication to a greater degree than did similarly sensitized cells from their heterozygous BALB/c littermates or C57BL/10ScN mice. The effectiveness of PC from sensitized nude mice for inhibiting viral replaction was even more marked when varying concentrations of PC were tested; only PC from nude mice were effective at PC/MEF ratios of 2 and 1. These results corroborate our previous studies which showed that EMCV replication was not affected by PC from mycobacteria-sensitized C57BL/10ScN mice at PC/MEF ratios of <5 (18).

Initially it appeared that the neutrophil might be the effector cell. However, hypaque-density gradients used to selectively enrich for the neutrophil showed that it was not important. The fraction harvested from the interface of the gradient which contained >1% neutrophils, and only 11% of the PC that were added to the gradient, inhibited viral replication as well as the unseparated PC (>90%). In contrast, the pellet, which was comprised of 41% neutrophils, was ineffective. Subsequent studies stressed the importance of the macrophage because our previous in vivo studies in C57BL/10ScN mice (10) have shown that silica, a purported macrophage toxin, abrogated resistance of mycobacteria-sensitized mice to a lethal challenge of EMCV. Furthermore, it has been shown by several investigators that nude mice have a high native resistance to experimental infection with several bacterial pathogens (2,3,15,16,23). To account for this increased resistance it has been suggested that macrophages of nude mice possess enhanced microbicidal activities (2,15,16,23). Furthermore, it has been shown that nude mice respond to the nonspecific adjuvant *Corynebacterium parvum* (5). It therefore seemed reasonable that nude mice might respond similarly to the mycobacteria, which could ultimately enhance the nonspecific antiviral activity of the macrophages.

Evidence that macrophages or macrophage-like cells were the effector cells was suggested by the nylon wool studies. High concentration of cells that were esterase positive and that phagocytosed latex adhered to the wool, and these cells were markedly effective in the inhibition of EMCV replication. In contrast, the nonadherent cells, which were $<1\%$ macrophages, did not inhibit replication. Furthermore, cells that did not adhere to plastic were ineffective. The nylon wool and hypaque-density gradient results also suggested that the PC were comprised of a small subpopulation of effector cells; enrichment techniques which yielded only 11 and 12% of the cells that were added to the gradient and the nylon wool were as effective in inhibiting EMCV replication as the unseparated PC.

To more clearly define the possibility that a macrophage or a macrophage-like cell was the effector cell, we turned our attention to a technique that is known to eliminate phagocytic cells (6,20). It was found that the nonphagocytic cells which remained after treatment with carbonyl iron and magnetic force did not inhibit EMCV replication; only 6% of these cells were phagocytic and $<1\%$ were esterase positive. In contrast, the untreated cells, which were comprised of 70% latex phagocytic cells, were markedly effective. Thus, it appeared that the effector cell from mycobacteria-sensitized nude mice that inhibited EMCV replication adhered to plastic and nylon wool, banded at the interface of hypaque-density gradients, and was removed by magnetic force after phagocytosis of protein-coated carbonyl iron. These data, in conjunction with the latex phagocytosis studies and esterase stains, suggested that the effector cell was a macrophage or a macrophage-like cell.

The mechanism(s) by which this cell inhibited EMCV replication is not known. It has been shown, however, that an (INF-γ) interferon that was released from PC of mycobacteria-sensitized C57BL/10ScN mice was of major importance in the nonspecific inhibition of EMCV replication in MEF monolayers (18). An antiviral substance with similar properties also has been detected in supernatant fluids prepared from PC cultures from mycobacteria-sensitized nude mice. Interestingly, others have shown that immune (INF-γ) interferon can be induced by phytohemagglutinin from nude mouse spleen cells (22). Whether the INF-γ released from PC of mycobacteria-sensitized nude mice has a significant role in the in vitro inhibition of EMCV replication is under investigation. Additional characterization of the INF-γ interferon, as well as the macrophage-like effector cell, also is in progress.

Acknowledgments

The authors thank Helen Blahnik for excellent secretarial assistance, and Scott Stewart and Rod Parker for care of the experimental animals.

References

1. Braun, D.G., B. Kindred, and E.B. Jacobson. Streptococcal group A carbohydrate antibodies in mice. Evidence for strain differences in magnitude and restriction of the response and for thymus dependence. Eu. J. Immunol. 32:138–143, 1972.
2. Cheers, C., and R. Waller. Activated macrophages in congenitally athymic "nude" mice. J. Immunol. 115:844–847, 1975.
3. Emmerling, R.H., H. Finger, and J. Bockemuhl. *Listeria monocytogenes* infection in nude mice. Infect. Immun. 12:437–439, 1975.
4. Gershwin, M.E., R.M. Ikeda, T.G. Kawakami, and R.B. Owens. Immunobiology of heterotransplanted tumors in nude mice. J. Natl. Cancer Inst. (USA) 58:1455–1461, 1977.
5. Goldberg, N.C., and J.C. Salomon. Thymoindependent antigenic stimulation in nude mice: Response to polyvinylprrolidone and adjuvant effects of *Corynebacterium parvum* and LHI. Intl. Arch. Allergy Appl. Immunol. 57:31–36, 1978.
6. Greenberg, A.H., L. Shen, and G. Medley. Characteristics of the effector cells mediating cytotoxicity against antibody-coated target cells. I. Phagocytic and non-phagocytic effector cell activity against erythrocyte and tumor target cells in a ^{51}Cr release cytotoxicity assay and (125I) IUDR growth inhibition assay. Immunology 29:719–729, 1975.
7. Julius, M.H., E. Simpson, and L.W. Herzenberg. A rapid method for the isolation of functional thymus-derived murine lymphocytes. Eu. J. Immunol. 3:645–649, 1973.
8. Koski, I.R., D.G. Poplack, and R.M. Blaese. A nonspecific esterase stain for the identification of monocytes and macrophages, pp. 359–362. In B. R. Bloom and J. R. David (eds.), *In vitro* methods in cell-mediated and tumor immunity. New York: Academic Press, 1976.
9. Lodmell, D.L., and L.C. Ewalt. Enhanced resistance against encephalomyocarditis virus infection in mice, induced by a nonviable *Mycobacterium tuberculosis* oil-droplet vaccine. Infect. Immun. 19:225–230, 1978.
10. Lodmell, D.L., and L.C. Ewalt. Induction of enhanced resistance against encephalomyocarditis virus infection of mice by nonviable *Mycobacterium tuberculosis*: Mechanisms of protection. Infect. Immun. 22:740–745, 1978.
11. Lodmell, D.L., and A.L. Notkins. Cellular immunity to herpes simplex virus mediated by interferon. J. Exptl. Med. 140:764–778, 1974.
12. Mahmoud, A.A.F. Purification of human granulocytes and production of the corresponding monospecific antisera, pp. 387–394. In B.R. Bloom and J.R. David (eds.), *In vitro* methods in cell mediated and tumor immunity. New York: Academic Press, 1976.
13. Manning, D.D., N.D. Reed, and C.F. Shaffer. Maintenance of skin xenografts of widely divergent phylogenetic origin on congenitally athymic (nude) mice. J. Exptl. Med. 138:488–494, 1973.
14. Milich, D.R., and M.E. Gershwin. T cell differentiation and the congenitally athymic (nude) mouse. Dev. Comp. Immunol. 1:289–298, 1977.
15. Mogensen, S.C., and H. Kerzel Anderson. Role of activated macrophages in resistance of congenitally athymic nude mice to hepatitis induced by herpes simplex virus type 2. Infect. Immun. 19:792–798, 1978.
16. Nickol, A.D., and P.F. Bonventre. Anomalous high native resistance of athymic mice to bacterial pathogens. Infect. Immun. 18:636–645, 1977.
17. Pantelouris, E.M., and P.A. Flisch. Responses of athymic (nude) mice to sheep red blood cells. Eu. J. Immunol. 2:236–239, 1972.
18. Pusateri, A.M., L.C. Ewalt, and D.L. Lodmell. Nonspecific inhibition of encephalomyo-

carditis virus replication by a type II interferon released from unstimulated cells of *Mycobacterium tuberculosis*-sensitized mice. J. Immunol. 124:1277–1283, 1980.
19. Rygaard, J., and C.O. Povlsen. Heterotransplantation of a human malignant tumor to "nude" mice. Acta Pathol. Microbiol. Scand. 77:758–760, 1969.
20. Sanderson, C.J., I.A. Clark, and G.A. Taylor. Different effector cell types in antibody-dependent cell-mediated cytotoxicity. Nature (London) 253:376–377, 1975.
21. Smialowicz, R.J., and J.H. Schwab. Inhibition of macrophage phagocytic activity by group A streptococcal cell walls. Infect. Immun. 20:258–261, 1978.
22. Wietzerbin, J., S. Stefanos, R. Falcoff, M. Lucero, L. Catinot, and E. Falcoff. Immune interferon induced by phytohemagglutinin in nude mouse spleen cells. Infect. Immun. 21:966–972, 1978.
23. Zinkernagel, R.M., and R.V. Blanden. Macrophage activation in mice lacking thymus-derived (T) cells. Experientia 31:591–593, 1975.

General Discussion

KRUEGER: Do immunized nudes show in vitro responses to mycobacteria; i.e., is there uptake of [^{3}H]thymidine by spleen cells exposed to mycobacteria? It has been shown that mycobacteria can be a B-cell mitogen and cause lymphokin release—perhaps this gives rise to activated macrophages and type II interferon.

LODMELL: We have not tested whether there is [^{3}H]thymidine uptake of cells from mycobacteria-sensitized nude mice exposed to mycobacteria. The addition of nonviable mycobacteria to unsensitized cells from nude mice does not elicit interferon production, whereas, interferon titers in cell cultures of mycobacteria-sensitized nude mice are enhanced by the addition of nonviable mycobacteria.

11

Friend Leukemia Virus Infection of C57BL/10 Nude Mice

Radmila B. Raikow,* James P. OKunewick, Ruby F. Meredith, and Kathleen C. Magliere

Cancer Research Unit, Allegheny General Hospital, 320 E. North Avenue, Pittsburgh, Pennsylvania 15212.

Abstract

Nude and normal C57BL/10 mice were injected with 100 SED units of Friend leukemia virus, a dose several times larger than that normally required to cause a 100% incidence of leukemia in virus-sensitive mice (1 SED 50/14 causes spleen enlargement in 50% of virus-sensitive SJL/J mice in 14 days). Although neither the nude C57BL/10 nor their normal littermates developed any signs of the disease, virus replication apparently took place in the nude mice after FLV injection. Plasma taken from these mice 120 days after FLV injection was capable of causing typical Friend erythroleukemia when injected into susceptible SJL/J mice. In contrast, plasma taken from normal C57BL/10 mice after FLV injection did not induce leukemia when injected into susceptible SJL/J mice. Furthermore, immunosuppression of normal C57BL/10 mice by cyclophosphamide injection prior to FLV infection resulted in the production of Friend virus in these immunosuppressed normal C57BL/10 mice. Virus replication was detectable by the in vitro XC assay in mice given cyclophosphamide followed by an injection of FLV, but not in normal C57BL/10 mice given cyclophosphamide alone with no subsequent FLV injection. It therefore appears that whereas normal C57BL/10 mice are capable of eliminating Friend virus, immunosuppressed C57BL/10 mice are not capable of eliminating Friend virus or preventing its replication. This appears to be the case whether the immunosuppression occurs as a result of the congenital athymic condition or is chemically induced with cyclophosphamide. These results suggest that the ability of C57BL/10 mice to repress replication of virus and to eliminate virus from the plasma is a function separate from the one governing the ability to resist leukemia transformation of its cells when injected with FLV.

---

* To whom correspondence should be addressed.
© 1982 Gustav Fischer New York, Inc.
Proceedings of the Third International Workshop on Nude Mice.

Introduction

The C57BL mouse strains are highly resistant to Friend leukemia virus (FLV) (13,28). This resistance is influenced by many genes (20), some of which affect events at the target level (10,12,19,28), while others work via the immune system (2–4,18). The most important factor influencing the virus resistance of C57BL mice appears to be their homozygosity for a resistance allele at the Fv2 locus, since mice with a sensitive allele at this locus against the C57BL genetic background develop leukemia after FLV injection (24). This locus affects the replication of the spleen focus-forming component of FLV (SFFV) (19) which is thought to be responsible for erythroleukemia development (9). However, the SFFV component is defective and requires a second, helper component, the lymphatic leukemia virus (LLV) for its maturation (2,10).

Stutman and Dupuy (28) reported that antilymphocyte serum, which was shown independently to be immunosuppressive, did not increase the negligible SFFV replication observed in C57BL mice. However, a small transient increase of infectious virus was observed in C57BL/6 mice after FLV injection, and this increase was temporally correlated with a virus-caused immunosuppression (21). Other reports confirm the importance of the immune system in C57BL virus resistance, viz., the facts that leukemia is inducible by FLV in neonatal C57BL/6 mice (3), and that treatment of C57BL/6 mice with ^{89}Sr, which destroys the bone marrow, renders even C57BL adults susceptible to Friend virus leukemogenesis (3,18). We have tested the effect of immunosuppression by cyclophosphamide and by the congenitally athymic condition in C57BL/10 mice on FLV infection. Some evidence for the presence of FLV antigen-directed, spleen cell-mediated cytotoxicity in normal and nude C57BL/10 mice is also reported. Our data indicate that T-cell-mediated immune functions are important in the elimination of injected virus in C57BL/10 mice. When virus is not eliminated in these mice it can replicate, apparently without causing cell transformation.

Materials and Methods

Animals. Normal C57BL/10J mice and SJL/J mice were purchased from Jackson Laboratories (Bar Harbor, Maine). A nucleus of C57BL/10 ScN mice carrying the athymic (nude) was obtained from Dr. Carl Hansen, National Cancer Institute (NCI), Bethesda, Maryland. Heterozygous females were bred to homozygous males. All mice were maintained in plastic cages with filter tops and were given autoclaved food (Purina No. 5010) and autoclaved acidified water. The diet of mice in the nude colony was also supplemented with approximately 10% Purina Puppy Chow. The cages, sawdust, nestlets (Ancare Corp., Manhasset, New York), food, and water for the nude mice were autoclaved and kept wrapped until use. The nude mice were handled with sterile gloves in a room where no other mice were kept. All mice were housed in an air-conditioned room with a 12-h light cycle and were used for experiments when they were 10 weeks old or older.

Virus. Our FLV stock was obtained from NCI in 1968. This original stock, stored at $-60°C$, was passaged one time in female SJL/J mice before use: Blood was collected from the vena cava of ether-anesthetized mice that had spleens enlarged at least fivefold from FLV injection initiated about 14 days previously. The plasma was diluted 1:1 (v/v) with 0.05 M Na citrate and stored in small aliquots at $-60°C$. The virus preparation was titered in SJL/J mice by a spleen enlargement assay as described by Chirigos et al. (7).

Cyclophosphamide treatment. Cyclophosphamide (Mead-Johnson) was dissolved in sterile saline immediately before use and was injected at 225 mg/kg of body weight, intraperitoneally (IP).

Cytotoxicity assay. Effector cells were isolated from mice that were injected wtih FLV in saline or with saline only 11 days before sacrifice. Their spleens were disrupted in McCoy's medium (GIBCO) with 5 mM EDTA and sieved through two layers of sterile gauze. Five ml of the resulting cell suspension (made 5×10^8 cells per ml) were layered on 5 ml of Ficoll-hypaque and centrifuged as described by Boyum (5). Lymphocytes collected from the interface were washed twice with McCoy's medium and resuspended in McCoy's medium with 10% fetal calf serum (Sterile Systems, Inc., Logan, Utah). The target cells were AD755 cells kindly supplied by Dr. J.J. Collins, Duke University Medical Center (Durham, North Carolina). These are C57BL/6 tumor cells that have been shown to display FLV antigens (8). Target cells were labeled with ^{51}Cr for 1 h. Approximately 10^5 target cells in 1 ml McCoy's medium plus 300 μC ^{51}Cr at specific activity >300 mC/mg (New England Nuclear, Boston, Massachusetts) were used. The labeled cells were washed three times to remove unincorporated chromium with McCoy's medium plus 10% fetal calf serum. Labeled target cell suspension (0.05 ml, 4×10^4 cells/ml) and 0.05 ml of effector cell suspension were added per well of Lindbro microtiter plates. Effector target ratios of 250, 25, and 2.5:1 were incubated for 4 or 22 h with gentle rocking (6). The counts released into the supernatant after incubation were collected using the Flow Titertek Supernatant Collection System (15) and counted on a gamma counter. Total releasable counts were estimated by the amount of soluble label in wells subjected to three successive freezing and thawing cycles.

UV–XC assay. Cells of the SC line (14) were plated at 3×10^4 cells/cm^2 in plastic tissue culture petri dishes, allowed to attach, treated with medium containing DEAE–dextran (2.0 μg/ml) for 1 h, and exposed to plasma being tested for the presence of virus (0.1 ml per 7.5 cm^2 petri dish, which is just enough to cover the cells, was used). After 10 min medium was added to a depth of about 0.5 cm. After 3 days the SC cells were inactivated by ultraviolet (UV) light (27) and XC cells (27) were added at 8×10^5 cells/cm^2. After 2 days of incubation the cells were fixed and stained with 1% crystal violet in 50% ethanol diluted 1:1 with 10% formalin. The total number of syncitia per plate was scored. This assay quantitates the LLV component of FLV (11,16).

Results

Effect of Immunosuppression by Cyclophosphamide in Normal C57BL/10 Mice

Normal 10- to 12-week-old female C57BL/10 mice were injected with 50 SED FLV on day 0, with 225 mg/kg cyclophosphamide (Cy) on day 5 and with another 50 SED FLV on day 7. Control mice received Cy alone or FLV alone. Approximately 60% of the mice that received Cy, either alone or in combination with FLV, died by day 90 (Figure 11-1). The cause of these deaths is not clear. The mice died in an emaciated state, indicating the possibility of generalized infection. However, none of the mice developed leukemia, as was ascertained by periodic peripheral white blood cell counts and by the condition of their spleens at death. The plasma of these mice was also periodically monitored for the presence of XC-positive virus. The results of these assays (Table 11-1) indicated that the titer of XC-positive virus in the plasma of immunosuppressed C57BL/10 mice that were inoculated with FLV increased more than 100-fold by 3 months after the infection.

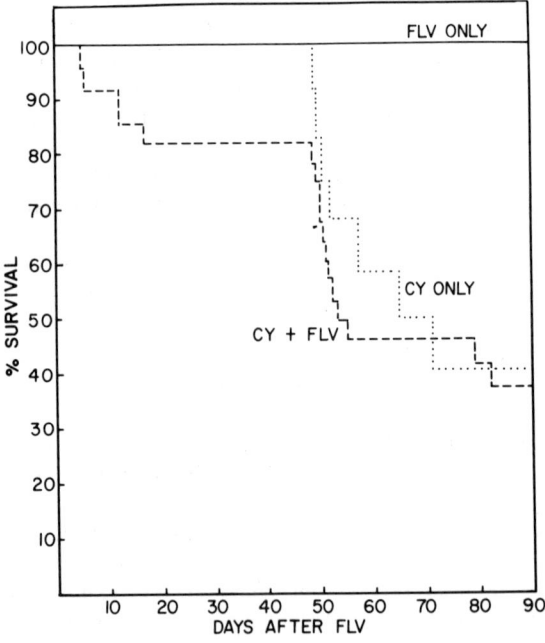

Figure 11-1. Survival of normal C57BL/10J mice given 100 SED of FLV either alone or in conjunction with an immunosuppressive dose of cyclophosphamide. Twelve-week-old, normal, female C57BL/10J mice were injected with FLV (50 SED) on day −7, with cyclophosphamide (225 mg/kg of body weight) on day −2, and again with FLV (50 SED) on day 0. Control groups received virus or cyclophosphamide only. There were 10–15 mice per group.

Table 11-1. Average number of syncitia produced per 7.5 cm^2 petri plate by 0.1 ml of plasma from C57BL/10 mice injected with 100 SED FLV either alone or after an immunosuppressive dose of Cy (control mice that received neither FLV nor Cy or that received Cy alone had no significant syncitia above background)

		No. of syncitia above background	
	Days after virus	FLV only	Cy + FLV
Experiment 1	1	12	93
	41	26	928
	90	20	>10,000
Experiment 2	92	0	962

FLV Infection in Nude C57BL/10 Mice

Six-month-old male C57BL/10 nude mice and control normal C57BL/10 mice, male or female, of various ages from 10 weeks to 6 months were used. The control mice were either part of the nude mouse colony or part of our regular stock purchased from Jackson labs. All mice were injected with 100 SED FLV on day 0. On days 80 and 120 blood was collected under ether anesthesia either by heart puncture or from the vena cava. Then, 0.1 ml of the plasma thus obtained was injected IP into virus-susceptible SJL/J mice, which were subsequently monitored for leukemia manifested by elevated peripheral white blood cell counts and splenomegaly.

Plasma from normal C57BL/10 mice did not cause leukemia in SJL/J mice in any of the tests. On the other hand, all SJL/J mice injected with plasma obtained from nude C57BL/10 mice did develop typical erythroleukemia (Table 11-2). The injected FLV therefore appeared to survive and probably to multiply in nude C57BL/10 mice just as was apparently the case in Cy-immunosuppressed

Table 11-2. Leukemia development in assay SJL/J mice injected with 0.1 ml of plasma collected from either normal or nude C57BL/10 mice

Source of plasma	Peripheral WBC[a] on day 25 after infection	Day of death	Spleen weight at death (g)
Nude C57BL/10 mice Plasma collected 80 days after FLV injection	106,529	34	1.3690
Plasma collected 120 days after FLV injection	117,209	27	1.5859
Normal C57BL/10 mice Plasma collected 80 days after FLV injection	13,794	128 (sacrificed)	Normal (<0.2)
Plasma colected on day 120 after FLV injection	10,403	121 (sacrificed)	Normal (<0.2)

[a] WBC, white blood cells/ml of blood $\times 10^{-3}$.

C57BL/10 mice (see above). In contrast to the chemically suppressed C57BL/10 mice, however, the athymic mice remained in good condition and appeared to maintain normal lifespans despite their viremia. Whether this was because of their more isolated environment remains to be determined.

Cytotoxicity as Measured by the ^{51}Cr Release Assay

We also tested whether FLV injection increased spleen cell-mediated cytotoxicity against cultured C57BL target cells that carry FLV antigens. The results (Figure 11-2) indicated that such FLV injection did increase specific lysis of these targets. The background spontaneous release, which was 7.4% after 4 h of incubation and 23.5% after 22 h of incubation in this experiment, was subtracted from each point, and each point is the mean of measurements made from quadruplicate vessels. As can be seen in Figure 11-2 the percentage of total counts released by splenocytes from mice that were exposed to FLV increased with increasing effector to target ratios up to nearly 50%, whereas the percentage

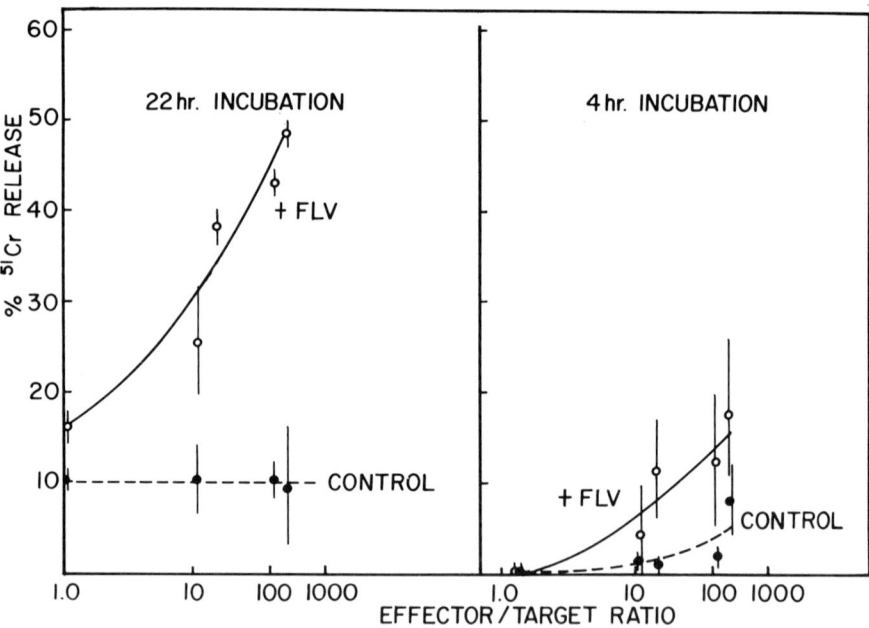

Figure 11-2. Cytotoxicity of spleen cells from normal C57BL/10J mice with and without FLV against AD755 cells in vitro. Spleen cells isolated from normal 12-week-old, female C57BL/10J mice that were injected with either 10 SED FLV in saline or with saline alone 11 days before sacrifice were incubated with ^{51}Cr-labeled AD755 cells. Means ± 1 SE of the percentage of the total releasable counts found in the supernatant of quadruplicate wells are given. The background spontaneous lysis, which was never more than 25%, was subtracted from each value.

Table 11-3. Percentage total releasable counts found in supernatants when effector and target cells were incubated for 4 h[a]

Source of effector cells	Normal control	Normal + virus	Nude + virus
Percentage total releasable counts	11.2	26.1	22.8

[a] Background lysis after 4 h was 10.1% and was subtracted from each value. Ratio of effectors to targets was 250:1. See Materials and Methods for further details.

of total counts released by splenocytes from mice that never received virus never rose above 10%.

Preliminary results (Table 11-3) indicate that splenocytes from FLV-injected nude mice are also capable of destroying target cells that display FLV antigens.

Discussion

Our results indicate that immunosuppression by cyclophosphamide, which is primarily a T-cell inhibition (26), or immunodeficiency caused by the congenitally athymic state allows FLV replication in C57BL/10 mice, in which viral replication is normally very limited (28). On the other hand, Stutman and Dupuy (28) reported that immunosuppression by rabbit antithymocyte serum did not influence Friend virus replication in C57BL/10 mice. This apparent discrepency may result from the type of assays for virus titer employed by these authors and by us. The in vivo spleen focus formation assay used by Stutman and Dupuy measures the SFFV component of FLV (1), whereas the in vitro XC assay used by us measures the LLV component of FLV (27). If the Fv2rr genotype of C57BL mice limits SFFV production at the cellular level, immunosuppression would not be expected to increase replication of this component. On the other hand, the LLV component may be eliminated in normal C57BL mice by an immune mechanism, possibly one that is antibody dependent and so involves T-cell helper functions. In support of this hypothesis, antibody production directed against leukemia virus (25) or endogenous virus (23) has been reported in several mouse strains. In the latter study (23) it was also demonstrated that nude mice were deficient in their ability to form such antibodies. The LLV replication therefore may increase with decreasing T-cell function. Our data also show that plasma from FLV-infected nude C57BL/10 mice causes typical erythroleukemia symptoms in SJL/J mice. Therefore, at least some SFFV must be present in these preparations. It has been shown that the SFFV titer can be very small compared to the LLV in preparations that are highly effective in producing erythroleukemia in susceptible mice (11). Very low SFFV proportions may have been beyond the level of detection by the spleen focus-forming assay utilized by Stutman and Dupuy (28).

Although SFFV replication is C57BL mice is very limited because of the Fv2rr genotype, a few appropriate FLV target cells apparently are transformed after FLV infection. That this is so is indicated by recent reports (3,18) that bone marrow destruction by ^{89}Sr allows Friend erythroleukemia development in

C57BL mice. Evidently a non-T-cell component is involved in the elimination of transformed cells that display appropriate cell surface antigens. Our preliminary results with cell-mediated cytotoxicity in vitro, directed against C57BL targets with FLV antigens, indicate that such cytotoxic cells do indeed exist in normal C57BL mice infected with FLV and probably also in nude C57BL mice. Macrophages have also been shown to destroy cells transformed by Rauscher leukemia virus (16). The existance of a non-T antileukemia cell, along with the fact that virus replication can occur without causing transformation (11,22), would explain why—despite the viremia produced by FLV infection in Cy-immunosuppressed or congenitally athymic C57BL/10 mice—the virus did not produce erythroleukemia in these mice.

References

1. Axelrad, A.A., and R.A. Steeves. Assay for Friend leukemia virus: Rapid quantification method based on enumeration of macroscopic spleen foci in mice. Virology 24:513–518, 1964.
2. Bendinelli, M. Role of Friend-associated lymphatic leukemia virus in immunization against Friend leukemia complex. Experientia 33:455–456, 1976.
3. Bennett, M., E.E. Baker, J.W. Eastcott, V. Kumar, and D. Yonkosky. Selective elimination of marrow precursors with the bone-seeking isotope ^{89}Sr: Implications for hemopoiesis, lymphopoiesis, viral leukemogenesis and infection. J. Reticuloendothel. Soc. 20: 71–87, 1976.
4. Blank, K.J., H.A. Freedman, and G. Lilly. T-Lymphocyte response to Friend virus-induced tumor cell lines of strains congenic at H-2. Nature (London) 260:250–252, 1976.
5. Boyum, A. Isolation of mononuclear cells and granulocytes from human blood by combining centrifugation and sedimentation at 1 g. Scand. J. Clin. Lab. Invest. (Suppl. 97) 21:9–109, 1968.
6. Canty, T.G., and J.R. Wunderlich. Quantitative in vitro assay of cytotoxic cellular immunity. J. Natl. Cancer Inst. (USA) 45:761–782, 1970.
7. Chirigos, M.A., E.D. Schwalb, and D. Scott. Friend leukemia virus spleen focus assay: Relationship of spleen foci to viremia splenomegaly, and survival time. Cancer Res. 27: 2249–2254, 1967.
8. Collins, J.J., G. Roloson, D.E. Haagensen, Jr., P.J. Fischinger, S.A. Wells, Jr., W. Holder, and D.P. Bolognesi. Immunologic control of the ascites form of murine adenocarcinoma 755. II. Tumor immunity associated with a Friend-Moloney-Rauscher-Type virus. J. Natl. Cancer Inst. (USA) 60:141–151, 1978.
9. Dietz, M.S.P., S.P. Fouchey, C. Longley, M.A. Rich, and P. Furmanski. Spontaneous regression of Friend virus-induced erythroleukemia I. The role of the helper murine leukemia virus component. J. Exptl. Med. 145:594–606, 1977.
10. Eckner, R.J. Helper-dependent properties of Friend spleen focus-forming virus: Effect of the Fv-1 gene on the late stages in virus synthesis. J. Virol. 12:523–533, 1973.
11. Eckner, R.J. Continuous replication of Friend virus complex (spleen focus-forming virus-lymphatic leukemia-inducing virus) in mouse embryo fibroblasts. Retention of leukemogenicity and loss of immunosuppressive properties. J. Exptl. Med. 142:936–948, 1975.
12. Eckner, R.J., V. Kumar, and M. Bennett. Immunogenetic analysis of the mechanism of induction of Friend virus leukemia. Transplant. Proc. 7:173–184, 1975.
13. Fieldsteel, H.A., P.J. Dawson, and W.L. Bostick. Quantitative aspects of Friend leukemia virus in various murine hosts. Proc. Soc. Exptl. Biol. Med. 108:826–829, 1961.
14. Hartley, J.W., and W.P. Rowe. Clonal cell lines from a feral mouse embryo which lack host-range restrictions for murine leukemia viruses. Virology 65:128–134, 1975.
15. Hirschberg, H., H. Skare, and E. Thorsby. Cell-mediated lympholysis: CML. A microplate technique requiring few target cells and employing a new method of supernatant collection. J. Immunol. Meth. 16:131–141, 1977.

16. Klement, V., W.P. Rowe, K.W. Hartley, and W.E. Pugh. Mixed culture cytopathogenicity: A new test for growth of murine leukemia viruses in tissue culture. Proc. Natl. Acad. Sci. (USA) 63:733–758, 1969.
17. Knyszynski, A., and D. Danon. The role of macrophages in defense against the development of Rauscher virus leukemia. J. Reticuloendothel. Soc. 22:341–348, 1977.
18. Kumar, V., L. Goldschmidt, J.W. Eastcott, and M. Bennett. Mechanisms of genetic resistance to Friend virus leukemia in mice IV. Identification of a gene (Fv-3) regulating immunosuppression in vitro, and its distinction from Fv-2 and genes regulating marrow allograft reactivity. J. Exptl. Med. 147:422–433, 1978.
19. Lilly, F. Fv2, identification and location of a second gene governing the spleen focus response to Friend leukemia virus. J. Natl. Cancer Inst. (USA) 45:163–169, 1972.
20. Meredith, R.F., and J.P. OKunewick. Genetic influence in murine viral leukemogenesis. Biomedicine 24:374–380, 1976.
21. Mortensen, R.F., W.S. Ceglowski, and H. Friedman. Leukemia virus induced immunosuppression X. Depression of T-cell-mediated cytotoxicity after infection of mice with Friend leukemia virus. J. Immunol. 112:2077–2086, 1974.
22. Nagao, K., Y. Kodama, K. Hamada, and K. Yoroko. Relationship between organotropism and leukemogenicity of type C RNA viruses as demonstrated in NIH Swiss mice inoculated at birth with Gross murine leukemia virus. J. Natl. Cancer Inst. (USA) 60:855–859, 1978.
23. Nowinski, R.C., and T. Doyle. Decreased immunity to viral antigens and increased expression of endogenous leukemia viruses in athymic (nude) mice. Virology 77:429–432, 1977.
24. Odaka, T. Inheritance of susceptibility to Friend mouse leukemia virus V. Introduction of a gene responsible for susceptibility in the genetic complement of resistant mice. J. Virol. 3:543–548, 1969.
25. OKunewick, J.P., E.L. Philips, and B. Brozovich. Effect of antiserum on transplantable hematopoietic colony-forming units during Rauscher leukemia development. Am. J. Hematol. 1:443–452, 1976.
26. Poulter, L.W., and J.L. Turk. Specificity of cellular immune reactivity to virus induced tumours. Nature (New Biol.) 5:17–18, 1972.
27. Rowe, W.P., W.E. Pugh, and J.W. Hartley. Plaque assay techniques for murine leukemia viruses. Virology 42:1136–1139, 1970.
28. Stutman, O., and J.M. Dupuy. Resistance to Friend leukemia virus in mice: Effect of immunosuppression. J. Natl. Cancer Inst. (USA) 49:1283–1293, 1972.

12

Characterization of a Rat Mutant (rnu^{nz}) Showing Similarities to the Nude Mouse

Michael V. Berridge,* L. Jane McNeilage,† Barbara F. Heslop,† and Tom E. Miller‡

Wellington Cancer and Medical Research Institute,* Clinical School of Medicine, Wellington, New Zealand, Department of Surgery,† University of Otago Medical School, Dunedin, New Zealand, Department of Medicine,‡ University of Auckland School of Medicine, Auckland Hospital, Park Road, Auckland 3, New Zealand.

Abstract

The spontaneous occurrence of an autosomal recessive mutant (rnu^{nz}) devoid of hair in a colony of outbred albino rats has led to characterization of the "nude" syndrome in an alternative species to the mouse. Survival of nudes past weaning is dependent on conditions of husbandry in the nude colony. Both male and female nudes have proved to be fertile. Homozygotes are essentially hairless except for stunted vibrissae and show abnormal thymus gland development. A putative thymus rudiment is described which was clearly distinct from other lymph nodelike structures in the anterior mediastinal region. Lymph nodes and spleens were largely devoid of lymphoid cells, particularly in thymus-dependent areas, although partial reconstitution was observed following thymus grafting. Skin grafts from histoincompatible BS ($RT1^1$) rats were accepted by a homozygous nude for at least 10 weeks, whereas a phenotypically normal littermate rejected similar grafts within 10 days. The absence of functional T lymphocytes was also indicated by lack of response of peripheral blood lymphocytes to the T-cell mitogens concanavalin A (ConA) and phytohemagglutinin (PHA), and by the absence of cells sensitive to alloantiserum against rat T lymphocytes. Although total blood leukocytes from a nude rat were within the normal range established with normal animals, differential counting of leukocytes showed a fourfold elevation of neutrophils and 2.5-fold reduction of lymphocytes. Phenotypically normal heterozygotes gave values intermediate between nude and normal rats. The nude gene is being placed on both DA and AS backgrounds. The rnu^{nz} gene has recently been shown to be allelic with the rnu gene with respect to hairlessness and thymus deficiency, although distinct phenotypic differences in the extent of residual hair growth were apparent between the two mutations.

---

* To whom correspondence should be addressed.
© 1982 Gustav Fischer New York, Inc.
Proceedings of the Third International Workshop on Nude Mice.

Introduction

The thymus plays a critical role in the development of the immune response. This was demonstrated by Miller (13), who showed that neonatal thymectomy resulted in severely reduced adaptive immunity and poor antibody responses to certain antigens (14). A hairless mouse mutant first observed about the same time (10) was later shown to lack a thymus (15) and thus provided an excellent experimental tool not only for studying the role of the thymus in the development of the immune response but also for growing foreign cells, including tumors from a variety of sources, in an in vivo environment (6).

During the past 3 years, three further nude mutations have been described, one in AKR/J mice, which has been shown to be allelic with the original mutation to nude (3), and two in outbred rat colonies (1,4), which have also been shown to occur at the same gene locus (M.F.W. Festing, personal communication). This contribution describes our experiences with the rnu^{nz} rat mutant [NB: With the demonstration of the allelic nature of the two rat mutations the designation of the New Zealand rat mutant has now been altered from $nznu$ (1,2) to rnu^{nz}.]

Materials and Methods

Animals held at Wellington, New Zealand, were raised under conventional conditions in a room with other rat stock. At Dunedin, heterozygotes and nude rats were maintained isolated from other rat stock as described previously (12). Cages, bedding, and most food were sterilized. Vitamin supplements (11) were employed with irradiated (2500 R) food. The Specific Pathogen Free (SPF) nude rat colony at Auckland was derived by hysterectomy from a single pregnant heterozygous female. This colony was barrier maintained and checked monthly for all category 4 rat and mouse pathogens.

Most nude rats born at Dunedin were subcutaneously grafted with a thymus within 24 h of birth using a laminar-flow cabinet and aseptic technique. Nude progeny from albino heterozygotes were grafted with thymuses from phenotypically normal littermates, whereas nude progeny of DA × nude crosses received DA thymuses.

The techniques of hemagglutination, skin grafting, lymphocyte transformation, and cytotoxicity have been described previously (2).

Results

Occurrence and Breeding

The spontaneous occurrence of a hairless rat in a colony of outbred albino animals held at Victoria University of Wellington in 1976 has led to the establishment of both conventional and SPF colonies producing hairless progeny (1)

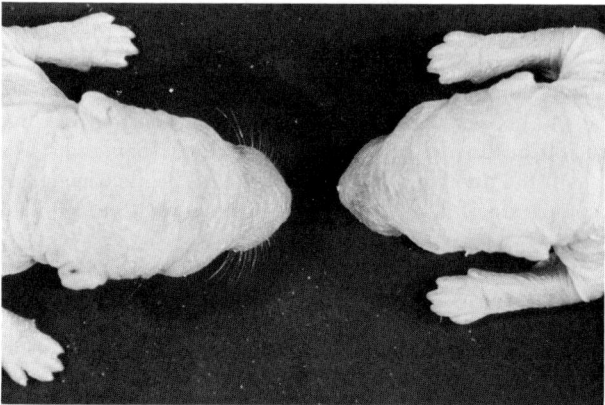

Figure 12-1. Normal (left) and nude (right) neonatal littermates within a few hours of birth showing the absence of vibrissae on the nude.

(Figure 12-1). The incidence of phenotypically abnormal animals and the breeding pattern established hairlessness as the homozygous state of an autosomal recessive gene mutation. Thus, combined results from conventional colonies held in Wellington (1) and Dunedin (12) show the incidence of hairless animals in litters to heterozygous parents to be 85 in 349 neonates, a figure not significantly different from the 25% expected for an autosomal recessive trait $\chi^2 = 0.077$ $(0.8 < P < 0.9)$. Survival past weaning appears to be dependent on the husbandry in the colony but under the best conventional conditions with thymus grafting was about 30%. Thus, animal breeding conditions at Dunedin (see Materials and Methods) resulted in a threefold increase in survival of homozygotes past weaning when compared with the Wellington colony. A similar improvement in the percentage of nudes attaining maturity was also observed. Figures from the SPF colony held in Auckland showed 11 nudes weaned out of 30 born, a figure not significantly different from the Dunedin experience (13 weaned out of 43 nudes born). Thus, it appears that thymus grafting of nude rats within 24 h of birth gives initial survival results similar to those obtained under SPF conditions. To date only eight nude rats raised conventionally have survived past 2 months and none past 6.5 months. In the SPF colony 27 nude have reached adulthood, of which 56% have been female. The average age of nude males which survived weaning was 5.5 months and that of nude females, 12 months.

Nude females are fertile but experience problems rearing their young. In the SPF colony at Auckland, nude females have sometimes needed assistance during birth and litters have had to be fostered immediately postpartum. At Dunedin nude females have not required assistance during birth but have been incapable of raising more than two offspring at a time. This may be caused by lactation problems, for the young reared by a nude mother are considerably smaller at weaning than their fostered littermates. In the SPF colony five out of six nude males have also been shown to be fertile but the average litter size was very small (two compared with nine for rnu^{nz}/rnu^{nz} female $\times$ $rnu^{nz}/+$ male and 12 for heterozygous matings).

The nude gene is being transferred to the DA background by repeated backcrossing of heterozygotes to DA rats. Heterozygotes are identified by crossing with rnu^{nz}/rnu^{nz} rats prior to crossing to the background strain. The survival figures of homozygous nudes in this colony are inferior to those of the albino colony because the heterozygous mothers have not accepted the nudes well. Of 86 nudes born, 80 have died preweaning, four died soon after weaning, while two are alive at 2 months. The rnu^{nz} gene is also being transferred to the AS background; comparable figures for these crosses are not yet available.

Pathology Findings

Because of high mortality among nude rats, only a few animals have been subjected to detailed pathological examination. Most of the deaths have occurred prior to weaning, and the young have been cannibalized. This has precluded adequate examination. The thymus in normal young rats is a large, whitish, bilobed structure, which occupies a substantial part of the anterior mediastinum. No such structure is evident in New Zealand nude rats, in which the anterior mediastinum has a strangely empty gross appearance. Very careful inspection of the area, however, has always revealed the presence of one or more tiny encapsulated portions of light brownish tissue. These have been slightly easier to find in the seven animals that have been thymus grafted subcutaneously than in the three rats that have not been grafted. It should be pointed out, however, that the ages of the animals examined have ranged from 1 week to 6 months, and differences in the size of these structures may merely reflect differences in age. It seems possible that at least some of the specimens removed from the thymic region have in fact been lymph nodes. The fact that the material from the thymic area has been larger in the recipients of subcutaneous thymus grafts accords with the observation that peripheral lymph nodes have also appeared larger in thymus-grafted animals.

The histology of the tissue in the thymic region has fallen into three categories:

1. In one of the thymus-grafted animals that had survived for over 6 months, a definite lymph node was found. This had enormous cortical germinal centers, some cells in the T-dependent areas (although fewer than normal), and a large number of plasma cells in the medulla, which was otherwise empty (Figure 12-2);
2. Structures consisting predominantly of large pale cells with copious cytoplasm and large vesicular nuclei. No keratin was evident and there were a few lymphoid cells. At first sight these gave the impression of being predominantly epithelial structures but the fact that a structure of identical appearance was seen in a superficial cervical lymph node at some distance from the area in which one might expect thymic remnants suggested that the thoracic structure was a lymph node. The presence of a subcapsular sinus supported this interpretation (Figure 12-3).
3. In two animals (one thymus grafted, one ungrafted) structures unlike either of the above were identified. These were highly vascular polycystic structures. Some of vessels were venules and a few were possibly lymphatic channels. There were considerable numbers of large pale cells, which

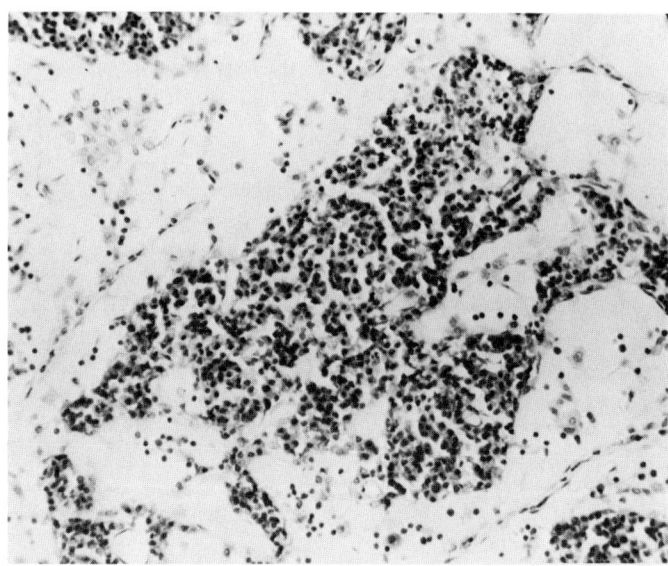

Figure 12-2. Medulla of lymph node from anterior mediastinum of 6-month-old thymus-grafted nude female. Apart from the substantial collections of plasma cells, the medulla is very poorly cellular. H&E ×208.

could possibly have been epithelial cells. Large quantities of extracellular material were present, but there was no identifiable keratin. In both cases a number of blast cells were present, together with lymphoid cells. In some cases the epithelial cells were arranged in acini so that the specimen bore a superficial resemblance to the parathyroid. On inspection under oil immersion, however, the cells had none of the characteristics of parathyroid epithelium. This structure has tentatively been designed a thymic rudiment and is shown in Figure 12-4. It is probably analogous to the thymic rudiment described in the mouse (8,9,16–18).

The lymph nodes in ungrafted nudes have consisted almost entirely of large cells with pale cytoplasm and vesicular nucleus. These are presumably cells of the reticuloendothelial system. There have been few lymphocytes, not only in the thymus-dependent areas but also in the cortex. In the thymus-grafted animals the cortex has often been well developed, with large germinal centers. The medulla of grafted animals has sometimes contained large numbers of plasma cells (Figure 12-2), although very few other cells have been seen. Examination of the spleen from ungrafted animals has shown marked depletion of T lymphocytes in thymus-dependent areas; characteristically the absence of perivascular lymphocyte cuffing in the nude spleen is seen in Figure 12-5. In thymus-grafted animals the spleen has contained a few germinal centers, and there have been plasma cells scattered throughout the red pulp. In none of the animals so far examined has thymus grafting brought about complete restitution of the peripheral lymphoid tissue.

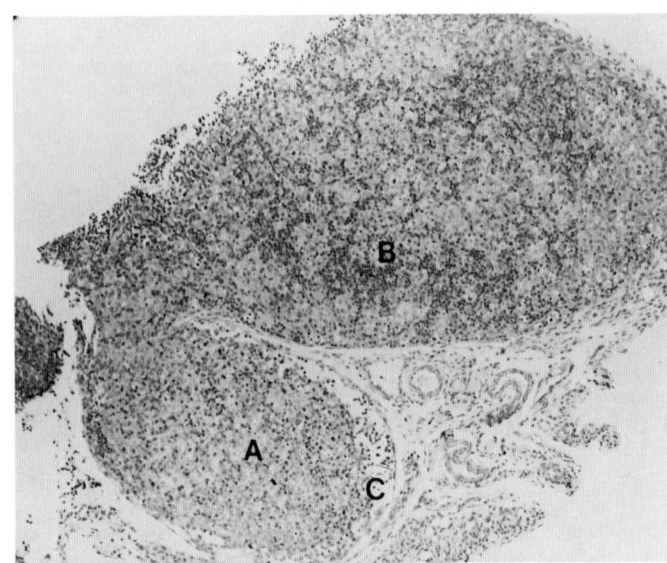

Figure 12-3. Superficial cervical nodes from a 21-day-old thymus-grafted nude rat. All the cervical nodes in this animal appeared similar, consisting of large pale cells (A), with some lymphoid cells (B). A subcapsular sinus is evident at C. A similar structure was present in the anterior mediastinum. H&E ×95.

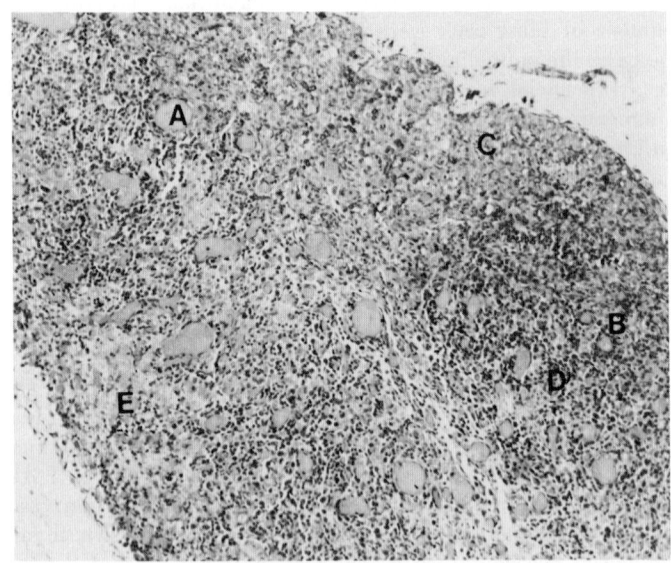

Figure 12-4. Structure removed from the anterior mediastinum of a 4-week-old thymus-grafted nude rat. There are numerous vascular channels (A), some epithelial lined acinar structures (B), numerous large pale cells (C), and some lymphoid cells (D). The specimen contains much extracellular material, best seen at E. H&E ×95.

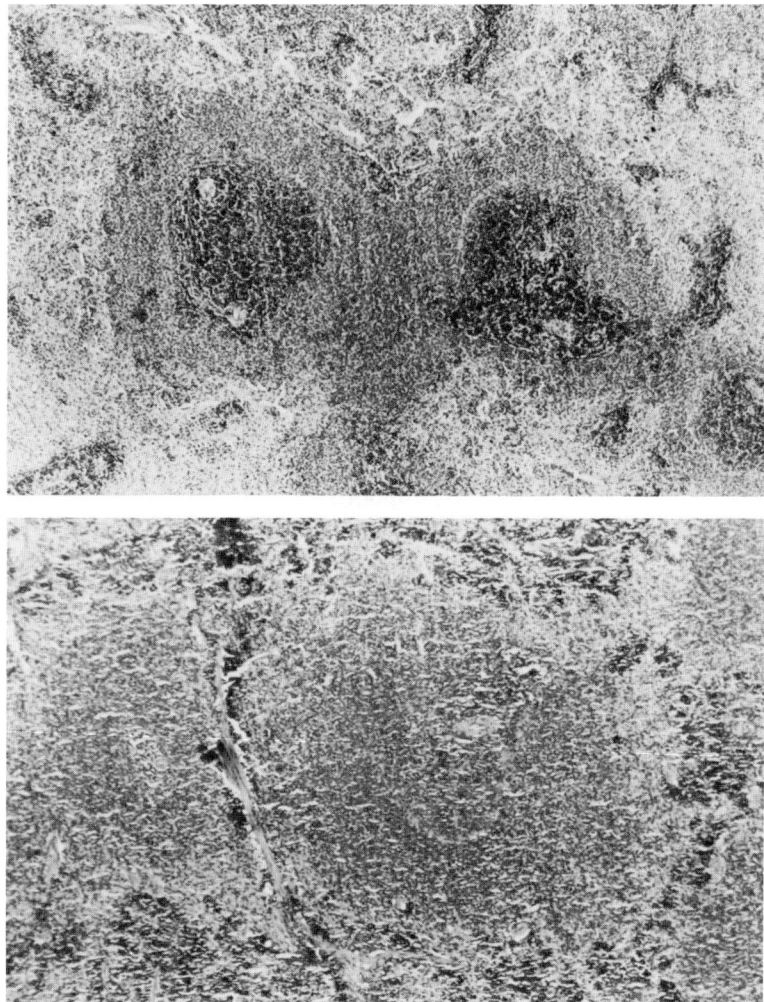

Figure 12-5. Normal spleen (top) and nude spleen (bottom) showing the absence of periarteriolar cuffing by T lymphocytes in the nude. H&E ×26.

Immunologic Characteristics

Studies on the immunology of the nude rat rnu^{nz} have been limited by the lack of availability of mature individuals and because most mutants have been thymus grafted soon after birth. However, limited investigations on a single animal have been described (2) and are here summarized.

1. Blood leukocytes from a nude rat that had been subjected to skin grafting 6 weeks previously were within the normal range established for normal rats from the same colony and for heterozygotes. However, differential

leukocyte analysis revealed only 34% lymphocytes in the nude rat compared with 77% ± 7.4% for normal animals and 58% ± 12% for heterozygotes. Neutrophils were elevated fourfold in the nude compared with normal rats.
2. Grafting of histoincompatible BS (RT1¹) skin onto a nude rat resulted in prolific growth of black hair on the graft and graft acceptance for at least 10 weeks posttransplantation. A normal littermate rejected its graft within 10 days.
3. Peripheral blood lymphocytes from the nude rat failed to respond to ConA or PHA, whereas phenotypically normal rats gave dose-dependent responses. No response to lipopolysaccharides (LPS) was observed in either normal or nude rats.
4. Peripheral blood lymphocytes from a nude rat were insensitive to alloantiserum against the PtaAl lymphocyte marker in a complement-dependent cytotoxic assay, whereas lymphocytes from normal rats were killed.

Discussion

The recent description of two independent mutations resulting in hairless athymic rats (1,4) which exhibit phenotypic characteristics similar to those of nude mice (2,5) should broaden the scope for investigation of various aspects of immunologic phenomena. Although the method of choice for breeding nude rats is obviously under germ-free or SPF conditions, some success has been achieved in raising thymus-grafted nude rats under clean conventional conditions. However, high mortality of nude neonates even when fostered to selected mothers results in few nudes reaching maturity (14–25% in the albino colony and about 2% in the colony backcrossed to DA rats).

Comparison of the histology of the thymic region of the nude rat with similar studies in the nude mouse (8,9,16,17) reveal certain similarities. For example in both nude rats and mice (8) mediastinal lymph nodes showing abnormal hyperplasia have been observed in which cells resembling thymic epithelia are apparent. The putative thymic rudiment presently described in both ungrafted and thymus-grafted nude rats does not resemble lymph node tissue but shows analogy with a similar structure observed in the nude mouse (8,9,16). Both structures are polycystic, are associated with acini of epithelial cells, and have a lymphoid cell component.

Although some of the observations described in this contribution were made with animals grafted subcutaneously with a thymus, the putative dysgenetic thymic rudiment nevertheless bears a close morphological resemblance to analogous structures in the nude mouse. These observations are of a preliminary nature and rest on light microscopic examination. More detailed analysis of these structures, including electron microscopy and analysis of cell surface markers on the epithelial cells and associated lymphocytes, will be required before definitive comments can be made concerning the histopathology of this rat mutant. The lack of effect of the rnu^{nz} mutation on blood leukocytes is more similar to the effect observed in streaker mice (19) than that in nude mice (15). For example, total blood leukocytes in the nude rat were within the normal range established for heterozygous and normal animals, whereas the percentage lymphocytes was depressed in the nude rat.

Initial plans to further characterize the immunology and pathology of the nude rat will place emphasis on the study of T-cell markers, T-cell-dependent immunologic reactions, and natural killer cell activity, which is thought to be responsible for the relatively low incidence of spontaneous neoplasms in nude mice (7). Once the nude gene has been placed on a known genetic background (both AS and DA are in progress) genetic studies of immunologic parameters will be initiated.

Acknowledgments

This work was supported by the Malaghan Research Fellowship to M.V.B. and by the Medical Research Council of New Zealand. We thank Richard Moore, Anne Marie McCarthy, and Vernon Jansen for proficient care of the nude colonies, Pamela Salmon for histology, and Dr. A.F. Baradi for discussions on the histology.

References

1. Berridge, M.V., R. Moore, B.F. Helsop, and L.J. McNeilage. Another nude rat (nznu). Rat News Lett. 4:23–26, 1978.
2. Berridge, M.V., N. Okech, L.J. McNeilage, B.F. Helsop, and R. Moore. Rat mutant (nznu) showing "nude" characteristics. Transplantation 27:410–413, 1979.
3. Eicher, E.M. Remutations. Mouse News Lett. 54:40, 1976.
4. Festing, M.F.W. More on the nude rat. Rat News Lett. 2:14–16, 1977.
5. Festing, M.F.W., D. May, T.A. Connors, D. Lovell, and S. Sparrow. An athymic nude mutation in the rat. Nature (London) 274:365–366, 1978.
6. Fogh, J., and B.C. Giovanella (eds.). The nude mouse in experimental and clinical research. New York: Academic Press, 1978.
7. Herberman, R.B. Natural cell-mediated cytotoxicity in nude mice, pp. 135–166. In J. Fogh and B.S. Giovanella (eds.), The nude mouse in experimental and clinical research. New York: Academic Press, 1978.
8. Holub, M., P. Rossmann, and B. Mandi. The dysgenetic thymic complex of the nude mouse. Folia Biol. (Praha) 24:416–418, 1978.
9. Holub, M., P. Rossmann, H. Tlaskalova, and H. Vidmarova. Thymus rudiment of the athymic nude mouse. Nature (London) 256: 491–493, 1975.
10. Isaacson, J.H., and B.M. Cattanach. Report. Mouse News Lett. 27:31, 1962.
11. Ley, F.J., J. Bleby, M.E. Coates, and J.S. Paterson. Sterilization of laboratory animal diets using gamma irradiation. Lab. Anim. 3:221–254, 1969.
12. McNeilage, L.J., B.F. Helsop, and M.V. Berridge. Nude (athymic) rats maintained in conventional conditions: A progress report. Proc. Univ. Otago Med. School 57:47–48, 1979.
13. Miller, J.F.A.P. Immunological function of the thymus. Lancet ii:748, 1961.
14. Miller, J.F.A.P. Effect of thymic ablation and replacement, pp. 436–460. In R.A. Good and A.E. Gabrielsen (eds.), The thymus in immunobiology. New York: Hoeber-Harper, 1964.
15. Pantelouris, E.M. Absence of thymus in a mouse mutant. Nature (London) 217:370–371, 1968.
16. Pantelouris, E.M., and J. Hair. Thymic dysgenesis in nude nu/nu mice. J. Embryol. Exptl. Morphol. 24:615–623, 1970.

17. Rychter, Z., M. Holub, and R. Vaněček. Topical and quantitative analysis of the thymus region in the nude mouse. Folia Biol. (Praha) 24:414–415, 1978.
18. Rygaard, J. Thymus and self, pp. 65–67. New York: John Wiley and Sons, 1973.
19. Shultz, L.D., H. Heiniger, and E.M. Eicher. Immunopathology of streaker mice: A remutation to nude in the AKR strain, pp. 211–222. *In* M.E. Gershwin and E.I. Cooper (eds.), Animal models of comparative and developmental aspects of immunity and disease. New York: Pergamon, 1978.

General Discussion

SORDAT: You mentioned the presence of germinal centers (GC) within the lymphoreticular tissue of the nude rat. Have any differences in number, size, or morphology of these GC been observed among the conventionally raised, "clean–conventional," and protected colonies?

BERRIDGE: Germinal centers were observed in the spleens and in the cortex of lymph nodes of thymus-grafted nude rats but not in ungrafted nudes maintained under "clean–conventional" conditions. No systematic studies have been done to compare the morphology of lymphoid organs between the three nude rat colonies described.

SORDAT: Could the lactation performance by a female nude rat be modified by a heterozygous thymus graft?

BERRIDGE: Under conventional conditions we have not been able to produce breeding female nudes except by thymus grafting. In the SPF colony, nude females have not been thymus grafted. Thus, although we do not know the effect of thymus grafting on lactation performance of nude females, it is possible that one effect of thymus grafting is to improve lactation.

MITCHELL: Do thymus grafts succeed in nude rats; i.e., is there any indication of active rejection or inhibition of thymus tissue?

BERRIDGE: Thymus grafting greatly improves the survival and breeding performance of nude rats and so may be considered to succeed functionally. In addition grafted nude rats show repopulation of T-dependent areas of lymphoid organs, including lymph nodes, spleen and perhaps rudimentary thymus tissue. Whether the graft itself survives is not known.

HOLLAND: What was the cause of death in SPF nude rats?

BERRIDGE: One of the primary causes of death in SPF nude rats appeared to be respiratory infection associated with bodily wasting. The exact nature of the infection has not been determined.

HOLLAND: One of the slides you showed us of a thymic rudiment bore marked resemblance to a nodule of ectopic thyroid tissue. In other congenital syndromes involving thymus dysgenesis, such as the DiGeorge syndrome, the thyroid frequently is involved; therefore, have you considered the possibility that this tissue is in fact of thyroid rather than thymic origin?

BERRIDGE: The putative thymic rudiment that we have observed appears to be morphologically distinct from the parathyroid. Characterization of the lymphoid cells in the rudiment is being undertaken in order to further characterize this organ.

13

Immunologic Studies and Growth of Human Tumors in the Athymic Rat

M. Joseph Colston, A. Howard Fieldsteel,*
R. Denise Lancaster,† and Peter J. Dawson‡

Life Sciences Division, SRI International, 333 Ravenswood Avenue, Menlo Park, California 94025, St. George's Hospital Medical School,† London SW17 ORE, United Kingdom, and Laboratory of Surgical Pathology,‡ University of Chicago, Chicago, Illinois 60637.

Abstract

This contribution describes a preliminary study of the immunologic status of the athymic rat. Histologically, lymphoid tissue from these animals was highly abnormal, with characteristic depletion in the thymus-dependent areas. Cultured splenic lymphocytes from athymic rats showed no response to concanavalin A, whereas low but significant responses were detectable in neonatally thymectomized rats. The athymic rats were able to accept skin allografts (Wistar/Furth rats) and xenografts (CBA mouse tail) and to support the growth of human tumor cell lines, although growth of some of the tumors was followed by complete regression. Athymic rats were also found to develop an enhanced and disseminated infection when inoculated in the foot pad with the human leprosy bacillus *Mycobacterium leprae*, an organism which, when inoculated into normal rats or mice, produces a limited, localized infection.

Introduction

An athymic, hairless mutation (*rnu*) of rats was recently described (5). One of the interesting properties of these rats is their ability to survive under conventional animal housing conditions; it was reported that under such conditions, postweaning losses of athymic rats were negligible and animals survived for more than 17 months (5). This is in marked contrast to athymic nude mice which rarely survive longer than 3-4 months when they are maintained under conventional conditions (4,8). This greater robustness of the athymic rat sug-

---

* To whom correspondence should be addressed.
© 1982 Gustav Fischer New York, Inc.
Proceedings of the Third International Workshop on Nude Mice.

gests that it may be of great value for biomedical research and emphasizes the importance of establishing its immunologic status.

The nude mouse has been shown to have no functional T lymphocytes as assessed by lack of in vitro response to T-cell mitogens (16) and its ability to accept skin allografts and xenografts (11,12), to support the growth of human tumors (7,13,14), and to be more susceptible than normal mice to a number of infectious agents (1). In addition, histological examination of lymphoid tissues of nude mice has revealed characteristic depletion of thymus-dependent areas (3). This contribution presents preliminary results of a study on: (a) the in vitro response to concanavalin A of splenic lymphocytes of athymic and neonatally thymectomized rats; (b) the ability of the athymic rat to accept skin allografts and xenografts and to support growth of human tumor cells; and (c) the infection resulting from the inoculation of athymic rats with the human leprosy bacillus, *Mycobacterium leprae*. This organism is known to produce an enhanced and disseminated infection in nude mice (2). The results of our histological examination of lymphoid tissue from athymic rats are also reported herein.

Materials and Methods

Animals. From the Laboratory Animal Centre, Carshalton, U.K., we obtained a small breeding nucleus of rats heterozygous for the *rnu* gene and have now established a colony of athymic rats. The mating system used is the homozygous recessive male crossed with the heterozygous female, commonly used for the breeding of nude mice (8). We are also transferring the gene for athymia onto the Lewis background by the cross–intercross system.

Athymic rats were usually 2–3 months old at the initiation of the experiments. Controls were either age- and sex-matched heterozygous littermates, or Lewis rats, as specified. Neonatally thymectomized Lewis rats (NTLR) were thymectomized within 16 h of birth as described previously (6).

Lymphocyte stimulation test. Spleens were removed aseptically and finely minced in RPMI 1640 medium. Erythrocytes were sedimented by centrifugation. Lymphocytes were washed twice and cultured at a concentration of 3×10^6 cells per milliliter in RPMI 1640 plus 10% fetal calf serum plus 25 μg/ml concanavalin A. Cultures were incubated for 48 h at 37°C in 5% CO_2 and 1 μCi of [^{3}H]thymidine was then added. After a further 18-h incubation, the lymphocytes were washed three times with buffered saline and finally with 5% trichloroacetic acid (TCA). TCA precipitates were collected on 25-mm Millipore filters, washed with five volumes of TCA, and counted in a Searle Mark III liquid scintillation counter.

Skin allografts and xenografts. Donor skin was prepared from the backs of Wistar/Furth rats and the tails of CBA mice. Two grafts, one allograft and one xenograft, were transplanted onto the dorsolateral area of the thorax of each recipient rat. The grafts were examined daily and the mean survival time (MST) was determined.

Human tumor cell lines. Cultured human tumor cells were grown in either Eagle's minimum essential medium or Dulbecco's modification of Eagle's me-

Immunological and Oncological Studies in Athymic Rats

dium, supplemented with 10% fetal bovine serum. Cultured cells were harvested by trypsinization and approximately 10^7 cells were inoculated subcutaneously between the scapulae of each rat. Each cell line was inoculated into four to seven athymic rats and two controls. When tumor growth was judged to have reached a maximum, one rat from each group was sacrificed and coded samples of the tumor, lymph nodes, lung, spleen, and kidney were examined histologically.

Infection with *M. leprae*. Eight athymic rats and eight heterozygote controls were inoculated in the left hind foot pad (LHP) with 10^5 *M. leprae* derived from mouse foot pad passage and in the right hind foot pad with 10^7 irradiation-killed *M. leprae*. Growth of *M. leprae* in the LFP and other organs was monitored by standard methods (15).

Results

The Response of Cultured Splenic Lymphocytes to Concanavalin A

The response to the T-cell mitogen concanavalin A of splenic lymphocytes of (a) normal Lewis rats, (b) NTLR, and (c) athymic rats is shown in Figure 13-1. The results are expressed as the lymphocyte stimulation index—that is, the ratio of counts of quadruplicate cultures in the presence of concanavalin A to those in the absence of concanavalin A.

The lymphocytes from athymic rats showed no significant response, with stimulation indices ranging from 0.6 to 1.5. The mean stimulation index in the

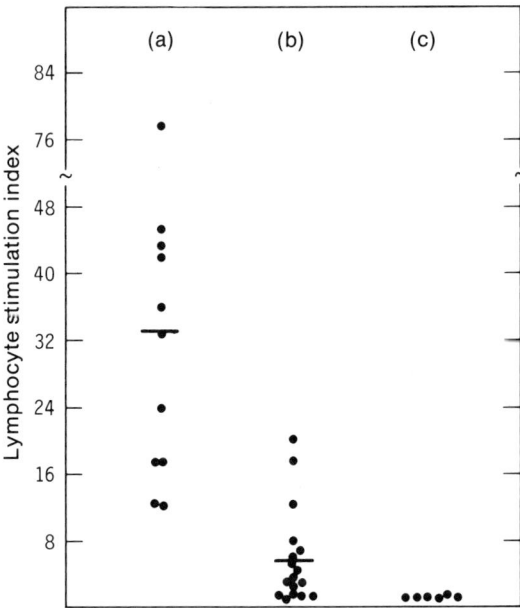

Figure 13-1. Response of cultured splenic lymphocytes to concanavalin A. (a) normal Lewis rat; (b) NTLR; (c) athymic rats.

Table 13-1. Survival time of skin allografts and xenografts on athymic rats

Recipient animal	Mean survival time (days)	
	Xenografts (Wistar/furth rat)	Xenografts (CBA mouse tail)
Control rats	9.5, 11, 11.5, 12.5, 13, 13.5, 15; $\bar{X} = 12.9$	7, 8, 9.5, 9.5, 10.5, 12, 12, 14, 15; $\bar{X} = 10.8$
Athymic rats	>42,[a] >117,[a] >130[a]	>42,[a] >117,[a] >130[a]

[a] Rats killed with grafts still intact.

control group was 33.3. The response in NTLR was lower than in controls, but significant responses were detectable in some of the rats; stimulation indices in this group ranged from 0.9 to 20.1 (mean 5.8).

Histological Examination of Lymphoid Tissue

Serial blocks of the thymus region from adult athymic rats have failed to reveal any thymic remnants. Lymph nodes exhibited marked paracortical depletion, reduction in the width of the cortex, and a marked reduction in the number of germinal centers, although a few remained. It is interesting to note that except for the medastinal nodes, there was extreme plasmacytopoiesis of the medullary canals.

Spleens were smaller than normal. There was lymphoid depletion of the perivascular sheath; Malpighian bodies appeared normal. The gut-associated lymphoid tissue was well developed but lacked germinal centers.

Survival Time of Skin Allografts and Xenografts on Athymic Rats

To determine whether athymic rats were capable of accepting skin allografts and xenografts, skin grafts from Wistar/Furth rats and CBA mouse tails were prepared and transplanted onto the backs of three athymic rats and nine heterozygote littermates.

The mean survival time (MST) for allografts and xenografts on heterozygote littermates (control rats) was 12.9 and 10.8 days, respectively (Table 13-1). The three athymic rats were killed at 42, 117, and 130 days after transplantation of the grafts; in each case, both the allograft and the xenograft were still intact. Therefore, athymic rats, like athymic mice (11,12), appear to be capable of accepting allografts and xenografts indefinitely.

Growth of Human Tumor Cell Lines in Athymic Rats

Five cell lines, all of which grew as monolayers, were inoculated subcutaneously into four to seven athymic rats and two heterozygous littermates. Details of the cell lines are shown in Table 13-2. Although the melanoma has been classified as fibroblastic by the Naval Biological Research Laboratory, the cells were not typically fibroblastic but were spindle shaped, with a tendency to pile up in culture.

Table 13-2. Human tumor cell lines inoculated into nude rats

Cell line	Passage Level	Tumor type	Cell type
A549	24	Lung carcinoma[a]	Epithelial
HSO695T	76	Melanoma[a]	Fibroblastic
HSO766T	10	Pancreas carcinoma[a]	Epithelial
HSO700T	13	Metastatic colon, intestine, or pancreas carcinoma[a]	Epithelial
Sk-CO-1	19	Colon carcinoma[b]	Epithelial

[a] Obtained from Dr. W. A. Nelson-Rees, NBRL, Oakland, California.
[b] Obtained from Dr. J.H. Pincus, SRI International, Menlo Park, California.

Only one cell line, SK-CO-1, failed to produce visible tumors in athymic rats, although minute tumors the size of grains of sand could be palpated from the ninth day after inoculation.

Cell line HSO766T, a metastatic pancreas carcinoma, produced visible tumors by 23 days. However, regression began at between 35 and 44 days, and the tumors had completely disappeared by 52 days. One of the tumors was removed when growth was at a maximum; it was characterized histologically as an anaplastic large-cell tumor consistent with a poorly differentiated carcinoma.

The pattern of growth of cell line A549, a lung carcinoma, was much the same as that for HSO766T. In most of the athymic rats, visible tumors had developed by 20 days and progressed until approximately 40 days, when regression started to occur. By 70 days most of the tumors were no longer palpable. Histologically, the tumor was classified as a poorly differentiated, mucin-secreting adenocarcinoma.

Cell line HSO700T—a metastatic colon, intestine, or pancreas carcinoma—produced visible tumors between 20 and 40 days after inoculation; tumor in three of the athymic rats regressed, but those in three others were progressive.

HSO695T, a malignant melanoma, was the only cell line to produce progressive tumors in all athymic rats. One tumor removed 77 days after inoculation weighed 51 grams. Histologically, it was described as a polygonal cell tumor with clear cytoplasm, round or oval nuclei, and some pigment-containing cells—consistent with a melanoma.

Growth of M. leprae *in Athymic Rats Following Foot Pad Inoculation*

Athymic and heterozygote control rats were inoculated in the LFP with 10^5 viable *M. leprae* and in the RFP with 10^7 irradiation-killed *M. leprae*. This dose of dead organisms is known to act as an immunogen in normal rats, preventing the growth of viable organisms in the contralateral foot pad.

Growth of *M. leprae* in the LFP of athymic rats (Table 13-3) was detectable in the first rat harvested; at 125 days, the number of organisms had increased 39-fold. There was no such increase in a control rat harvested at 130 days. By 164 days, 7.7×10^7 *M. leprae* were harvested from the LFP of the athymic rat and small numbers of organisms could be detected in liver and spleen. The

Table 13-3. Growth of *M. leprae* in athymic rats following foot pad inoculation

Animal	Days after inoculation	Acid-fast bacilli (AFB) harvested					
		LFP	Liver	Spleen	Ears	Tongue	Tail
Athymic rat	125	3.9×10^6	—	—	—	—	—
Control rat	130	1.7×10^5	—	—	—	—	—
Athymic rat	137	2.7×10^7	—	—	—	—	—
Control rat	157	2.1×10^5	—	—	—	—	—
Athymic rat	164	7.7×10^7	3.0×10^{4a}	1.8×10^{4a}	—	—	—
Athymic rat	291	6.7×10^7	1×10^{5a}	2×10^{5a}	8×10^3	3.6×10^4	1.7×10^6

^a Number of organisms per gram of tissue.

athymic rat harvested at 291 days had 6.7×10^7 *M. leprae* in the LFP. It is difficult to say whether this represents a "plateauing" of the infection, because at this stage we are comparing growth in only two animals. In any event, dissemination of the infection to peripheral sites, especially the tail, was evident in the athymic rat harvested at 291 days; such dissemination is never seen in normal rats infected with *M. leprae*.

Discussion

Clearly, the athymic rat is an important new model for biomedical research. In this study we were unable to detect any in vitro response to the T-cell mitogen concanavalin A, whereas low but significant responses were detectable in neonatally thymectomized rats.

The athymic rats were able to accept skin allografts and xenografts indefinitely and to support the growth of human tumor cell lines. The complete regression of at least two of the tumors is interesting and confirms the report by Festing et al. (5) of growth and then regression of a mouse plasma cell tumor and a human colon carcinoma. Reports of growth followed by regression of human tumor cell lines in nude mice have been few, although this has been reported to occur in a significant number of human tumors surgically transplanted into nude mice (14). This growth followed by regression implies that natural, cell-mediated cytotoxic mechanisms, thought to be important in controlling tumor development in nude mice (9), are also important in the athymic rat.

The enhanced infection seen in athymic rats inoculated with *M. leprae* suggests that, as with the nude mouse (2,10), these animals are highly susceptible to infection with this organism and may prove to be a useful model of multibacillary leprosy. The leprosy infection is extremely chronic and the long survival of athymic rats, even under conventional conditions, makes them particularly valuable. NTLR are currently being used in our laboratory as a model of multibacillary leprosy and fulfill most of our requirements. One of their major drawbacks, however, is variability in their degree of immunosuppression. This is illustrated by the response of splenic lymphocytes of NTLR to concanavalin A reported herein. These inadequately immunosuppressed animals have to be screened out of our experimental protocols, but even among the remaining NTLR there is individual variability. Clearly, an athymic rat that exhibits good survival might be expected to be much more uniform in its response to *M. leprae* infection, particularly when the gene for athymia has been transferred to an inbred-strain background.

The severely depleted specific cellular immune mechanisms of the athymic rat make it an important model for studies involving tumors from humans and infectious agents, particularly where longevity is an important factor. However, the ability of the athymic rat to survive under conventional animal housing conditions and the observation that some of the human tumors, when transplanted into athymic rats, grew and then regressed suggest that non-T-cell-mediated cellular immune mechanisms may be important to the athymic rat.

Acknowledgments

This work was supported by the United States–Japan Cooperative Medical Science Program, National Institute of Allergy and Infectious Diseases (Grant R22 AI-08417), by a grant from the Chemotherapy of Leprosy (THELEP) component of the UN DP/World Bank/WHO Special Programme for Research and Training in Tropical Diseases, and by an Internal Research and Development grant from SRI International.

M.J. Colston is supported by a Victor Heiser Fellowship for Leprosy Research.

Cell lines HSO766T, A549, HSO700T, and HSO695T were obtained from Dr. W.A. Nelsen-Rees and were produced with support from the National Cancer Institute, Biological Carcinogenesis Branch, Division of Cancer Cause and Prevention, under the auspices of the Office of Naval Research and the regents of the University of California. Cell line SK-CO-1 was obtained from Dr. J.H. Pincus of SRI International.

References

1. Armstrong, D., and P. Walzer. Experimental infections in the nude mouse, pp. 477–489. In J. Fogh and B.C. Giovanelli (eds.), The nude mouse in experimental and clinical research. New York: Academic Press, 1978.
2. Colston, M.J., and G.R.F. Hilson. Growth of *Mycobacterium leprae* and *M. marinum* in congenitally athymic (nude) mice. Nature (London) 262:399–410, 1976.
3. De Sousa, M.A.B., D.M.V. Parrott, and R.M. Pantelouris. The lymphoid tissues in mice with congenital aplasia of the thymus. Clin. Exptl. Immunol. 4:637–644, 1969.
4. Eaton, G.J., H.C. Outzen, R.P. Custer, and F.N. Johnson. Husbandry of the "nude" mouse in conventional and germfree environments. Lab. Anim. Sci. 25:309–314, 1975.
5. Festing, M.F.W., D. May, T. A. Connors, D. Lovell, and S. Sparrow. An athymic nude mutation in the rat. Nature (London) 365–366, 1978.
6. Fieldsteel, A.H., and A.H. McIntosh. Effect of neonatal thymectomy and antithymocytic serum on susceptibility of rats to *Mycobacterium leprae* infection. Proc. Soc. Exptl. Biol. Med. 138:408–413, 1971.
7. Giovanella, B.C., J.S. Stehlin, and L.J. Williams. Development of invasive tumors in the "nude" mouse after injection of cultured human melanoma cells. J. Natl. Cancer Inst. (USA) 48:1531–1533, 1972.

8. Gullino, P.M., R.D. Ediger, B. Giovanella, B. Merchant, H.C. Outzen, Jr., N.D. Reed, and H.H. Wortis. (Committee on Care and Use of the "Nude" Mouse). Guide for the care and use of the nude (thymus-deficient) mouse in biomedical research. Inst. Lab. Anim. Resources News 19:M5–M20, 1976.
9. Herberman, R.B. Natural cell-mediated cytotoxicity in nude mice, pp. 135–166. *In* J. Fogh and B.C. Giovanella (eds.), The nude mouse in experimental and clinical research. New York: Academic Press, 1978.
10. Kohsaka, K., T. Mori, and T. Ito. Lepromatoid lesion developed in nude mouse inoculated with *Mycobacterium leprae*. Animal transmission of leprosy. La Lepro (Jpn. J. Leprosy) 45:177–187, 1976.
11. Manning, D.D., N.D. Reed, and C.F. Shafter. Maintenance of skin xenografts of widely divergent phylogenetic origin on congenitally athymic (nude) mice. J. Exptl. Med. 138: 488–494, 1973.
12. Pennycuik, P.R. Unresponsiveness of nude mice to skin allografts. Transplantation 11: 417–418, 1971.
13. Povlsen, C.O., and J. Rygaard. Heterotransplantation of human adenocarcinomas of the colon and rectum to the mouse mutant nude. A study of nine consecutive transplantations. Acta Pathol. Microbiol. Scand. (A) 79:159–169, 1971.
14. Sharkey, F.E., J.M. Fogh, S.I. Hajdu, P.J. Fitzgerald, and J. Fogh. Experience in surgical pathology with human tumor growth in the nude mouse, pp. 187–214. *In* J. Fogh and B.C. Giovanella (eds.), The nude mouse in experimental and clinical research. New York: Academic Press, 1978.
15. Shepard, C.C. and D.H. McRae. A method for counting acid-fast bacteria. Intl. J. Lep. 36:78–82, 1968.
16. Thurman, G.B., B.B. Silver, J.A. Hooper, B.C. Giovanella, and A.L. Goldstein. *In vitro* mitogenic responses of spleen cells and ultrastructural studies of lymph nodes from nude mice following thymosin administration *in vivo*, pp. 105–117. *In* J. Rygaard, and C.O. Povlsen (eds.), Proceedings of the first international workshop on the nude mouse. Stuttgart: Gustav Fischer Verlag, 1974.

General Discussion

FESTING: Can you tell me the diameter of the skin allo- and xenografts transplanted to the nude rats?

FIELDSTEEL: They were 10 mm in diameter and were placed on the back—not the tail—of the rats.

14

Induction of Lymphatic Tissue in the Nude Mouse Dysgenetic Thymus

Miroslav Holub,* Z. Rychter,† and A. Machoninová

Institute for Clinical and Experimental Medicine, 146 22 Prague 4, Czechoslovakia, and Department of Histology,† Medical School, Charles University, Prague 2, Czechoslovakia.

Abstract

Levamisole induces a fast neoformation of dense lymphatic tissue containing lymphocytes with the Thy-1.2 surface marker in the dysgenetic thymus complex of nude mice. The effect is well marked 5 days after the start of levamisole treatment and can be observed as late as 2 months after its termination. On day 5 after the start of levamisole treatment the number of bone marrow colony-forming units (CFUs') of nu/nu and nu/+ mice also increases and blood leukopenia is temporarily corrected. There is no observable short-term effect of levamisole on the nu/+ thymus which is de norma smaller and has a reduced thymocyte population compared to a +/+ thymus. It is argued that the dysgenetic thymus microenvironment can be induced to support differentiation of T cells from local mesechymal precursors and that levamisole affects, directly or indirectly, basically defective mesenchymal cell centers in the bone marrow and thymus.

Introduction

Since it has been established that the dysgenetic thymus of the nude mouse does not support immigration and differentiation of prethymic precursor cells, although these cells are present in the nude mouse bone marrow (13,17), little attention has been given to the state of the dysgenetic thymus during immunostimulation of the nude mouse. Neither is there a consensus on the relation of thymus dysgenesis to the lower stem cell potential of mice bearing the *nu* gene (2). A mesenchymal disorder may be a direct effect of the *nu* gene; it may influence not only the bone marrow cellular relations but also, possibly through

---

* To whom correspondence should be addressed.
© 1982 Gustav Fischer New York, Inc.
Proceedings of the Third International Workshop in Nude Mice.

humoral factors and direct invasion of the thymic anlage, the development of the thymus itself (4). Therefore, we examined the influence of levamisole, which was found to enhance T-cell differentiation in nude mice (15) and may have a positive effect on the maturation, recruitment, and replenishment of the circulating pools of other white blood cells (1), on the state of the bone marrow stem cell potential, circulating white blood cell counts, and thymus of *nu* gene-bearing mice.

Materials and Methods

Mice were reared under SPF conditions in two closed colonies, one with a BALB/c background (ninth backcross generation), the other with C57BL/10Sn background (third backcross generation). The *nu/+* mice were obtained from *nu/nu* × *+/+* matings. Because we have a high male/female ratio among our *nu/nu* mice, males were used in most experiments; some data on thymus cellularity and bone marrow stem cell potential were obtained in females from a colony with a CBA/J background (cross–intercross breeding) which meanwhile has been discontinued.

Levamisole (solution R 12564), 20 mg/ml, obtained from Janssen Pharmaceutica (Beerse, Belgium) was diluted in pyrogen-free saline to contain in 0.2 ml a dose of 25 mg of the salt per kilogram of body weight. In most experiments single subcutaneous (SC) injections were applied for 3 consecutive days.

White blood cell counts and thymus cellularities were estimated in a Bürker chamber. Thy-1.2 marker on cells teased from the thymus or spleen was detected by the cytotoxic assay (7) and the bone marrow colony-forming units were measured by the Till and McCulloch assay using 850-rad irradiated littermates as femoral bone marrow cell recipients.

For morphometric, cytologic, and histological evaluation, dysgenetic thymus complexes were isolated by microdissection under a stereomicroscope. Paraffin embedded, hemotoxylin–eosin, or methyl-green-pyronin stained frontal sections were used for histological analysis. Diffusion chamber cultures were performed in carefully sealed Millipore filter envelopes (0.22 μm porosity). Diffusion chambers were implanted intraperitoneally (IP) to congeneic recipients under thiopental anesthesia. Cultures of dysgenetic thymuses with bone marrow cells and bone marrow cell transfer experiments were performed in the mice of BALB/c background.

Results

Heterozygous (*nu/+*) mice had white blood cell count and bone marrow stem cell potential comparable to *nu/nu* mice, i.e., significantly lower than *+/+* mice (Table 14-1). The thymus of *nu/+* mice was histologically normal; however, there was a lower content of lymphocytes especially in the medulla, where the number of epithelial cells per unit area was proportionally increased. The projection area outlined by a drawing device attached to the stereomicroscope

Induction of Lymphatic Tissue in Nude Mice

Table 14-1. White blood cells and colony-forming units in $+/+$, $nu/+$, and nu/nu mice

CBA/J females 2 months	n	WBCC[a]	CFU/s[b]
$+/+$	5	7590 ± 361[c]	30.8 ± 2.96
$nu/+$	5	2920 ± 277[d]	18.4 ± 2.23[d]
nu/nu	9	2160 ± 347[d]	21.1 ± 2.98[d]

[a] Number of leukocytes per cubic millimeter of tail vein blood.
[b] Number of colony-forming units in 10^5 bone marrow cells as determined in the Till-McCulloch assay (3).
[c] Mean $\pm$ SEM.
[d] Significant difference from $+/+$ ($P > 0.01$).

was used to compare the size and topographic position of the thymuses. The crossing points of internal thoracic vessels and phrenic nerves were used as invariant points. It was found (Figure 14-1) that the nu/nu dysgenetic thymus occupies the caput thymi area of the $+/+$ thymus and the $nu/+$ thymus the corpus thymi area. The $nu/+$ thymus was invariably smaller than the $+/+$ thymus. Also the thymus weight and cell index were significantly lower in $nu/+$ mice than in $+/+$ mice of the same sex and age (Table 14-2 and 14-6).

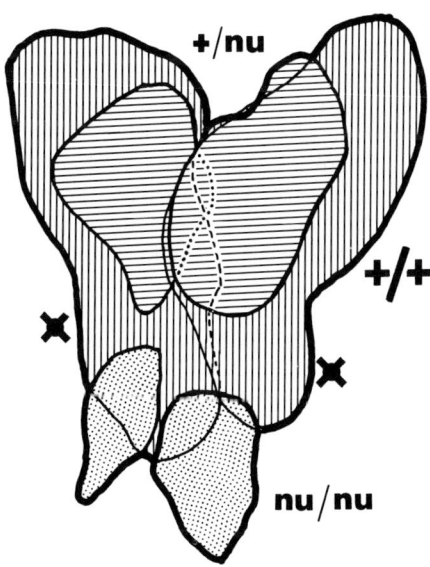

Figure 14-1. Projection area of the dysgenetic thymus (nu/nu) and thymuses of $nu/+$ and $+/+$ mice. ×, crossing points of phrenic nerves with internal thoracic vessels. Single specimen outlines obtained with a drawing device attached to the stereomicroscope were superimposed separately for the right and left sides; according to the number of individuals in each series, the outlines with a significantly low probability of occurrence were excluded and the "mean" position was estimated (2).

Table 14-2. White blood cell counts, thymus weights, and cell indices of $+/+$ and $nu/+$ mice

CBA/J females 3 months	n	WBCC[a]	Thymus weight (mg/g)	Thymus cell index (cells $\times 10^{-6}$/body weight in grams)
$+/+$	7	5360–8480	2.01 ± 0.18[b]	1.62 ± 0.29[b]
$nu/+$	6	2120–3720	1.26 ± 0.21[c]	0.55 ± 0.09[c]

[a] Number of leukocytes per cubic millimeter of tail vein blood.
[b] Mean $\pm$ SEM.
[c] Significant difference from $+/+$ ($P < 0.01$).

Levamisole (25 mg/kg) administered on days 1, 2 and 3 provoked a marked increase of the bone marrow stem cell potential of nu/nu and $nu/+$ mice on day 5 (Table 14-3). Levamisole elicited no such response in the $+/+$ mice, in one experiment CFU content decreased in the pooled femoral bone marrows from two $+/+$ donors (Table 14-3). The total number of cells flushed out from these femurs was 18×10^6 cells per mouse. Compared with 13.5 to 15×10^6 cell count per mouse found in other bone marrow donors this number suggests that the decrease of CFU may have resulted from the dilution of stem cells by maturing cell stages. Alternatively, levamisole may have increased the stem cell export from the bone marrow.

The white blood cell counts of nu/nu and $nu/+$ mice increased on day 5 after the beginning of levamisole application but returned to the initial values on day 12 (Table 14-4).

In the nu/nu dysgenetic thymus complex multiple loose lymphocytic infiltrates in the connective tissue around the cysts and vessels were present on days 3 and 4 after the beginning of levamisole application. From day 5 on dense lymphocytic accumulations were formed in the reticular framework of large cells with pale nuclei; they were partly demarcated by fibroblasts (Figure 14-2). Sometimes round acini of large, pale nucleated cells intermixed with lymphocytes or

Table 14-3. Colony-forming units in bone marrow cells in nontreated mice and mice treated with levamisole on day 1, 2, and 3

C57BL/10Sn males 4–6 weeks	CFU/s on day 5[a]	
	Not treated	Levamisole treated
$+/+$	28.34[b]	16.67[b]
	26.11[b]	24.10[b]
$nu/+$	17.84	25.26
	18.80	23.54
nu/nu	19.38	38.49
	19.00	24.20

[a] Number of colony-forming units in 10^5 bone marrow cells; pools of bone marrow cells from two donors.
[b] Two separate experiments.

Table 14-4. White blood cell counts in mice treated with levamisole on day 1, 2, and 3

C57BL/10Sn Males 4–6 weeks	n	WBCC[a]		
		Before treatment	After levamisole treatment	
			Day 5	Day 12
+/+	8	7820 ± 1318[b]	9160 ± 3165	7760 ± 1318
nu/+	6	3180 ± 488	7660 ± 1375[c]	2700 ± 551
nu/nu	10	1687 ± 296	4644 ± 1987[c]	2184 ± 877

[a] Number of leukocytes per cubic millimeter of tail vein blood.
[b] Mean ± SEM.
[c] Increase significant ($P < 0.001$).

rosettes of one large cell surrounded by lymphocytes could be distinguished. Lymphocytes permeated the cylindrical epithelium of some cysts. Numerous blood capillaries appeared in the dense lymphatic areas. In some dysgenetic thymi the arrangement of single dense lymphatic areas resembled a section of normal thymus cortex, well demarcated by fibroblasts on the periphery (Figure

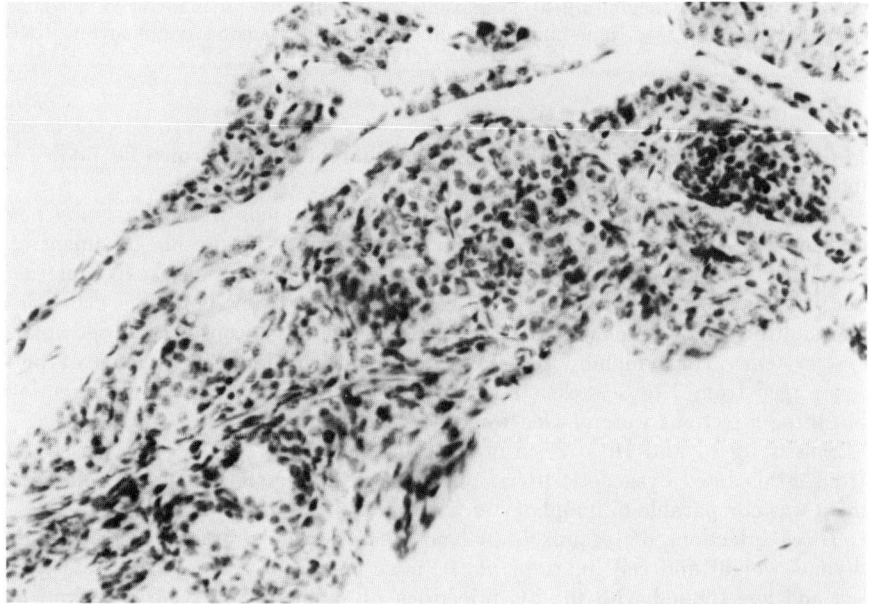

Figure 14-2. Central area of the nu/nu dysgenetic thymus complex on day 5 after the beginning of levamisole treatment (nu/nu male aged 6 weeks, frontal section). Lymphoid cells are mostly assembled in lobuli, partly demarcated by fibroblasts. Lymphocytes permeate the cyst-lining epithelium (bottom right). The lymphatic tissue contains numerous vessels. Secondary small cysts already can be seen in some lymphatic lobuli. H&E, ×330.

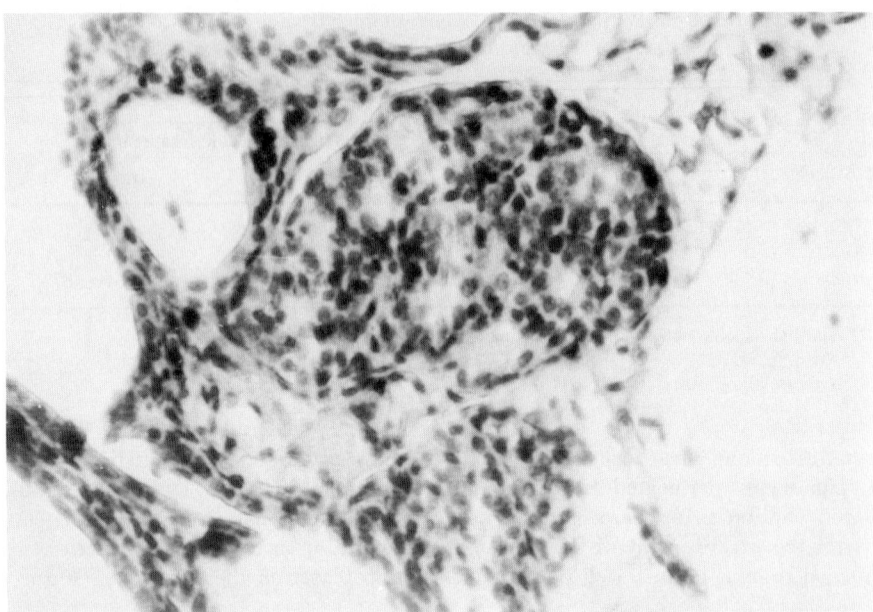

Figure 14-3. A well-demarcated lymphatic lobulus in the *nu/nu* dysgenetic thymus on day 6 after the beginning of levamisole treatment (*nu/nu* male aged 6 weeks, frontal section). Large, pale-nucleated cells can be seen among lymphocytes. H&E, ×522.

14-3). There was a marked reduction of undifferentiated or brown fat bodies in the area.

This lymphoid transformation of the dysgenetic thymus was a persisting phenomenon and could be found even 2 months after the levamisole treatment; at this late stage, the entire dysgenetic thymus complex was markedly enlarged and contained giant cysts lined with flat epithelium, a few epithelial cell acini, and multiple small lymphatic lobuli with connective tissue capsules and secondary cyst systems. The lymphoid transformation could be induced in mice of both sexes aged from 3 to 8 weeks. It was visible after a single levamisole injection, but three injections were needed for its full development.

Sensitivity to anti-Thy-1.2 serum + complement of lymphocytes teased out from the isolated dysgenetic thymi on day 6 after the start of levamisole treatment was comparable to lymphocytes from a *nu/+* thymus (Table 14-5).

Three injections of levamisole had no demonstrable short-term effect on the thymus weight and cell index of *nu/+* mice compared with mice of the same sex and age treated with the SC injections of saline (Table 14-6). Levamisole even decreased the thymus cell index of *+/+* mice (Table 14-6).

When bone marrow cells obtained by flushing the femurs of 6-week-old female BALB/c mice with Hanks solution were injected in the amount of 2 to 2.7 × 10^6 cells in 0.2 ml IV plus 11.4 to 13.5 × 10^6 in 0.5 ml IP into congeneic *nu/nu* recipients (6-week-old females), they provoked an increase in white blood cell counts lasting from day 2 until day 12. The proportion of blood

Table 14-5. Cytotoxic assay with the anti-Thy 1.2 serum using cells from nontreated and levamisole-treated mice

C57BL/10Sn males 3 weeks		Anti-Thy. 1.2 serum $+C^a$ (%)	Normal AKR serum $+C^a$ (%)
Levamisole treated day 6			
nu/nu	dysgen. thymus	8	97
	spleen	92	84
nu/+	thymus	10	90
Nontreated			
nu/nu	dysgen. thymus	16	94
	spleen	76	92
nu/+	thymus	5	95

^{a} Results expressed as percentage of viable cells in trypan blue exclusion test.

lymphocytes was increased. In the dysgenetic thymus area only small nodules and loose lymphocytic infiltrates were found on day 5; on day 12 there was no difference between the dysgenetic thymuses of the bone marrow cell-injected and control nu/nu mice.

Isolated dysgenetic thymuses from 8-week-old BALB/c nu/nu males were cultured for 5–6 days in diffusion chambers in peritoneal cavities of nu/+ littermates. Bone marrow cells (8 to 9 × 10⁴ cells in 0.1 ml of Hanks solution) from congeneic nu/nu donors or +/+ donors of the same age and sex were added to some chambers. The dysgenetic thymus donors or the bone marrow cell donors or diffusion chamber recipients were treated with 25 mg/kg leva-

Table 14-6. White blood cell counts, thymus weights, and cell indices in levamisole-treated nu/+ and +/+ mice

C57BL/10Sn^a males 20 days	n	WBCC^b (mean)	Day 6 after treatment		
			WBCC^c (mean)	Thymus weight (mg/g) (mean ± SEM)	Thymus cell index (cells × 10⁻⁶/body weight in grams) (mean ± SEM)
nu/+ levamisole	5	3540	6330	3.80 ± 0.92	1.78 ± 1.25
nu/+ saline	5	3180	4670	4.63 ± 0.89	2.70 ± 1.06^d
+/+ levamisole	5	9060	9100	4.44 ± 0.98	2.94 ± 0.95^d
+/+ saline	5	8600	7600	5.72 ± 0.89	5.33 ± 1.49

^{a} Mice were injected subcutaneously with levamisole or saline on days 1, 2, and 3.
^{b} Number of leukocytes per cubic millimeter of tail vein blood before treatment.
^{c} Number of leukocytes per cubic millimeter of tail vein blood on day 6 after treatment.
^{d} Significant difference from +/+ controls (last line) ($P < 0.05$).

misole SC or IP. The only positive findings (lymphoid transformation of the dysgenetic thymus complex) were obtained in diffusion chambers with dysgenetic thymuses from donors treated with levamisole 24 h before killing. The presence of bone marrow cells had no effect.

Discussion

It has been shown that mice bearing the *nu* gene (*nu/nu* and *nu/+*) have a reduced bone marrow stem cell potential as confirmed both by the CFUs' assay and in diffusion chamber cultures (3). We have suggested, therefore, that this stem cell defect may be a primary consequence of a mesenchymal disorder caused by the *nu* gene (3). In the *nu/+* hybrid, the bone marrow stem cell defect obviously affects the thymus size and reduces its lymphocytic population. Levamisole treatment stimulated the *nu/nu* and *nu/+* bone marrow stem cell potential, possibly temporarily, as suggested by the return of the white blood cell counts to normal *nu/nu* and *nu/+* values on day 12 after levamisole application.

On the other hand, levamisole has a prompt and lasting effect on the *nu/nu* dysgenetic thymus, inducing there a conspicuous lymphoid transformation. Multiple aggregates of dense lymphatic tissue are formed, reminiscent of the solitary lymphatic rudiment found in all dysgenetic thymus complexes of non-treated *nu/nu* mice (6,7), but occupying an area about three to four times larger in the frontal section (11); the aggregates contain T lymphocytes and fuse into well-demarcated lobuli which replace the undifferentiated and brown fat bodies. Unless we accept a very special effect of levamisole on the blood-borne precursor cells, the most likely explanation of the lymphoid transformation is a stimulation of local, intrinsic T-cell precursors, as suggested also by the diffusion chamber and bone marrow transfer experiments. Thymus repopulation by blood-borne prethymic cells starts as late as 10 days after a lethal irradiation in normal mice (16) and after 2–3 weeks in *nu/nu* mice with an implanted allogeneic thymus (10). Intrinsic thymic precursor cells (stem cells or transit cells) were disclosed in normal thymuses upon lethal irradiation (9); these cells are solely responsible for the thymocyte differentiation between 2 and 8 days after the destruction of the preexisting thymocyte pool (9). The local mesenchymal precursors in the *nu/nu* dysgenetic thymus complex may be the settled and dedifferentiated progeny of the large blastic cells (5) that migrate in the embryo but do not find a proper inducing microenvironment in the dysgenetic thymus. Consequently, we have to assume that levamisole treatment affects also the dysgenetic thymus undifferentiated epithelial cells (6). "Epithelial structures" in the *nu/nu* dysgenetic thymus which do respond to humoral factors from a normal epithelial thymus graft were described (8). We have found a marked enlargement of epithelial acini in the *nu/nu* dysgenetic thymus after orthotopic transplantation (cranially to the normal thymus) to congeneic euthymic recipients (6). Levamisole effects are currently likened to thymic humoral factors (15). On the other hand, if levamisole induces T-cell differentiation and activation directly or indirectly (15), it may as well have an effect on thymic epithelia bearing the same Thy-1 surface marker (14). A few thymus-specific epithelial cells containing thymosin-like material were found even in nontreated nu/nu dysgenetic thymuses (12).

Levamisole had no short-term effect on the $nu/+$ thymus weight and cellularity. In $+/+$ mice there was even a decrease of the thymic cell index on day 6 after levamisole. Again, the bone marrow stem cell stimulation had no immediate relevance to the thymocyte pool in the $nu/+$ thymus. In normal, well-balanced centers of cell differentiations (thymus, bone marrow) enhancement of the cell maturation and traffic may dominate the overall picture after levamisole application.

In the basically defective mesenchymal and epithelial cell centers levamisole seems to stimulate, directly or indirectly, the available undifferentiated primitive cells. These cells must then be present in the nu/nu dysgenetic thymus complex.

References

1. Amery, W.K. The mechanism of action of levamisole: Immune restoration through enhanced cell maturation. J. Reticuloendothel. Soc. 24:187–193, 1978.
2. Barták, M., M. Bokorová, Z. Rychter, and M. Holub. The thymus of the $nu/+$ hybrid. Folia Biol. (Praha) 24:419–420, 1978.
3. Dolenská, S., M. Holub, and B. Mándi. Bone marrow stem cell potential in mice bearing the nu gene. Folia Biol. (Praha) 24:421–423, 1978.
4. Groscurth, P., and G. Kistler. Histogenese des Immunosystems der "nude" Maus. I. Pränatale Entwicklung des Thymus: Eine lichtmikroskopische Studie. Beitr. Pathol. 154:109–124, 1975.
5. Habu, S., and N. Tamaoki. Thymocyte differentiation from precursor cells in the embryonic thymus of nude and normal mice, pp. 197–206. In T. Nomura, N. Ohsawa, N. Tamaoki, and K. Fujiwara (eds.), Proceedings of the second international workshop on nude mice. Tokyo: University of Tokyo Press, 1977.
6. Holub, M., P. Rossmann, and B. Mándi. The dysgenetic thymic complex of the nude mouse. Folia Biol. (Praha) 24:416–418, 1978.
7. Holub, M., P. Rossmann, H. Tlaskalová, and H. Vidmarová. Thymus rudiment of the athymic nude mouse. Nature (London) 256:491–493, 1975.
8. Hong, R., H. Schulte-Wissermann, E. Jarrett-Toth, S.D. Horowitz, and D.D. Manning. Transplantation of cultured thymic fragments. II. Results in nude mice. J. Exptl. Med. 149:398–415, 1979.
9. Kadish, J.L., and R.S. Basch. Thymic regeneration after lethal irradiation: Evidence for an intra-thymic radioresistant T cell precursor. J. Immunol. 114:452–458, 1975.
10. Loor, F., and B. Kindred. Differentiation of T-cell precursors in nude mice demonstrated by immunofluorescence of T-cell membrane markers. J. Exptl. Med. 138:1044–1055, 1975.
11. Machoninová, A., Z. Rychter, M. Holub, L. Korčáková, and B. Mándi. Influence of thymosin and levamisole on dysgenetic thymus of the nude mouse. Folia Biol. (Praha) 24:424–425, 1978.
12. Mándi, B., M. Holub, P. Rossmann, B. Csaba, T. Glant, and E. Ölveti. Detection of thymosin 5 in calf and mouse thymus and in nude mouse dysgenetic thymus. Folia Biol. (Praha) 25:49–55, 1979.
13. Pritchard, H., and H.S. Micklem. Haemopoietic stem cells and progenitors of functional T-lymphocytes in the bone marrow of "nude" mice. Clin. Exptl. Immunol. 14:597–607, 1973.
14. Raedler, A., A. Arndt, E. Raedler, D. Jablonski, and H.-G. Thiele. Evidence for the presence of Thy-1 on cultured thymic epithelial cells of mice and rats. Eur. J. Immunol. 8:728–730, 1978.
15. Renoux, G., and M. Renoux. Thymus-like activities of sulphur derivatives on T-cell differentiation. J. Exptl. Med. 145:466–471, 1977.

16. Takada, A., and Y. Takada. Proliferation of donor marrow and thymus cells in the myeloid and lymphoid organs of irradiated syngeneic host mice. J. Exptl. Med. 137: 543–546, 1973.
17. Wortis, H.H., S. Nehlsen, and J.J. Owen. Abnormal development of the thymus in "nude" mice. J. Exptl. Med. 134:681–692, 1971.

15

Induction of a T-Cell-like Response in Athymic Mice

Gillian Beattie,* Joseph Lipsick, Robert A. Lannom, Steven Baird, Nathan O. Kaplan, and Abraham G. Osler

Department of Chemistry, Medicine, and Pathology, University of California at San Diego, La Jolla, California 92093.

Abstract

Our athymic mouse colony has on occasion yielded *nu/nu* mice capable of mounting an antibody response to sheep erythrocytes (SRBC), an antigen known to be T dependent in mice. These mice also appeared ill and many were wasting. Autopsies revealed all animals to be truly athymic but infected with the pinworms *Aspiculuris tetraptera* or *Syphacia obvelata* or both. The antiserum produced by the pinworm infected mice in response to immunization with SRBC did not cross-react with burro erythrocytes (BRBC), suggesting that the immune activation was not polyclonal. The antibody titer of this antiserum to SRBC was not altered after absorption of the serum with pinworms, suggesting that the response was not induced by or directed at the worms themselves. Infected mice treated with piperazine, an antihelminthic drug, were still able to mount a response to T-dependent antigen 3 months after the infection had disappeared. Immunofluorescence studies of mesenteric lymph nodes and spleens from pinworm-infected mice showed an increase in both T and B cells and the appearance of germinal centers and plasma cells. In vitro proliferation responses of spleen cells of infected mice showed a higher thymidine incorporation than did the cells of uninfected mice, and this incorporation was either inhibited or not affected by concanavalin A (ConA) and phytohemagglutinin (PHA) at concentrations that stimulate *nu/+* spleen cells.

Introduction

Natural infection of conventional mouse colonies with the pinworms *Aspiculuris tetraptera* and *Syphacia obvelata* is very common occurrence (13). Although not so common in pathogen-free colonies, such as our Athymic Mouse Research, infection with nematode parasites is still possible.

* To whom correspondence should be addressed.
© 1982 Gustav Fischer New York, Inc.
Proceedings of the Third International Workshop on Nude Mice.

There is evidence that athymic mice become more heavily infected with pinworms with time than do conventional mice and that resistance to this infection is thymus dependent (2).

It has also been shown that tumor growth in rats infected with the nematode *Nippostrongylus brasiliensis* can be enhanced or suppressed, depending on the timing of the tumor cell inoculum in relation to the parasitic infection (3). In view of these findings, it is important to evaluate the effect pinworm infection has on the immune response of athymic mice.

Materials and Methods

Mice. BALB/c heterozygous ($nu/+$) and homozygous (nu/nu) mice were bred under pathogen-free conditions in the Athymic Mouse Research Center of this laboratory. The mice originated from breeding stock on the BALB/c genetic background purchased from Bomholdtsgaard Ltd., 8680, Ry, Denmark. The colony is composed of descendants of one homozygous male and two heterozygous females.

Immune response. Mice were inoculated intraperitoneally (IP) with 2×10^8 SRBC or BRBC or 4×10^8 chicken erythrocytes (CRBC) (all from Colorado Serum Laboratories). Six days later, the serum agglutination titer was determined with the particular erythrocyte, starting with a 1:20 serum dilution.

Fecal smear. Mice were assigned separate cages containing no bedding and the 24-h fecal sample was examined for eggs (5).

Antihelminthic treatment. Piperazine hexahydrate was sterilized by filtration and added to the sterile drinking water at 1.3 g/liter.

Mitogen assay. Spleens were teased in RPMI 1640 (GIBCO, Grand Island, New York) into single-cell suspensions. Cells were then washed and incubated at cell concentrations of 1.25×10^6 cells/ml in 200 μl RPMI 1640 supplemented with 5% fetal calf serum (GIBCO), 5×10^{-5} M 2-mercaptoethanol (Sigma Corp., St. Louis, Missouri), and 1 μg PHA per milliliter (Burroughs Wellcome, Research Triangle Park, North Carolina) or 2 μg ConA per milliliter (Sigma Corp., St. Louis, Missouri). After 72 h of incubation at 37°C, 5% CO_2, humidified atmosphere, the cells were pulsed with [^{3}H]thymidine (5 μCi/ml, New England Nuclear, Boston, Massachusetts) for 4 h. Samples were then harvested by filtration, treated with trichloracetic acid (TCA) for precipitation of DNA, and then dried on filters with ethanol. The samples were then counted in a liquid scintillation counter.

Immunofluorescence studies. Anti-T antiserum was prepared in rabbits using mouse thymocytes, absorbed with the cloned mouse tissue culture line RAW 307 (a non-T, non-B lymphoblastoid line), and then further absorbed with normal non-pinworm-infected nude mouse spleen cells.

Anti-B cell serum was prepared by the pertussis method. Serum from mice immunized with Lilly (Indianapolis, Indiana) pertussis vaccine was incubated with pertussis vaccine and washed; then the complex was used to immunize rabbits to produce antiserum specific for mouse IgM, IgG, and IgA. For staining, slides were incubated with these antisera at 1:50 dilution at 37°C. for 30 min. After washing with phosphate buffered saline (PBS), fluorescinated antiserum

Table 15-1. Correlation of pinworm infestation and SRBC antibody response[a]

No. of mice	Antibody response[b]	Pinworm infestation	Diseased or stressed appearance
2	1:160	+	+
5	1:80	+	+
9	1:40	+	+
8	1:20	+	+
4	0	0	+

[a] Mice were inoculated IP with 2×10^8 SRBC and serum was assayed 6 days later.
[b] Highest dilution of serum yielding agglutination of the T-dependent antigen SRBC. The $nu/+$ response was $= 1:300$; the healthy nu/nu response was $= 0$.

(goat antirabbit IgG, Cappel Labs, Cochranville, Pennsylvania) was added to the slides at 1:8 dilution and incubated and washed as before. They were examined using a Zeiss photomicroscope with appropriate fluorescence filters.

Results

The correlation of pinworm infection with SRBC antibody responses is shown in Table 15-1. Of 28 mice selected for stressed appearance, 24 (86%), showed a positive SRBC agglutination at a serum dilution of 1:20 or higher. All 24 of these mice were infected with pinworms as shown by autopsy. The data showing that the induction of an SRBC antibody response was attributable to the pinworm infection are in Table 15-2. Healthy, uninfected, SRBC agglutinin-negative mice were intubated with live pinworms taken from the cecum of an infected

Table 15-2. SRBC antibody response of nude mice following pinworm infection[a]

	Mouse no.				
	1	2	3	4	5
Day 0					
Fecal smear, No. of eggs	0	0	0	0	0
Antibody response to SRBC	0	0	0	0	0
Day 21					
Fecal smear, No. of eggs	13	19	13	24	170
Day 35					
Antibody response to SRBC[b]	1:20	1:40	0	1:20	1:20

[a] Mice intubated day 1 with pinworms in PBS.
[b] Highest serum dilution that agglutinated SRBC. Mice were immunized on day 29.

mouse. After 21 days these mice showed heavy pinworm infection and when assayed at day 35, four-fifths showed a positive response in the SRBC agglutinin assay.

Specificity of Response

In order to determine whether the type of response elicited by immunization with SRBC resulted from a cross-reactivity with pinworm antigens, antisera from mice immunized with SRBC were absorbed overnight at 4°C with live pinworms. The agglutination titer of the antiserum to SRBC was not altered by the absorption, suggesting that the response to SRBC immunization was specific. Also, antisera to SRBC did not agglutinate BRBC, another T-dependent antigen with which antibodies to SRBC do not cross-react, suggesting that the B-cell activation was not polyclonal. Assays of the antibody response of sera from infected mice immunized with SRBC before treatment with piperazine hexahydrate (1.3 g/liter in the drinking water) and with CRBC afterwards indicated that the ability to mount an antibody response was still present 3 months after effective treatment of the infection (Table 15-3).

Immunofluorescence Studies

Figures 15-1 through 15-4 show an increase in both T and B cells in the mesenteric lymph node of infected mice when stained with anti-T or anti-B

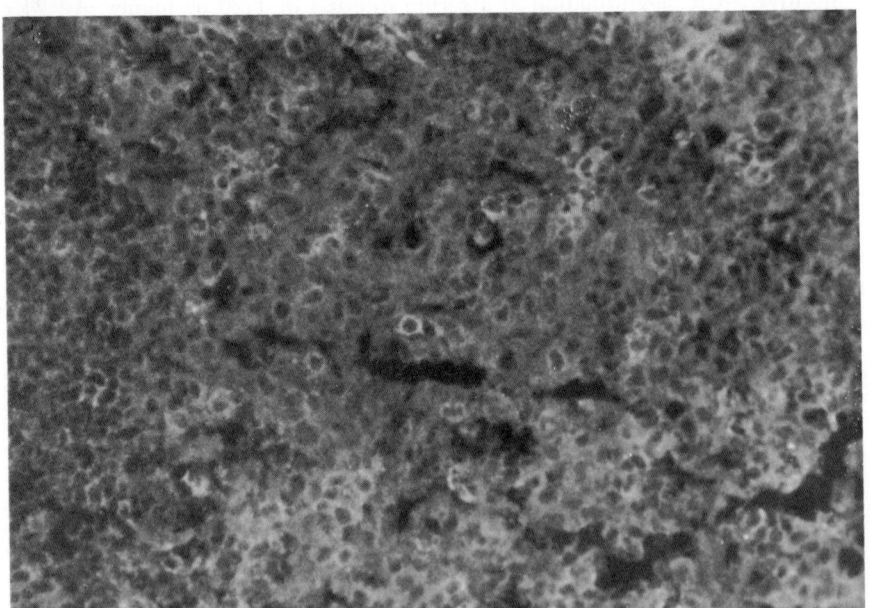

Figure 15-1. Anti-T-cell staining of a healthy nude mesenteric lymph node: The area shown is diffuse cortex, a T-dependent area. Only a few single cells are positively stained by immunofluorescence; most cells are negative.

Table 15-3. Effect of piperazine treatment in pinworm infection and antibody response

	Pre-Piperazine treatment	3 months post-Piperazine treatment
Percentage of mice with egg-positive fecal smear ($n = 18$)[a]	100	0
Percentage of mice with positive T-cell-dependent response ($n = 18$)	100 using SRBC as antigen	77 using CRBC as antigen

[a] n, number of mice.

sera, and fluorescent anti-Ig. There is patchy increased cellularity in the diffuse cortex of the lymph node area. A few germinal centers are also present. In the spleen the increase in cellularity is in the periarteriolar region and in occasional germinal centers (data not shown). The diffuse cortex of lymph nodes and the periarteriolar white sheath of spleen have been shown to be T-cell domains in normal animals. The formation of germinal centers has been shown to require both T and B cells under the experimental conditions examined so far.

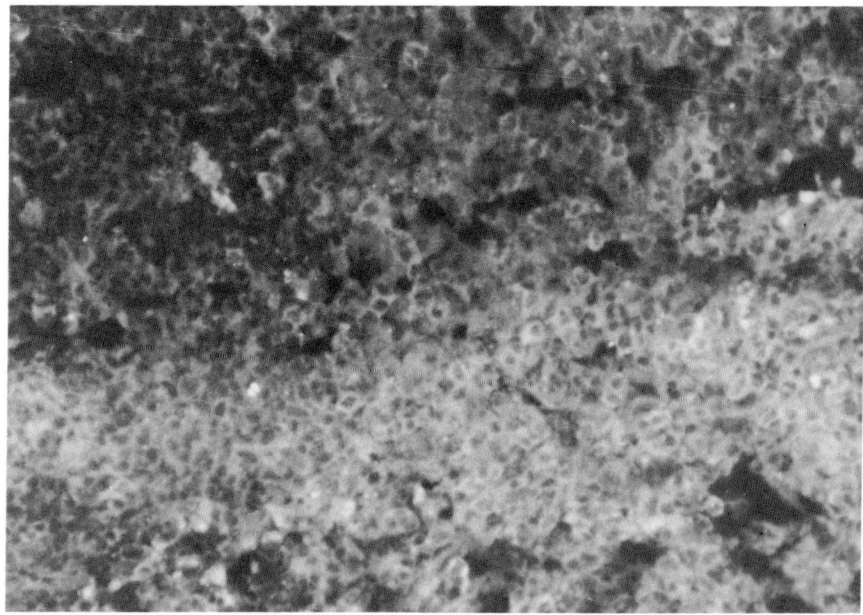

Figure 15-2. Anti-T-cell staining of lymph node from a pinworm-infected nude: The area shown is similar to that in Figure 15-1. Note the increase in cellularity compared to Figure 15-1, and a proliferating population of positive cells, rather than single positive cells as in Figure 15-1.

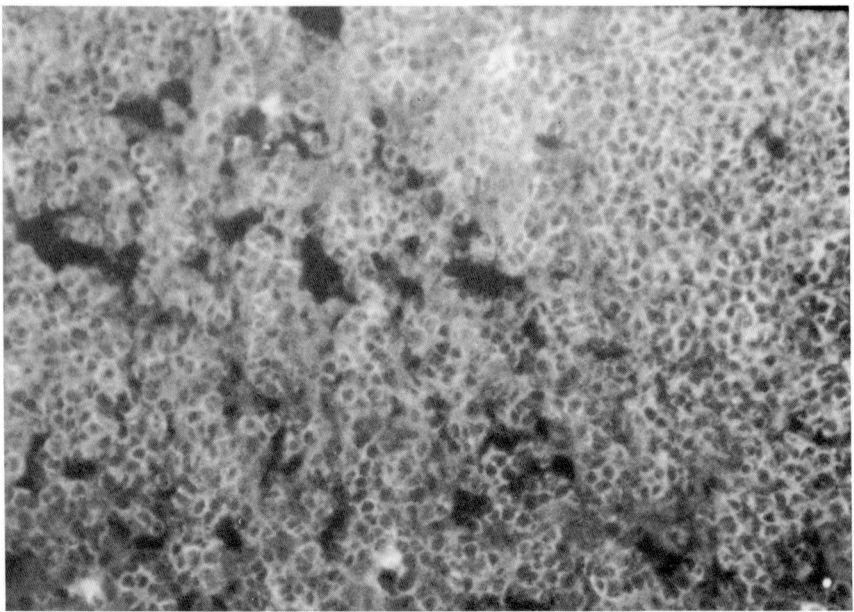

Figure 15-3. Anti-B-cell staining of a healthy nude mesenteric lymph node: follicle and diffuse cortex. Most cells stain positive.

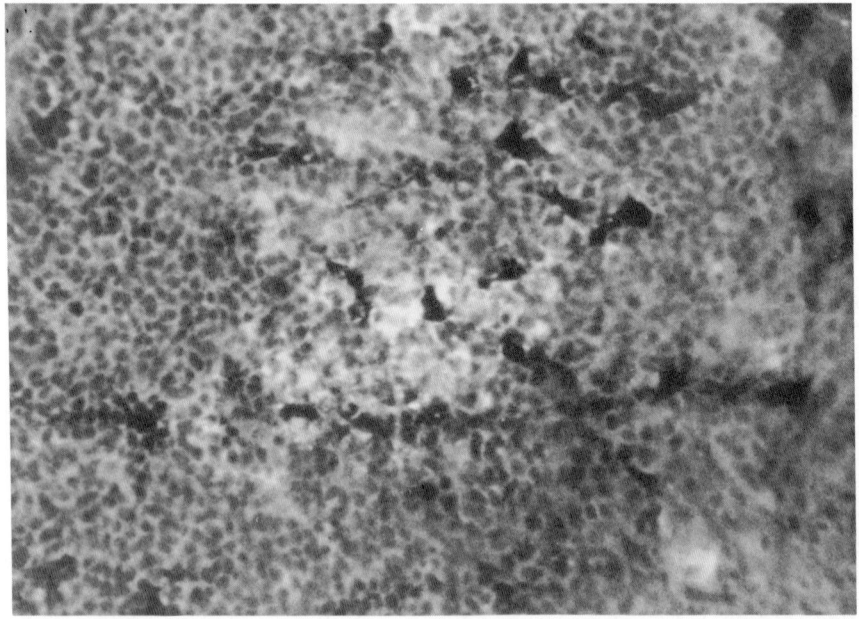

Figure 15-4. Anti-B-cell staining of lymph node from a pinworm-infected nude: follicle, diffuse cortex, and a germinal center. Note the brightly fluorescing plasma cells of the germinal center, not usually seen in nudes.

Mitogen Assay

In vitro proliferation of spleen cells from uninfected and infected SRBC agglutinin-positive athymic mice and their heterozygous siblings was examined (Table 15-4). It is apparent that the infected athymic mice showed a much higher [^{3}H]thymidine uptake in the absence of mitogens than either the healthy athymic or the heterozygous controls. Whereas the T-cell-stimulating mitogens ConA and PHA caused the expected stimulation of heterozygous spleen cells, they were either ineffective or inhibitory on both healthy and infected athymic mouse spleen cells.

Discussion

Infection of athymic BALB/c mice with the pinworms *S. obvelata* and *A. tetraptera* changes the immune status of these mice. Three measurable differences were observed. First, the infected nude mice produced specific SRBC agglutinins after in vivo challenge at a high level never seen in healthy nude mice (6) (Table 15-1). This new capability, however, could be induced in healthy nude mice by controlled induction of infection with the worms (Table 15-2). Further, the immune response was specific in that SRBC-specific antisera from these mice did not agglutinate BRBC. The antibody response did not appear to be attributable to the worms themselves since absorption with worms did not change the anti-SRBC titer. In addition, the ability to mount an antibody response remained intact for at least 3 months in mice cured of pinworm infection as measured by later challenge with non-cross-reacting CRBC (Table 15-3).

Second, nude mice infected with these worms showed an increased number of both B & T cells in spleen and lymph nodes relative to healthy nude controls. The presence of germinal centers and plasma cells, not normally seen in nude mice (1), was also noted in the infected animals.

Third, nude mice infected with these worms and showing positive SRBC agglutination responses also had a greater background rate of proliferation of

Table 15-4. In vitro proliferation responses[a]

Mouse	No mitogen	2 µg ConA per milliliter	1 µg PHA per milliliter
Healthy *nu/nu*	1,847	1,708	684
Healthy *nu/nu*	2,600	2,108	1,464
Healthy *nu/+*	1,860	101,444	72,167
Infected *nu/nu*	4,208	4,445	1,386
Infected *nu/nu*	4,839	2,336	909
Infected *nu/nu*	5,220	4,018	1,519
Infected *nu/nu*	5,226	5,341	1,753
Infected *nu/nu*	8,184	8,225	6,138
Infected *nu/nu*	6,342	7,538	4,163

[a] Data are the average of triplicate cultures in cpm [^{3}H]thymidine per well. Assays were carried out with 2.5×10^5 cells per 200 µl per well.

spleen cells in vitro relative to healthy nude or $nu/+$ controls. These proliferating cell populations, however, did not increase their rate of proliferation in response to the T-cell mitogens ConA or PHA at levels which greatly stimulated $nu/+$ control spleen cells (Table 15-4).

The presence of positive agglutinin responses to xenogenic T-dependent RBC antigens and the presence of increased numbers of B and T cells as detected by immunofluorescence, in both spleen and lymph nodes in pinworm-infected animals suggest induction of a T-cell population which can "help" the B cells known to be present in nude mice (11). Other substances were injected in vivo in an attempt to provoke the induction of agglutinin responses, including dibutyryl adenosine 3',5'-cyclic monophosphate, muramyl dipeptide, lipopolysaccharides (LPS), levamisole, diethyl dithiocarbamic acid, and thymopoietin pentapeptide (4,7,10,14). All were without effect. Stressing the mice with heat or cold also failed to induce such responses. If such a T-cell population is induced by the pinworms it does not appear to be directed against pinworm antigens but is more likely to be polyclonal or nonspecific.

On the other hand, the unusually high in vitro rate of proliferation of spleen cells seen in infected nude mice relative to healthy nu/nu or $nu/+$ controls was not further stimulated by T-cell mitogens. This would suggest that this proliferating population is either not T-cell-like or is maximally stimulated. T-cell lymphoma lines growing maximally in tissue culture are known to be inhibited by T-cell mitogens (8) and this could be true of all maximally stimulated T cells, such as may be present in these infected nude mice.

In view of the possibly conflicting evidence concerning the presence of functional T cells in these animals given by immunofluorescence and in vitro proliferation studies, the key question is whether the agglutinin response is T dependent. In vitro studies are now underway to determine the nature of the in vivo anti-SRBC response and to further characterize the proliferating spleen cells in these pinworm-infected nude mice. It is hoped that B- and T-cell-specific antisera will allow us to differentiate the possibility of polyclonal induction of helper T cells by the worms from a possible polyclonal induction of B-cells by the worms. Simultaneous induction of a Thy-1-positive, but nonfunctional cell population in spleen and lymph nodes has been seen with many proposed thymic hormones in vitro (12) and hepatitis infections in vivo (9). If we can demonstrate that T-antigen-bearing functional cells can be induced in a mouse with no thymus, then elucidation of the mechanism and comparison to the function of the normal thymus should prove fascinating.

Acknowledgments

This work was supported in part by grants from the U.S. National Institutes of Health, Public Health Services [CA 11683-09, CA 23052-01, GM 07198-05 (J.L.), and AI 15798-02 (A.J.O.)], the American Cancer Society (BC-60), and the National Foundation/March of Dimes.

References

1. de Sousa, M.A.B., H. Pritchard, and D.M.V. Parrott. An analysis of some morphological features of the lympoid system in nude mice, pp. 119–127. *In* J. Rygaard and C.O. Povlsen (eds.), Proceedings of the first international workshop on nude mice. Stuttgart: Gustav Fischer, 1974.
2. Jacobson, R.H., and N.D. Reed. The Thymus dependency of resistance to pinworm infection in mice. J. Parasitol. 60:976–979, 1974.
3. Keller, R., B.M. Ogilvie and E. Simpson. Tumor growth in nematode infected animals. Lancet i:678–680, 1971.
4. Leclerc, C., E. Bourgeois, and L. Chedid. Enhancement by muramyl dipeptide of in vitro nude mice responses to a T dependent antigen. Immunol. Commun. 8 (1):55–64, 1979.
5. McQuay, R.M. Medical Helminthology and Entomology, pp. 917–979. *In* I. Davidsohn and J.B. Henry (eds.), Todd and Sanfords Clinical Diagnosis by Laboratory Methods, Philadelphia: W.B. Saunders, 1969.
6. Mitchell, G.F. T cell Modification of B cell responses to antigens in mice, pp. 97–116. *In* M.D. Cooper (ed.), Contemporary topics in immunobiology, vol. 3. New York: Plenum Press, 1974.
7. Renoux, G., and M. Renoux. Thymus-like activities of sulfur derivatives on T-cell differentiation. J. Exptl. Med. 145:466–471, 1977.
8. Ralph, P., and I. Nakoinz. Inhibitory effects of lectins and lymphocyte mitogens on murine lymphomas and myelomas. J. Natl. Cancer Inst. (USA) 51:883–890, 1973.
9. Scheid, M.P., G. Goldstein, and E.A. Boyse. Differentiation of T cells in nude mice. Science 190:1211, 1975.
10. Scheid, M.P., M.K. Hoffmann, K. Komuro, U. Hämmerling, J. Abbott, E.A. Boyse, G.H. Cohen, J.A. Hooper, R.S. Schulof, and A.L. Goldstein. Differentiation of T cells induced by preparations from thymus and by non-thymic agents. J. Exptl. Med. 138:1027–1032, 1973.
11. Sprent, J. Migration and life span of circulating B lymphocytes of nude mice, pp. 11–23. *In* J. Rygaard and C.O. Povlsen (eds.), Proceedings of the first international workshop on nude mice. Stuttgart: Gustav Fischer, 1974.
12. Stutman, O. Intrathymic and extra thymic T cell maturation. Immunol. Rev. 42:139–182, 1978.
13. Taffs, L.F. Pinworm infections in laboratory rodents: A review. Lab. Anim. 10:1–13, 1971.
14. Weksler, M.E., J.B. Innes, and G. Goldstein. Immunological studies of aging IV. The contribution of thymic involution to the immune deficiencies of aging mice and reversal with thymopoetin. J. Exptl. Med. 148:996–1006, 1979.

General Discussion

KINDRED: You showed one germinal center. Was that an isolated example or was the frequency of germinal centers something like normal?

BEATTIE: We found several germinal centers in the mesenteric lymph nodes and spleens of infected nudes; we never saw them in healthy nudes.

KINDRED: We also found a higher incidence of anti-SRBC responses in sick nudes —but no secondary response. Did you look for a secondary response?

BEATTIE: No, not yet.

16

Clearance of Enzymes by the Reticuloendothelial System in Euthymic and Athymic Mice

J. Gabriel Michael,* Linda DiPersio, Andreas P. Kyriazis, and Amadeo J. Pesce

University of Cincinnati Medical Center, Departments of Microbiology and Pathology, 231 Bethesda Avenue, Cincinnati, Ohio 45267.

Abstract

The nude gene is expressed by the absence of mature T lymphocytes, defective immune response, and immunoglobulin deficiency. We investigated whether this gene extends its influence to reticuloendothelial system (RES) function in athymic (nu/nu) or heterozygous (nu/+) animals. Human and mouse lactic dehydrogenase isoenzymes (LDH) were employed to test clearance activity of the mouse RES. This was studied using stimulated or suppressed RES. The stimulation was achieved by administration of *Corynebacterium parvum* and suppression by infection with lactic dehydrogenase virus (LDV). Extracts of human and mouse tumor cell lines served as sources of LDH. The rate of clearance of both human and mouse LDH in untreated athymic, heterozygous, and normal mice was found to be about equal. However, in *C. parvum*-stimulated normal mice the clearance rate of LDH isoenzymes was significantly increased: For human LDH the enhancement was about threefold. In athymic mice pretreatment with *C. parvum* had no stimulating effect and the clearance rate remained unchanged from the untreated animals. Interestingly, in heterozygous mice the clearance rate following *C. parvum* treatment increased only slightly. Infection with LDV totally suppressed clearance of all forms of LDH isoenzymes. In addition, even *C. parvum*-stimulated mice showed total suppression of clearance when infected with LDV. Our data clearly indicate that athymic and heterozygous mice in part are deficient in their response to nonspecific stimulation by *C. parvum* as reflected in RES clearance. Thus, the nude gene seems to be also expressed in the impairment of RES function.

---

* To whom correspondence should be addressed.
© 1982 Gustav Fischer New York, Inc.
Proceedings of the Third International Workshop on Nude Mice.

Introduction

It is generally accepted that phagocytic cells of athymic mice are fully functional. Indeed, several reports indicated that bactericidal and tumoricidal activity of macrophages from athymic mice are elevated above that observed in euthymic mice (4). Clearance by the reticuloendothelial system (RES) depends on the activity of the phagocytic cells. In the study reported here we attempted to determine whether RES functions equally well in athymic (nu/nu), heterozygous ($nu/+$), and normal ($+/+$) mice. The clearance of an enzyme, lactic dehydrogenase (LDH), which is known to be eliminated efficiently from mouse blood plasma by the phagocytic cells of RES, was investigated. RES clearance rates of the enzyme in untreated animals as well as in mice stimulated with *Corynebacterium parvum* were examined. The clearance rate of the enzyme was impaired by lactic dehydrogenase virus infection. The clearance rates of the enzyme in stimulated animals were found to differ significantly from group to group, indicating a regulatory role for T lymphocytes and their products in the elimination of foreign and endogenous materials from the blood circulation.

Materials and Methods

Animals. Congenitally athymic nude mice homozygous for the *nu* allele were bred in our laboratory from matings of BALB/c-*nu*/*nu* homozygous males and BALB/c-*nu*/+ heterozygous females. Heterozygous ($nu/+$) mice were obtained from matings of homozygous (nu/nu) males and homozygous ($+/+$) BALB/c females. BALB/c mice were purchased from Sprague-Dawley Corporation, Madison, Wisconsin.

Human tumor. The HEp-2 cell line, derived from a carcinoma of the human larynx, was obtained from American Type Culture Collection (CCL 23) and maintained in minimum essential medium (MEM) supplemented with 10% newborn calf serum (Grand Island Biological Company, Grand Island, New York), and 100 μg/ml each of penicillin and streptomycin.

Preparation of enzyme extracts. LDH enzyme extract was prepared by freeze–thawing of a 0.9% NaCl suspension of washed HEp-2 tumor cells containing 5×10^7 cells/ml. The extract was filtered through a 0.22 μm Swinnex filter (Millipore Corporation, Bedford, Massachusetts).

Collection of mouse plasma samples. For examination of plasma LDH levels, 0.05–0.1 ml of blood was drawn by orbital venipuncture with a heparinized Natelson capillary tube. Plasma samples were stored at −80°C until assayed for enzyme content. Each experimental group consisted of 5–10 mice.

Enzyme clearance. A formula devised by Biozzi, Benacerraf, and Halpern (1) was used. The rate of clearance (C) followed an experimental function of the time (T):

$$K = (\log C_1 - \log C_2)T.$$

LDH isoenzyme procedure. The LDH-I, Pol-E-Film system was used (Pfizer Diagnostics Division, Clifton, New Jersey). A standard LDH preparation was made by preparing sonicate from a suspension of HEp-2 cells. The assay was performed using the lactate to pyruvate conversion, and standard reference sera that contained 160 mIU of enzyme per milliliter were obtained from Hyland Division, Travenol Laboratories, Inc., Costa Mesa, California.

The LDH isoenzyme separation, identification, and quantification were done following the procedure described in our earlier publication (8). The amount of enzyme present was quantitated using a colorimetric densitometer (Helena Quickscan, Helena Laboratories, Beaumont, Texas). One microliter of standard extract was placed in duplicate on the agarose film and used as the standard. Bands four and five comprised 60% of the total color measured by densitometry. The amount of enzyme in human bands four and five present in mouse plasma was determined by comparing densitometer readings obtained from scanning all bands of the standard LDH preparation containing 100 mIU/ml.

Corynebacterium parvum stimulation. Killed bacteria, 0.7 mg per mouse, in 0.1 ml 0.9% NaCl were injected intravenously (IV) 7 days before the peritoneal cells were harvested. Bacteria were donated by Burroughs-Wellcome Company, Research Triangle, North Carolina. Animals stimulated with C. parvum as above were also employed in RES clearance assays.

Cytotoxicity assays. In vitro tumoricidal activity of peritoneal cells (PEC) was determined using the method of Norbury and Fidler (5). Briefly, target cells were labeled by growth in medium containing [^{125}I]-IUdR. Peritoneal cells were added to the tumor cells in a 10:1 effector–target cells ratio and the release of isotopically labeled DNA from the monolayer into the medium at 48 h was determined. Cytotoxicity assays represent the average ± standard deviation of six tests on each mouse. In addition, the percentage kill is shown. This was calculated by the following equation:

$$\frac{\text{test release} - \text{spontaneous release}}{\text{total available isotope} - \text{spontaneous}} \times 100$$

LDH virus. A stock preparation of the LDH virus (LDV) was obtained from American Type Culture Collection and used throughout these experiments. Mice were infected with the LDH agent by intraperitoneal injection of a $10^8 \times$ infective dose 50 (ID_{50}). The infection was ascertained by observing rapid elevation of LDH in mouse plasma 24 h after administration of the virus.

Results

Activation of Peritoneal Cells by C. parvum for Tumoricidal Killing

One of the characteristic properties of activated macrophages is their ability to kill tumor cells (10). Peritoneal cells from C. parvum-treated mice were examined for their cytotoxicity against HEp-2 tumor cells. As shown in Table 16-1, PEC from C. parvum-treated mice showed cytotoxic activity. This effect was evident with peritoneal cells from nude (nu/nu), heterozygous (nu/+), or homozygous BALB/c (+/+) mice. It can be therefore concluded that activation of peritoneal cells, presumably macrophages, is independent of T lymphocytes.

Table 16-1. Cytotoxicity of peritoneal cells ($[^{125}I]$-UdR release) for HEp-2 target cells[a]

Treatment	Mouse genotype	HEp-2 cells cpm ± S.D.	% killing
None	nu/nu	1669 ± 161	16
	nu/+	1730 ± 18	18
	+/+	1770 ± 81	14
C. parvum	nu/nu	3250 ± 97	71
	nu/+	2975 ± 90	61
	+/+	3642 ± 84	66

[a] For description of the treatment see Materials and Methods section.

Clearance of Human LDH Isoenzymes in Nude and BALB/c Mice

The fate of human LDH isoenzymes was followed after IV injection of HEp-2 cell extract. The isoenzymes in nude and BALB/c mice disappeared at about the same rate with an estimated half-life of 2.5 h. Thus the RES clearance rate in athymic mice seems to be equal to that in euthymic mice.

As can be noted from Figure 16-1, at about 10 h after the LDH extracts were injected only insignificant amounts of human LDH remained in mouse plasma. Therefore, in subsequent experiments concentration of human LDH isoenzymes was measured between 1 and 2 h after administration of the extract. Usually four to six LDH determinations were made during that time period.

Clearance of Human LDH Isoenzyme in C. parvum-Treated Mice

As shown earlier (Table 16-1) C. parvum treatment activated macrophages in nude mice as expressed in their tumoricidal activity and it was therefore of interest to determine whether this activation was also reflected in RES clearance of the LDH enzyme. As shown in Table 16-2 C. parvum treatment of BALB/c (+/+) mice resulted in elevated RES clearance rate of the isoenzymes. However, clearance of the enzymes by RES of C. parvum-injected nude mice remained the same as in untreated mice. Clearance rate in heterozygous animals was only slightly increased following C. parvum treatment.

Table 16-2. Clearance rate of LDH after C. parvum treatment

Treatment	Mouse genotype	K values
None	nu/nu	0.031 ± 0.005
	nu/+	0.040 ± 0.004
	+/+	0.035 ± 0.001
C. parvum	nu/nu	0.029 ± 0.005
	nu/+	0.051 ± 0.004
	+/+	0.130 ± 0.01

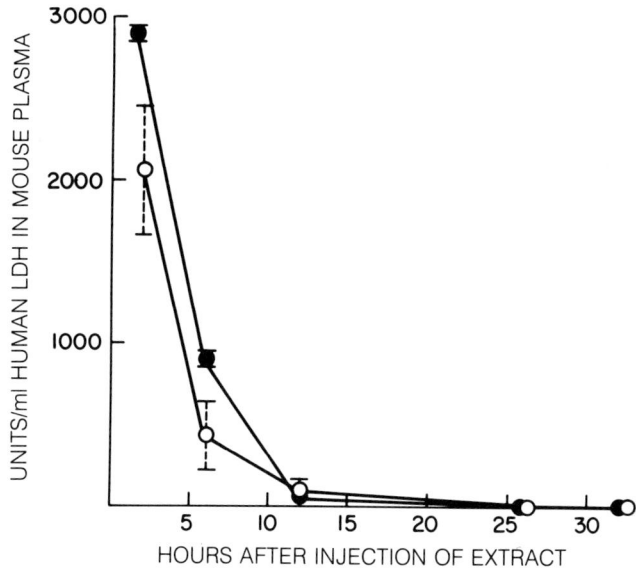

Figure 16-1. Human LDH in mouse plasma after IV injection of 0.2 ml of an extract of 1×10^7 HEp-2 cells. ●, nude mice; ○, BALB/c mice.

Human LDH Clearance in LDV-Infected Mice

Since stimulation with *C. parvum* only selectively influenced RES clearance of human LDH it was decided to examine the clearance following impairment of RES by LDH virus. Both untreated and *C. parvum*-treated mice were infected with the virus. As shown in Table 16-3, all groups of mice were about equally affected by the infection by exhibiting marked reduction in their ability to clear human LDH isoenzyme.

Table 16-3. Clearance rate following infection of mice with LDV

Treatment	Mouse genotype	K values
None	nu/nu	0.035 ± 0.002
	nu/+	0.044 ± 0.005
	+/+	0.047 ± 0.005
LDV	nu/nu	0.002
	nu/+	0.001
	+/+	0.002
LDV + *C. parvum*	nu/nu	0.004
	nu/+	0.005
	+/+	0.002

Discussion

Our data indicate that clearance of human LDH isoenzymes by RES in untreated athymic and euthymic mice proceeds at about the same rate. These data are not unexpected because the phagocytic cells of athymic mice are known to function well. More surprising are the data dealing with RES clearance in *C. parvum*-stimulated mice. Although peritoneal macrophages of athymic mice were activated for increased tumoricidal activity, RES clearance of these animals was not elevated following such stimulation.

Injection of *C. parvum* is known to produce in euthymic mice a very intense and prolonged stimulation of the phagocytic activity of RES. Histological studies reveal a marked proliferation of lymphohistiocytic elements which invade the liver and the spleen (2,10).

Among the effects of *C. parvum* are increased clearance of foreign materials from the circulation, increased resistance to various bacterial and protozoan infections, and increased antibody production. The contribution of thymus and T lymphocytes to the increased phagocytic activity is well known. Although T lymphocyte products activate macrophages, a direct stimulation of macrophages by *C. parvum* in the absence of T cells is possible. It therefore has to be concluded that in RES clearance T lymphocytes and their products play a significant but as yet undefined role.

The mechanism by which LDH virus impairs the normal clearance for certain enzymes is not fully understood. This impairment may be related to the persistent high level of viremia characteristic of this infection. Circulating virus particles are cleared from the plasma by the RES and consequently there may occur competitive inhibition of plasma enzyme clearance. In any event, our data support the contention put forward by Wakim and Fleisher (11) that the RES controls plasma enzyme levels by clearing endogenous enzymes from the blood. LDH virus infection in mice causes a similar pattern of plasma enzyme increases to that caused by RES blockade (6,9). These increases correlate with reductions in plasma enzyme clearance rates in infected mice. It is conceivable that elimination of enzyme activity from the plasma may result in part from direct inactivation in the plasma rather than from movement out of plasma. However, in clearance experiments using radioactively labeled enzymes it has been shown that the radioactivity and enzyme activity disappear from the plasma at a similar rate, suggesting that the enzyme is removed and not merely inactivated. It is assumed therefore that the enzymes are cleared in a manner similar to radioactively labeled proteins.

The heterozygous $(nu/+)$ mice responded poorly to *C. parvum* stimulation with minimal increase in RES clearance rate. Heterozygous mice are considered by many to have a normal thymus function and they are often used as normal controls. There is, however, the possibility that the *nu* gene may be expressed in these animals resulting in partially impaired immune systems. It has been reported that heterozygous mice have relatively low levels of IgG immunoglobulins in plasma similar to that found in athymic mice (7). Further investigations are needed to clarify this interesting observation.

References

1. Biozzi, G., B. Benacerraf, and B.N. Halpern. Quantitative study of the granulopoeitic activity of the RES. Br. J. Exptl. Med. 34:441–457, 1953.
2. Halpern, B.N., A.R. Prevot, G. Biozzi, C. Stiffel, D. Monton, J.C. Morand, Y. Bonthiller, and D. Decreusefond. Stimulation de l'activite phagocytaire du systeme reticuloendothelial provoquec par *Corynebacterium parvum*. J. Reticuloendothel. Soc. 1:77–96, 1963.
3. Massarrat, S. Enzyme kinetics, half-life and immunological properties of iodine 131-labelled transaminase in pig blood. Nature (London) 206:508–510, 1965.
4. Meltzer, M.W. Tumoricidal responses *in vitro* of peritoneal macrophages from conventionally housed and germ-free nude mice. Cell. Immunol. 22:176–181, 1976.
5. Norbury, K.D., and I.J. Fidler. In vitro tumor cell destruction by synergic mouse macrophages: Methods for assaying cytotoxicity. J. Immunol. Meth. 7:109–122, 1975.
6. Notkins, A.L., and C.H. Schecle. Impaired clearance of enzymes in mice infected with LDH agent. J. Natl. Cancer Inst. (USA) 33:741–750, 1964.
7. Okudaira, H., Y. Komagata, A. Ghoda, and K. Ishizaka. Thymus-independent and dependent aspects of immunoglobulin synthesis and specific antibody formation in nude mice, 167–175. *In* T. Nomura, N. Ohsawa, N. Tamaoki, and K. Fujiwara (eds.), Proceedings of the second international workshop on nude mice. Tokyo: University of Tokyo Press, 1977.
8. Pesce, A.J., H.C. Bubel, L. DiPersio, and J.G. Michael. Human lactic dehydrogenase as a marker for human tumor cells grown in athymic mice. Cancer Res. 37:1998–2003, 1977.
9. Rowson, K.E.K., and B.W.J. Mahy. Lactic dehydrogenase virus. New York: Springer-Verlag, 1975.
10. Scott, M.T. In vivo cortisone sensitivity of nonspecific anti-tumor activity of *Corynebacterium parvum* activated mouse peritoneal macrophages. J. Natl. Cancer Inst. (USA) 54:789–792, 1975.
11. Wakim, K.G., and G.A. Fleisher. The fate of enzymes in body fluids—An experimental study. IV. Relativity of the RES to activities and disappearance rate of various enzymes. J. Lab. Clin. Med. 61:107–115, 1963.

General Discussion

LODMELL: I have three questions: (1) Was the *C. parvum* viable? (2) What was the route of inoculation of *C. parvum*? (3) What target cells were used in the cytotoxicity assays?

MICHAEL: *C. parvum* bacteria were heat killed and injected IV into the mice. In cytotoxicity assays we used HEp-2 cell line as target cells.

GERSHWIN: What was the origin of the BALB/c mice used in this study?

MICHAEL: All our animals came from the same source—Sprague-Dawley.

GERSHWIN: What I would like to suggest is that the nudes (nu/nu) and $(nu/+)$ on a BALB/c background used in these experiments are significantly different from Sprague conventional BALB/c. A more appropriate control would be not to use BALB/c's as your source of $+/+$ mice but to mate heterozygotes and produce $+/+$ mice which you would have to identify by test crossing to nudes. There are certainly differences among colonies of BALB/c's.

MICHAEL: To insure genetic proximity between BALB/c $(+/+)$ and heterozygous animals $(nu/+)$, we mated BALB/c with nude homozygous (nu/nu) and the first generation progeny (F1) was used in our experiments.

RYGAARD: I have a comment to a comment. I believe Dr. Gershwin is right—that we should try to conform our homozygous $(+/+)$ controls with the nudes we are using, but I believe the way he suggested is a backward way of doing it because you have a nude that does not conform to standard BALB/c and then he suggests that you should produce new homozygotes $(+/+)$ that conform to the nudes.

I believe we should conform our nudes with the established inbred strains.

UNIDENTIFIED: Did you thymus graft your nude mice and then treat them with *C. parvum*?

MICHAEL: No, this was not done.

CUTLER: With particulate material it has been shown in many different ways that clearance is by the RES—with a soluble substance what evidence do you have that you are indeed looking at RES clearance?

MICHAEL: It has been shown by Wakim and his associates that clearance of LDH measured by reduction in radiolabel is very similar to our clearance, which was measured by loss of enzyme activity. Thus, we concluded that LDH is cleared by RES.

17

T Helper Cells That Differentiate in an Allogeneic Thymus and the Requirement for H-2 Compatibility for T-B Help

Berenice Kindred*

Institute for Immunology and Genetics, German Cancer Research Center, Im Neuenheimerfeld 280, D-6900 Heidelberg 1, West Germany.

Abstract

Nude mice can be reconstituted by an allogeneic, neonatal thymus graft and such mice possess all the T-cell functions that have been tested, although responses are usually lower than normal. In experiments using radiation chimeras, evidence has been produced that T cells that differentiate in an F1 thymus learn to recognize the second parent as self and will help B cells of the second parental haplotype. This is not the case when thymus-grafted nude mice are used. If BALB/c-*nu* are grafted with a neonatal C57BL/6 or F1 thymus, host precursor cells differentiate in the grafted thymus and migrate to the peripheral lymphoid organs. Spleen and lymph node cells from such mice can be treated with anti-C57 serum and complement and transferred to naive C57BL/6-*nu*, F1-*nu*, or BALB/c-*nu* for testing. The use of nude mice as secondary hosts permits the testing of the T cells in an environment where B cells and accessory cells are all of the same genotype and no contaminating radiation-resistant cells are present. In this test system, BALB/c-*nu* precursors that differentiate in a C57BL/6 or F1 thymus restore only BALB/c-*nu*. There is no restoration of C57BL/6-*nu* or F1-*nu*. The failure to restore F1-*nu* is consistent with the failure of parental T cells from normal donors to restore F1-*nu*. This is again different to results obtained with radiation chimeras, where parental T cells do cooperate with F1 B cells. It is suggested that these differences are caused by residual or radiation-resistant T cells or accessory cells which play an essential part in the initiation of the immune response.

---

* Present address: Max-Planck-Institut für Biologie, Abteilung Immunogenetik, Corrensstrasse 42, D-7400 Tübingen, West Germany.
© 1982 Gustav Fischer New York, Inc.
Proceedings of the Third International Workshop on Nude Mice.

Introduction

Many of the T-cell functions that are deficient in the nude mouse can be restored after grafting of a congenic or allogeneic neonatal thymus. It has been shown that the nude mouse has T-cell precursors which are able to enter the thymus graft, differentiate, and migrate to the periphery (12,16). At least partial recovery of normal histology of peripheral lymphoid organs has been described by de Sousa and Pritchard (2) and functional restoration of T-cell mitogen responsiveness (11), response to T-dependent antigens (5,11,15,17), skin graft rejection (4,5,11,14,16,17,23), and in vitro cytotoxicity for allogeneic cells (3) have all been demonstrated after grafting with a congenic or allogeneic neonatal thymus. Cytotoxic cells that are capable of killing virus-infected cells can develop in a congenic thymus graft (24) but the situation is not clear when an allogeneic graft is used. It has been reported, however, that in adult, thymectomized, x-irradiated, bone marrow-reconstituted (ATxBM) mice grafted with an allogeneic thymus no killers of virus-infected cells were generated (25).

Work with lethally irradiated bone marrow- (or fetal liver-) reconstituted F1 mice has yielded evidence that cytotoxic cells which kill virus-infected cells can learn, in the thymus, to recognize as self the H-2 haplotype of the second parent as well as that which is genetically determined. Further there are reports that this is also true of helper cells (6,19,22), although others could find no evidence for this (8,10).

The aim of the work presented here was to study helper T cells that differentiate in congenic, allogeneic, or F1 thymus grafts in nude mice to determine whether they are able to help B cells of host (i.e., T-cell precursor) type or thymus donor type or both—in other words, to find out whether helper T cells learn to recognize and help B cells of a different H-2 haplotype.

Materials and Methods

The mice used in this study were BALB/c, C57BL/6, BALB/c-*nu*, C57BL/6-*nu*, and F1-*nu* purchased from GL Bomholtgard (Ry, Denmark). The F1-*nu* were kindly bred by Dr. Carl Friis for this work.

Neonatal BALB/c, C57BL/6, or F1 thymus grafts were placed subcutaneously in BALB/c-*nu* and left for at least 6 weeks to allow precursor cells to enter the graft, differentiate and repopulate T-dependent areas of peripheral lymphoid organs. Spleens were then taken from these animals and transferred as shown in the scheme in Figure 17-1 (9). The cells were treated with anti-C57BL/6 serum and complement before transfer to remove cells that could have been introduced with the C57BL/6 or F1 grafts.

Cells from the same spleen cell suspension were injected intravenously into BALB/c-*nu*, C57BL/6-*nu*, or F1-*nu* hosts, which were then immunized with 0.2 ml of a 10% sheep red blood cell (SRBC) solution and boosted 1 week later with the same dose.

Secondary anti-SRBC responses were measured since nude mice may produce

Differentiation of Helper T Cells in Allogeneic Thymus

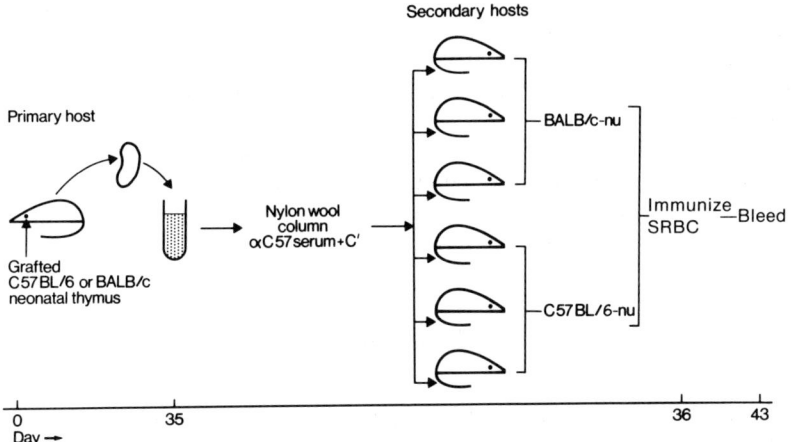

Figure 17-1. Scheme of transfer experiments. Spleen cells were taken from thymus-grafted BALB/c-*nu* at least 6 weeks after grafting. The cells were passed over nylon wool columns and treated with anti-C57BL/6 serum and complement before injection into BALB/c-*nu* or C57BL/6-*nu*. The recipients were then immunized with SRBC and boosted 7 days later. From Kindred (9).

variable primary responses without reconstitution. Antibodies were assayed by hemagglutination.

Results

As shown in Table 17-1, splenic T cells from BALB/c-*nu* thymus-grafted donors were able to restore a secondary anti-SRBC response in BALB/c-*nu* but not in C57BL/6-*nu* or F1-*nu*, regardless of the thymus in which the cells had differentiated. The reason for the very low response in the group that received cells from F1 → BALB/c-*nu* donors is not clear. The F1 → BALB/c-*nu* donors

Table 17-1. Secondary anti-SRBC responses in nude recipients of anti-C57BL/6 and complement-treated splenic T cells from BALB/c-*nu* donors bearing neonatal thymus grafts

	Transfer recipient					
	BALB/c-*nu*		C57BL/6-*nu*		F1-*nu*	
Thymus graft	No.	Response[a]	No.	Response	No.	Response
BALB/c	4	7.5	7	0	2	0
F1	3	3.0	3	0	3	0
C57BL/6	5	6.4	6	0	2	0
BALB/c thymus cells	5	6.5	5	0	5	0[b]

[a] Mean $\log^{-2}$ titer.
[b] Note that parental T cells did not restore F1.

had been grafted with thymuses 1 week later than the other donors because F1 newborns were not available at the same time but the grafts had been in place for 7 weeks before transfer and we had shown previously that 5 weeks after thymus grafting a normal number of T cells could be found in lymph nodes and spleen (12).

T cells from BALB/c-*nu* precursors restored neither C57BL/6-*nu* nor F1-*nu* recipients, even when they differentiated in an F1 thymus. In fact with regard to ability to promote a secondary anti-SRBC response, they were similar to normal BALB/c thymocytes.

Discussion

Chimeras have been classified by Zinkernagel (24) as (a) lymphohematopoietic, i.e., those in which the host and the thymus are of the same major histocompatibility complex (MHC) type but the lymphohematopoietic stem cells are from a different strain, and (b) thymus chimeras, i.e., those in which the host and stem cells are the same and the thymus is of a different MHC type. The chimeras used in this work are of the latter kind and transfer to naive nudes for testing has the advantage that T cells were primed in an environment where B cells, macrophages, and any other cells that might be involved were from the same animal and only T cells that had differentiated in the thymus graft were introduced (T cells that might have been transferred with the graft having been removed by treatment with appropriate antiserum and complement).

Under these conditions the thymus was necessary for the development of precursors to functional T cells but helper T cells did not learn to help a second strain either in an F1 or an allogeneic thymus. Helper T cells do differentiate in an allogeneic thymus but they reconstitute only the strain of the nude that provided the precursors. Reports to the contrary have been made by workers using lymphohematopoietic chimeras with priming in the chimera (6,19,22). The problems inherent in this system include allogeneic effects and graft versus host reactions and have recently been discussed by Katz et al. (8).

The failure of either normal parental thymus cells or parental T cells that have differentiated in an allogeneic or F1 thymus to reconstitute F1-*nu* emphasizes another point of difference. Several groups have shown that parental T cells cooperate perfectly well with F1 B cells (7,18,20,21); however, this work has been done using primed cells. Apart from the failure of parental T cells to restore F1-*nu* there are other indications that unprimed parental T and F1 B cells do not cooperate (1,13). Therefore it appears that the conditions for cell cooperation in initiating a response may be different from those required in a memory response. As work showing learning of restriction by helper T cells has routinely been performed using primed cells, this is likely to be another cause of discrepancies in such studies and that presented here.

The abnormalities inherent in the different kinds of chimeras, e.g., radiation-resistant cells in the irradiated F1 or possibly incomplete precursor repertoire in the nude, priming before or after transfer and possible artifacts in the test systems, particularly when an anti-SRBC response in vitro is used, can best be untangled by experiments varying only one of these components at a time. We are at present testing the effects of priming before and after transfer from nude chimeras and are planning to compare lymphohematopoietic and thymus chimeras using the same transfer schedule in order to clarify some of these problems.

References

1. Claman, H.N., E.A. Chaperon, and L.L. Hayes. Thymus-marrow immunocompetence. IV. The growth and immunocompetence of transferred marrow, thymus and spleen cells in parent and F_1 hybrid mice. Transplantation 7:87–98, 1969.
2. de Sousa, M.A.B., and H. Pritchard. The cellular basis of immunological recovery in nude mice after thymus grafting. Immunology 26:769–776, 1974.
3. Engers, H.D., B. Sordat, and C. Merenda. Functional reconstitution of nude mice. Generation of cytolytic T lymphocyte activity *in vitro* using spleen cells from thymus grafted nude mice. Manuscript in preparation.
4. Isaak, D.D. Fate of skin grafts from different inbred strains on nude mice bearing allogeneic thymus grafts. J. Reticuloendothel. Soc. 23:231–233, 1978.
5. Jutila, J.W., N.D. Reed, and D.D. Isaak. Studies on the immune response of congenitally athymic (nude) mice. Birth Defects, Orig. Art. Series XI:522–527, 1975.
6. Kappler, J.W., and P. Marrack. The role of H-2 linked genes in helper T cell function. IV Importance of T cell genotype and host environment in I-region and Ir gene expression. J. Exptl. Med. 148:1510–1522, 1978.
7. Katz, D.H., T. Hamaoka, M.E. Dorf, and B. Benacerraf. Cell interactions between histoincompatible T and B lymphocytes. II Failure of physiologic cooperative interactions between T and B lymphocytes from allogeneic donor strains in humoral response to hapten-protein conjugates. J. Exptl. Med. 137:1405–1418, 1973.
8. Katz, D.H., L.H. Katz, C.A. Bogowitz, and B.J. Skidmore. Adaptive differentiation of murine lymphocytes. II The thymic microenvironment does not restrict the cooperative partner cell preference of helper T cells differentiating in $F_1 \to F_1$ thymic chimeras. J. Exptl. Med. 149:1360–1370, 1979.
9. Kindred, B. Functional activity of T cells which differentiate from nude mouse precursors in a congenic or allogeneic thymus graft. Immunol. Rev. 42:60–75, 1978.
10. Kindred, B. Lymphocytes which differentiate in an allogeneic thymus. III Parental T cells whch differentiate in an allogeneic or semi-allogeneic thymus help only B cells of their own genotype. Cell. Immunol. 51:64–71, 1980.
11. Kindred, B., and F. Loor. Activity of host derived T cells which differentiate in nude mice grafted with coisogenic or allogeneic thymuses. J. Exptl. Med. 139:1215–1227, 1974.
12. Loor, F., and B. Kindred. Differentiation of T cell precursors in nude mice demonstrated by immunofluorescence of T-cell membrane markers. J. Exptl. Med. 138:1044–1055, 1973.
13. Marušić, M., and B. Kindred. Failure of parental T cells to restore T cell deficient F_1 mice. Manuscript in preparation.
14. Pantelouris, E.M. Observations on the immunobiology of nude mice. Immunology 20:247–252, 1971.
15. Pantelouris, E.M. Thymic involution and ageing: A hypothesis. Exptl. Gerontol. 7:73–81, 1972.
16. Pritchard, H., and H.S. Micklem. Haemopoetic stem cells and progenitors of functional T lymphocytes in the bone marrow of 'nude' mice. Clin. Exptl. Immunol. 14:597–607, 1973.
17. Radov, L.A., D.H. Sussdorf, and R.L. McCann. Relationship between age of allogeneic donor and immunological restoration of athymic (nude) mice. Immunology 29:977–988, 1975.
18. Sprent, J. Two subgroups of helper cells in F_1 hybrid mice revealed by negative selection to heterologous erythrocytes *in vivo*. J. Immunol. 121:1691–1695, 1978.

19. Sprent, J., and H. von Boehmer. T-helper function of parent → F_1 chimaeras. Presence of a separate T cell subgroup able to stimulate allogeneic B cells but not syngeneic B cells. J. Exptl. Med. 149:387–397, 1979.
20. Swierkosz, J.E., K. Rock, P. Marrack, and J.W. Kappler. The role of H-2 linked genes in helper T-cell function. II Isolation on antigen-pulsed macrophages of two separate populations of F_1 helper T cells each specific for antigen of one set of parental H-2 products. J. Exptl. Med. 147:554–570, 1978.
21. Waldman, H. Conditions determining the generation and expression of T helper cells. Immunol. Rev. 35:121–145, 1977.
22. Waldman, H., H. Pope, C. Bettles, and A.J.S. Davies. The influence of thymus on the development of MHC restrictions exhibited by helper T cells. Nature (London) 277: 137–138, 1979.
23. Wortis, H.H., S. Nehlsen, and J.J. Owen. Abnormal development of the thymus in "nude' mice. J. Exptl. Med. 134:681–692, 1971.
24. Zinkernagel, R.M. Thymus and lymphohematopoietic cells: Their role in T cell maturation, in selection of T cells' H-2-restriction-specificity and in H-2 linked Ir gene control. Immunol. Rev. 42:224–270, 1978.
25. Zinkernagel, R.M., G.N. Callahan, A. Althage, S. Cooper, P.A. Klein, and J. Klein. On the thymus in the differentiation of "H-2 self-recognition" by T cells: Evidence for dual recognition? J. Exptl. Med. 147:882–896, 1978.

General Discussion

MITCHELL: Would you like to speculate on why the results of your experiments with grafted nudes are different from those reported by Zinkernagel and others using radiation chimeras?

KINDRED: I think the differences partly result from the use in nudes of a whole neonatal thymus graft, while the thymus of the chimeras is irradiated. We have found irradiated thymuses very poor for reconstitution. I also think the nudes provide a more sensitive system and one free of the problem of residual cells not killed by the irradiation. When Zinkernagel uses nudes he gets similar results to ours.

MINATO: If you take BALB/c nude mice which have been grafted with C57BL/6 thymus are the T cells tolerant to C57BL/6?

KINDRED: Yes. They don't respond in MLR and cytotoxic cells are not generated in vitro. They usually don't reject C57BL/6 skin—occasionally they do, but if they do they don't reject a second graft. So I think the skin graft rejection is on the basis of a skin-specific antigen. There is a skin-specific antigen described in this combination which T cells wouldn't learn to recognize in the thymus.

18

Tolerance Induction for Alloantigens by Thymus Epithelium

Richard Hong,* Roger Klopp, Robert Struble, and Judith K. Manning†

Departments of Pediatrics and Medical Microbiology,† University of Wisconsin Center for Health Sciences, 600 Highland Avenue, Madison, Wisconsin 53792.

Abstract

Nude mice transplanted with cultured thymus fragments are fully reconstituted as assessed by numerous immunologic parameters. When an allogeneic graft is used, tolerance for those alloantigens is seen and skin grafts will remain in place for over a year. Allograft rejection capability for strains different from the nude recipient and the thymus donor is retained, however. Clonal deletion does not occur and reactivity against donor antigens can be demonstrated in vitro by mixed leukocyte culture and cell-mediated lympholysis. Preliminary data suggest that cellular suppression of the allograft rejection occurs. Humoral factors have not been sought as yet.

Introduction

In the past 3 years we have employed organ cultures of thymus glands as a reconstitution modality. The results to date have indicated that both syngeneic and allogeneic organs provide complete reconstitution of the congenitally athymic nude mice and a significant benefit to humans with combined immunodeficiency (2,3).

Nude mice, transplanted with allogeneic noncultured thymuses, will accept organ grafts of the identical strain as the thymus donor (second party) while possessing the capability for rejecting in a normal manner an allogeneic graft from a third party (i.e., neither host nor thymus donor strain). Grafts of both skin and heart follow this behavior (3,4,6,7,9,10).

---
* To whom correspondence should be addressed.
© 1982 Gustav Fischer New York, Inc.
Proceedings of the Third International Workshop on Nude Mice.

In the previous work by others, transplantation of intact thymus glands or gland fragments was necessarily associated with the transplantation of donor thymocytes. It is conceivable that the presence of donor lymphocytes at a time when thymic reconstitution was just beginning could result in tolerance by virtue of allogeneic exposure at a critical time in T-cell system development. Inasmuch as our cultured thymic transplants are essentially devoid of thymocytes, we investigated whether or not similar tolerance of second-party skin grafts would also be found.

Materials and Methods

Thymus glands were obtained from newborn C57BL/6 mice and cultured as described previously (3). Briefly, the glands are cut into small fragments and placed on stainless steel grids. The tissues are cultured in Ham's F-12 medium with 10% fetal calf serum, in a 5% CO_2 atmosphere for 8–9 days. Transplants are placed under the renal capsule.

Skin transplants were performed as described by Manning and Krueger (8) 2 months after the thymus implantation. Luxuriant hair growth was observed for all grafts and taken as a sign of acceptance.

Mixed leukocyte culture (MLC) and cell-mediated lympholysis (CML) was performed by the methods of Hartzman et al. (1) and Zarling et al. (11).

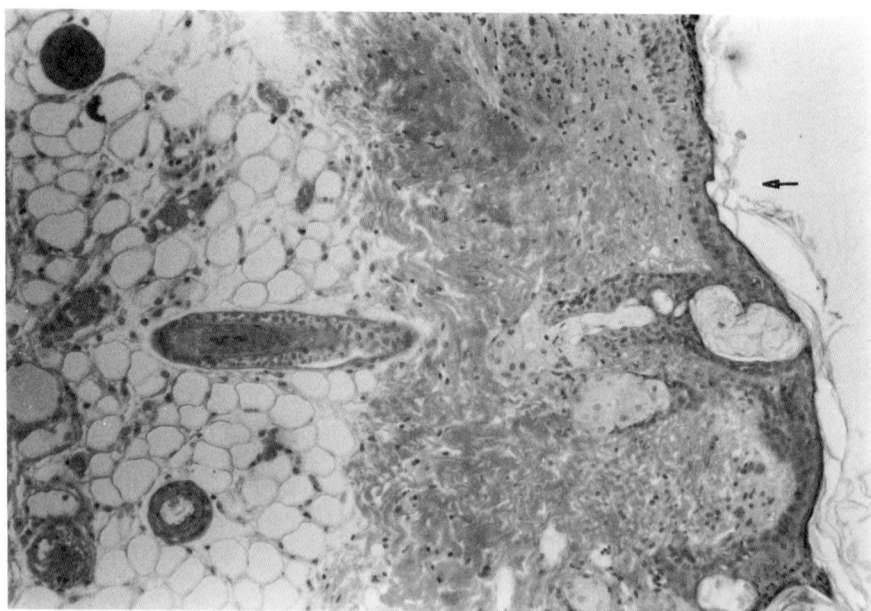

Figure 18-1. Junction of C57BL/6 skin (at left) and nude (BALB/c) at right indicated by arrow. Graft in place 8 months. Restoration of paniculus is seen. No inflammation present. ×100.

Tolerance Induction for Alloantigens by Thymus Epithelium   233

Table 18-1. Mixed leukocyte culture

Responder	Stimulator	Stimulation index
BALB/c (normal)	C57BL/6	33.1 ± 24.4 (6)[a]
BALB/c (nude) reconstituted and skin grafted with C57BL/6	C57BL/6	32.8 ± 34.2 (6)[a]

[a] Number in parentheses = number of mice.

Results and Discussion

To date, of 15 transplants, we have observed prolonged acceptance of second-party skin grafts in all, in some cases for periods in excess of 1 year. All transplants have been performed in nude mice on a BALB/c background (H-2d). Thymus transplants were derived from HaLCr (H-2q) and C57BL/6 (H-2b).

All grafts showed a luxuriant hair growth and histological examination of the tissues revealed complete normality with no evidence whatever of inflammatory responses (Figure 18-1).

The mechanism for this acceptance must involve one of two processes, clonal deletion or active suppression. To test the former hypothesis, mixed leukocyte cultures and cell-mediated lympholysis experiments were performed. The results of these studies are shown in Tables 18-1 and 18-2. In no case to date has there been any evidence of lack of reactivity of the reconstituted nudes against lymphocytes of the thymus donor.

Only preliminary data are available in regard to the presence of a suppressor population. Table 18-3 shows the result of MLCs performed in the presence of

Table 18-2. Cell-mediated lympholysis

Effector cell source	E/T ratio[a]				
	40	20	10	5	2.5
Normal BALB/c	15.6[b]	2.2	1.4	0.4	ND[c]
Reconstituted nude					
No. 1[d]	11.4	5.7	1.1	0.9	ND
2	8.7	2.3	1.5	1.0	ND
3	ND	26	23	23	ND
4	ND	44	27	22	ND
5	ND	8.6	ND	ND	11.0
6	ND	12.5	13.5	14	0

[a] E/T ratio = effector cell to target cell ratio.

[b] Percentage specific lysis = $\dfrac{\text{experimental cpm} - \text{spontaneous cpm}}{\text{maximal cpm} - \text{spontaneous cpm}} \times 100$.

[c] ND, not done.

[d] Reconstituted with C57BL/6 thymus. Targets were either C57BL/6 spleen cell or EL-4 targets.

Table 18-3. Lack of in vitro suppression by reconstituted nude spleen cells

SI ratio[a]
14.1/11.1 = 1.27
23.3/14.3 = 1.63
18.9/12.8 = 1.48
66.5/10.6[b] = 6.27
19.4/10.6[b] = 1.83
14.1/14.3 = 0.99
14.6/14.3 = 1.02

[a] Ratio of stimulation indices (SI) of normal BALB/c responders stimulated by C57BL/6 mixtures to which have been added spleen cells from reconstituted nudes (numerator) or normal BALB/c mice (denominator).

[b] Background proliferation (nude cells alone) was 10,562 cpm.

spleen cells obtained from a reconstituted nude mouse bearing a skin graft of the thymus donor. In no case was the response of normal BALB/c cells less in the presence of spleen cells from tolerant nudes. In some cases, the high background levels of spontaneous proliferation of the reconstituted nude mouse spleen cells cloud the interpretation of these data. There is no ready explanation for this phenomenon, and the animals appeared to be in excellent health. However, the animals were 18 months of age at testing and the skin grafts had been in place for nearly a year. It is possible that the high spontaneous proliferation resulted from aging or waning of thymic activity.

In cell transfer experiments, four normal BALB/c mice received 5×10^6 spleen cells from the C57BL/6 "tolerized" nudes. Three mice showed prolonged (>21 days) acceptance of C57BL/6 skin, but permanent tolerance has as yet not been shown. We believe that the results are uninformative and that larger cell doses must be tested before final conclusions are drawn.

The possibility of an antibody or humoral suppressor has not been tested for as yet.

In two instances, we transplanted skin from a C57BL/6 donor after the original graft had been in place 1 year. No second set rejection occurred; in fact the second graft was completely accepted for 2 months until the termination of the experiment. Although the nude recipient had been reconstituted 18 months previously, he was still able to reject vigorously a third-party (C3H) allograft (Figure 18-2).

It is clear that a lasting tolerance for donor H-2 antigens can be seen in nude mice reconstituted with allogeneic thymic epithelial cells. Since virtually no donor thymocytes are transplanted, the epithelium must in some way influence subsequent allogeneic responses. The phenomenon described here, that of inducing tolerance to alloantigens with thymus correction of the athymic nude, has been observed on many previous occasions (3,4,6,7,9,10). Heretofore, however, the thymocytes transplanted concomitantly with the thymus obscured the cellular basis of this acquisition of tolerance. Our lymphoid-depleted transplants strongly implicate the epithelial cells as the major if not only determinant of tolerance. These data, then, indicate that in addition to thymocyte differentiation another major function of thymic epithelium is the prevention of clonal expres-

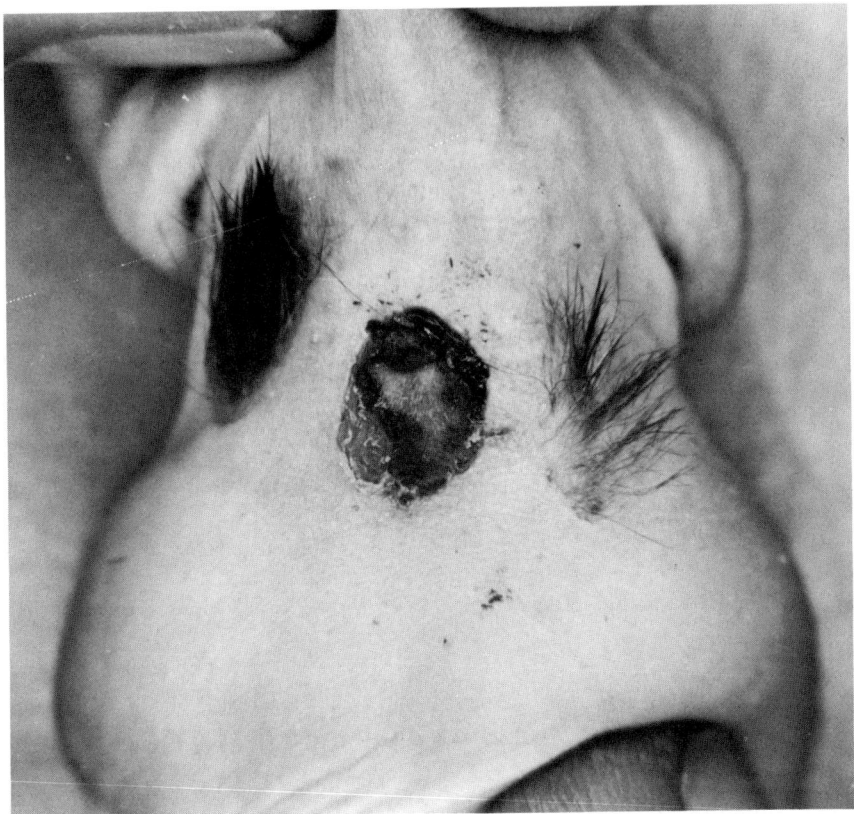

Figure 18-2. Nude (BALB/c) reconstituted with C57BL/6 CTF. Original C57BL/6 graft is on right and second C57BL/6 graft has been applied on left. Both grafts are in place and appear to be accepted, although hair growth of older graft is somewhat sparse. Acute rejection of third-party graft is seen in center.

sion reactive against thymic H-2 alloantigens. Naturally, in normals, thymic H-2 is self, and this study provides direct evidence for a role of thymic epithelium in self-tolerance, in general accord with the theory of Jerne (5).

Although these data provide insight into the cellular basis of the self-tolerance in normals, a paradoxical situation obtains in nudes. Since the congenitally athymic animals also retain tolerance for their own BALB/c histocompatibility transplantation one can surmise that self-tolerance can also be obtained by non-H-2 compatible thymic mechanisms. At least the allogeneic thymus epithelium does not permit or generate clones that would recognize and react with recipient alloantigens, even though they differ from those of the thymus. The acquisition of self-tolerance cannot uniquely require commonality of thymic epithelial and lymphocyte H-2 products. In fact, in the case of the donor antigens, clonal deletion by continuous antigen exposure probably occurs. In any event, no autoreactivity can be demonstrated.

Studies by Zinkernagel purport to show that genetic restriction of cytotoxic T-cell killing is controlled by the H-2 of the thymus and not the H-2 of the thymocyte precursor (13). If this is true, a corollary may be that another manifestation of genetic restriction is the development of postthymic cells that see thymic H-2 as self. However, preliminary work in our laboratory, using CTF rather than whole glands to reconstitute nudes, does not find genetic restriction to be controlled by the thymic epithelial H-2 (Hong, unpublished). Furthermore, Zinkernagel's studies predict that H-2 compatibility between thymocyte precursor and thymus epithelium is necessary to accomplish reconstitution (12). Fully allogeneic tissues have achieved reconstitution both in nude mice and humans; therefore, we do not believe that the thymic epithelium can confer upon its differentiated progeny the capacity to behave in interactions between H-2 gene products of various cell lines in any way other than would be predicted by its own H-2 genes.

Although our data are incomplete concerning the mechanisms for thymus-induced allogeneic tolerance, the prevailing evidence favors a suppressor effect. That a negative signal may produce long-lasting allograft tolerance has significant clinical implications. One can now ask whether the negative signal can override a positive signal. If so, a mechanism for tolerance induction in normals may be at hand.

Acknowledgments

This work was supported by the U.S. National Institutes of Health, Grant AI 14354.

References

1. Hartzman, R.J., M. Segall, M.L. Bach, and F.H. Bach. Histocompatibility matching. VI. Miniaturization of the mixed leukocyte culture test: A preliminary report. Transplantation 11:268–273, 1971.
2. Hong, R., H. Schulte-Wissermann, and S.D. Horowitz. Thymic transplantation for relief of immunodeficiency diseases. Surg. Clin. North Am. 59:299–312, 1979.
3. Hong, R., H. Schulte-Wissermann, E. Jarrett-Toth, S.D. Horowitz, and D.D. Manning. Transplantation of cultured thymic fragments II. Results in nude mice. J. Exptl. Med. 149:398–415, 1979.
4. Isaak, D.D. Fate of skin grafts from different inbred strains on nude mice bearing allogeneic thymus grafts. J. Reticuloendothel. Soc. 23:231–233, 1978.
5. Jerne, N.K. The somatic generation of immune recognition. Eu. J. Immunol. 1:1–9, 1971.
6. Kindred, B., and F. Loor. Activity of host-derived T cells which differentiate in nude mice grafted with coisogenic or allogeneic thymuses. J. Exptl. Med. 139:1215–1227, 1974.
7. Kindred, B., and B. Sordat. Lymphocytes which differentiate in an allogeneic thymus. II. Evidence for both central and peripheral mechanisms in tolerance to donor strain tissues. Eu. J. Immunol. 7:437–442, 1977.
8. Manning, D.D., and G.G. Krueger. Use of cyanoacrylate cement in skin grafting congenitally athymic (nude) mice. Transplantation 18:380–383, 1974.

9. Pritchard, H., and H.S. Micklem. Haemopoietic stem cells and progenitors of functional T-lymphocytes in bone marrow of "nude" mice. Clin. Exptl. Immunol. 14:597–607, 1973.
10. Splitter, G.A., T.C. McGuire, and W.C. Davis. The differentiation of bone marrow cells to functional lymphocytes following implantation of thymus grafts and thymic stroma in nude and ATxBM mice. Cell. Immunol. 34:93–103, 1977.
11. Zarling, J.M., M. McKeough, and F.H. Bach. A sensitive micromethod for generating and assaying allogeneically induced cytotoxic human lymphocytes. Transplantation 21:468–476, 1976.
12. Zinkernagel, R.M. Thymus function and reconstitution of immunodeficiency (letter). New Engl. J. Med. 298:222, 1978.
13. Zinkernagel, R.M., G.N. Callahan, A. Althage, S. Cooper, J.W. Streilin, and J. Klein. The lymphoreticular system in triggering virus plus self-specific cytotoxic T cells: Evidence for T help. J. Exptl. Med. 147:897–911, 1978.

General Discussion

KINDRED: I saw a manuscript recently where someone had compared whole neonatal thymus grafts with irradiated adult thymus grafts; with the irradiated adult thymus there was no reconstitution, but there was reconstitution with an irradiated F1 thymus or with a whole neonatal F1 thymus. With the whole neonatal F1 thymus there was tolerance to the second parent, whereas with the irradiated there wasn't, which would suggest that the lymphocytes in the thymus are playing a role or are responsible for the tolerance which is induced—rather than the thymus epithelium. Would you agree with this?

HONG: Well, of course we cannot state absolutely that our cultured tissue is totally devoid of lymphocytes. Certainly the number is several orders of magnitude less than in the intact gland. I am quite certain that any lymphocytes we do transplant are dead. If it is due to lymphocytes, it is a very small number of nonviable lymphocytes.

UNIDENTIFIED: Can you alter tolerance once it is established?

HONG: We have given C57BL/6 (donor) lymph node cells to reconstituted nude mice bearing donor skin. Reactions vary from decrease in size of graft (probably slow rejection), to loss of hair pigment and acute graft rejection. We have not followed the animals long enough to know whether the thymus is also rejected, but those studies are in progress.

CAPEL: Did you determine whether or not antibodies are formed against the haplotype of the thymus graft? This looks like the situation where enhancement of grafts in regular mice is induced by antibodies; in that situation, with kidney grafts you can get an enhanced period and during the enhanced period you can find MLC and CML reactivity. However, you can't find any induction of memory cells. Might it be that this phenomenon you describe is due to some humoral antibody?

HONG: No, we have not measured antibody responses.

MITCHELL: Will serum from the tolerant animals inhibit MLC reactions?

HONG: We have not tested this point as yet.

KINDRED: I have tried this. With an allogeneic neonatal thymus graft you get tolerance to grafts but serum from these mice did not inhibit the MLC.

MITCHELL: It strikes me that a suppressor phenomenon is indicated, yet you get positive MLC and CML. Now how does this suppressor activity act? If it is a cell it must have been in the MLC cultures.

KINDRED: I don't think it is known.

CAPEL: When did you take the serum? It is important at what time of the response you take the serum.

KINDRED: I can't remember, actually.

19

IgE Responses in Nude Mice Reconstituted with Cultured Thymic Fragments

Judith K. Manning* and Richard Hong

Departments of Medical Microbiology* and Pediatrics, University of Wisconsin Center for Health Sciences, Madison, Wisconsin 53792.

Abstract

Previous studies have shown that grafting neonatal cultured thymic fragments (CTF) into nude mice restores the ability of such animals to make approximately normal Ig levels of IgM, IgG, and IgA, to make IgM and IgG antibodies to SRBC, and to reject skin grafts. In this study we found that the IgE antibody response of nude mice grafted with CTF was also restored. Mice were immunized with a mixture of 10 µg ovalbumen (OVA) and 1 mg $Al(OH)_3$ gel intraperitoneally. The IgE antibody response to OVA was measured by the passive cutaneous anaphylaxis reaction performed in rats. These data imply that T-helper function for IgE responses is restored.

Introduction

Although nude mice cannot make IgE antibodies, it has been shown by Michael and Bernstein (6) that nude mice given adult thymocytes are able to make IgE antibodies. Additionally, the grafting of whole neonatal thymus glands into nudes also restores the ability to make IgE antibodies (2; Reed et al., Chapter 9, this volume). Previous studies in our laboratory have demonstrated that cultured thymic fragments (CTF), from which most of the lymphocytes have been depleted, are able to restore the ability of nude mice to make IgM and IgG antibodies to sheep red blood cells. In this study we explored the restoration of antibody-forming capacity further by examining the IgE antibody responses of nude mice given CTF.

---

* To whom correspondence should be addressed.
© 1982 Gustav Fischer New York, Inc.
Proceedings of the Third International Workshop on Nude Mice.

Materials and Methods

Nude mice (nu/nu) on a BALB/c background were purchased from Sprague-Dawley Co., Madison, Wisconsin.

The method used to culture thymus fragments was essentially as described previously (1). Thymus glands were removed from neonatal BALB/c mice, cut into approximately six pieces, placed on organ culture grids, and cultured in Ham's F-12 medium containing 10% fetal calf serum for 10 days. The medium was changed on days 2 and 7. After culturing, 10–15 fragments were routinely transplanted under the kidney capsule of BALB/c nude mice 4–6 weeks old.

At various times after implantation, usually 2 months, the mice were immunized by intraperitoneal injection of 10 μg ovalbumen (OVA) mixed with 1 mg Al(OH)$_3$ gel to induce IgE responses to OVA. At intervals after immunization, sera were analyzed for IgE antibodies to OVA by the passive cutaneous anaphylaxis (PCA) test performed in rats (7) as follows: Twofold dilutions of serum were injected intradermally into the skin of rats; 24–48 h later these rats were injected intravenously with OVA and Evans blue dye and killed 30 min later. Their skin was reflected, the underside examined, and the titer of the antiserum was recorded as the reciprocal of the last dilution showing a discrete, blue spot at least 4 mm in diameter. For the intradermal and intravenous injections the rats were sedated with Innovar-Vet (Pitman-Moore, Inc., Washington Crossing, New Jersey). A pool of antiserum, with the PCA titer predetermined by previous testing in several rats, was used as a standard in each rat.

Results and Discussion

Table 19-1 shows the IgE antiovalbumen responses of nude mice given neonatal CTF compared with ungrafted nude mice and normal BALB/c mice. Unrestored nudes did not make detectable IgE antibody, whereas nudes given CTF did and this response was greater than that of normal BALB/c. These results may be skewed because of the high response of a few animals. We cannot explain

Table 19-1. Effect of transplanting cultured thymic fragments on IgE production in nude mice

	Mean PCA titer	
	Day 14[a]	Day 21
Nude (3)[b]	<8[c]	<8
Nude + CTF (8)	1008	976
Normal BALB/c (13)	216	404

[a] Days after immunization with OVA + Al(OH)$_3$ gel.
[b] Number of mice.
[c] Results are expressed as a mean anti-OVA passive cutaneous anaphylaxis (PCA) titer.

Table 19-2. IgE responses in nude mice at various times after transplantation of cultured thymus fragments

Time after transplant	Mean PCA titer	
	Nude + CTF	Age-matched normal BALB/c
1 month	430[a] (8–1024)[b]	426 (256–512)
2 months	320 (128–512)	426 (256–512)
3 months	299 (128–512)	417 (128–512)

[a] Results are expressed as a mean anti-OVA passive cutaneous anaphylaxis (PCA) titer, 21 days after immunization with OVA + $Al(OH)_3$ gel.
[b] Range of the responses, three mice per group.

this. Possibly, these high responses result from differences in the amount and/or quality of the transplanted tissue and reflect a difference in regulatory capacity (3).

Preliminary studies indicate that IgE antibody is produced in grafted nudes for at least 3 months after immunization. In normal BALB/c mice, only one injection of OVA in $Al(OH)_3$ gel results in production of IgE antibodies for at least 6 months afterward (J.K. Manning, unpublished observations).

As shown in Table 19-2 the PCA titers were similar in nudes that had been grafted 1 month, 2 months or 3 months prior to immunization. In this experiment, the responses of age-matched BALB/c mice were similar to those of grafted nudes. Therefore, the ability to make IgE antibody was restored by 1 month after nude mice received CTF.

In the studies just described, nude mice were grafted with CTF from neonatal (1–3 days old) BALB/c mice. Studies in other laboratories (4,5,8) have shown that adult thymus tissue grafted into nude mice failed to restore to normal IgM and IgG antibody-forming capacity. We therefore examined the effect of giving CTF from adult (6 weeks old) BALB/c mice on the IgE antibody response. The mean PCA titer was <8 in three nude mice given adult CTF and 512 in three nude mice given neonatal CTF when assayed 21 days after immunization. These results agree with the earlier reports (4,5,8) in that nude mice grafted with adult CTF failed to make detectable IgE antibody to OVA.

In summary, the grafting of neonatal CTF from genetically similar mice enabled nude mice to produce IgE antibody responses. A probable explanation is that grafting with CTF restored T-helper function for IgE by maturation of the nude's T-cell precursors by the grafted thymic epithelium.

Acknowledgment

This work was supported by the U.S. National Institutes of Health, Grants AI 14354 and AI 10404.

References

1. Hong, R., H. Schulte-Wissermann, E. Jarrett-Toth, S.D. Horowitz, and D.D. Manning. Transplantation of cultured thymic fragments II. Results in nude mice. J. Exptl. Med. 149:398–415, 1979.
2. Hsu, C.K., and S.H. Hsu. Immunopathology of schistosomiasis in athymic mice. Nature (London) 262:397–399, 1976.
3. Ishizaka, K. Cellular events in the IgE antibody response. Adv. Immunol. 23:1–75, 1976.
4. Loor, F., and L-B Hagg. The restoration of the T-lymphoid system of nude mice: Lower efficiency of nonlymphoid, epithelial thymus grafts. Cell. Immunol. 29:200–209, 1977.
5. Loor F., and L-B. Hagg. Restoration of the T-lymphoid system of nude mice: different regeneration of neonatal and adult thymuses. Eur. J. Immunol. 7:278–282, 1977.
6. Michael, J.G., and I.L. Bernstein. Thymus dependence of reaginic antibody formation in mice. J. Immunol. 111:1600–1601, 1973.
7. Ovary, Z., S.S. Caizza, and S. Kojima. PCA reactions with mouse antibodies in mice and rats. Intl. Arch. Allergy Appl. Immunol. 48:16–19, 1975.
8. Radov, L.A., D.H. Sussdorf, and R.L. McCann. Relationship between age of allogeneic thymus donor and immunological restoration of athymic (nude) mice. Immunology 29:977–988, 1975.

General Discussion

OUTZEN: Nudes + CTF appeared to produce less IgE when tested 3 months after transplantation compared to those tested 1 month after transplantation. Is this difference due to experimental variation or could it be the result of a proliferation of a suppressor T-cell population?

MANNING: This is difficult to assess. When the mean PCA titers are compared there is very little difference between the groups. Comparing the ranges of the responses, it appears that nudes tested 1 month after grafting may be capable of giving a slightly higher response. However, there were only three animals in that group: the PCA titers were 8, 256, and 1024. I am unable to answer the question because of the wide variation in the results.

GAUTSCH: Have you tried to reconstitute the IgE response by grafting Millipore chamber-enclosed CTF into your nude mice? And if so, with what result?

MANNING: No, we have not done this.

MITCHELL: In a nude mouse grafted with *both* neonatal and adult thymuses, the positive reconstitutive effect of neonatal thymus is presumably dominant over the negative effect of adult thymus. Is this so?

MANNING: We have not used this particular experimental design.

20

Detection of IgE in Congenitally Athymic (Nude) Mice

Dean Roberts,* Toru Takenaka, and Joe M. Jones

National Center for Toxicological Research, Jefferson, Arkansas 72079.

Congenitally athymic (nu/nu) and immunologically intact ($nu/+$) mice were primed subcutaneously (SC) with tetanus toxoid (TT) plus $Al(OH)_3$ and challenged intraperitoneally (IP) with TT adsorbed on carbon. Tetanus antitoxin was measured using a toxin neutralization bioassay and tetanus-specific IgE was measured by passive cutaneous anaphylaxis (PCA). Radioimmunoassays were used to measure total IgE and tetanus-specific gammaglobulin, IgE, and IgM. Local peritoneal eosinophil responses and antibody responses were evaluated on days 0, 4, 7, 14, and 20 after challenge. High eosinophil responses were seen only in T-cell-competent mice that had received primary immunization. Fourteen days after challenge, $nu/+$ mice had a neutralizing titer of 90,000 and a PCA titer of 512. In contrast, nu/nu mice lacked detectable neutralizing antibodies (titer <300) and IgE (PCA titer <2). When total IgE was measured, however, nu/nu mice had significant levels. The presence of IgE of unknown specificity was confirmed by cutaneous anaphylaxis reactions following intradermal injection of monospecific anti-IgE but not following intradermal injection of anti-IgG or anti-IgA. Weanling nu/nu mice injected intradermally with immune mouse serum developed positive PCA reactions when injected intravenously (IV) 24 h later with TT or anti-IgE plus Evans blue. The appearance of IgE in nude mice was age related and confirmed by Ouchterlony analysis using monospecific antimouse IgE. The results indicate that although T-cell-deficient mice are unable to make a specific IgE response to TT, they have functional mast cells and are able to synthesize IgE.

---

* To whom correspondence should be addressed.
Proceedings of the Third International Workshop on Nude Mice.

21

Induced Oophoritis and Gastritis in Nude Mice: A New Approach to the Localized Type of Autoimmunity

Akinori Kojima,* Osamu Taguchi, and Yasuaki Nishizuka

Laboratory of Experimental Pathology, Aichi Cancer Center Research Institute, Chikusa, Nagoya 464, Japan.

Abstract

No spontaneous model of localized autoimmune disease has so far been available in the mouse. Neonatal thymectomy (nTx) induces specific oophoritis and gastritis in BALB/c-+/? mice, while BALB/c-*nu/nu* mice of the same colony do not develop either of these diseases. Oophoritis was found in 40% of 3-month-old nude females injected at 4 weeks of age with relatively low doses (10^5, 10^6) of peripheral lymphoid cells from normal male or neonatally ovariectomized female mice, but not from normal females. In contrast, normal peripheral lymphoid cells were generally ineffective in inducing gastritis. Female and male thymus cells (10^6, 10^7) were partially effective in inducing both diseases (20-30%). Spleen or lymph node cells (10^7) from nTx mice were more effective (70%) and oophoritis was induced following transfer of spleen cells from disease-developing nTx mice to nude mice (100%) but not to normal mice (0%). An indirect immunofluorescence test has shown that circulating antibodies (IgG) were directed against ooplasm or zona pellucida in oophoritis and to parietal cells in gastritis. These results suggest that tolerance is naturally established with both ovarian and stomach antigens at the level of peripheral T cells in normal female mice as a result of interaction between peripheral T cells and specific antigens. In normal males, tolerance is not established against specific ovarian antigens because of the absence of that organ. Although the specificity of autoimmune attack against a particular organ following neonatal thymectomy appears to be dependent on the genetic background, there is no clear relationship between the organ specificity and H-2 haplotypes of mice under study.

---

* To whom correspondence should be addressed.
© 1982 Gustav Fischer New York, Inc.
Proceedings of the Third International Workshop on Nude Mice.

Table 21-1. Autoimmune diseases (11a) and their mouse models

---

1. Diseases primarily involving blood cells (hemolytic anemia)
   NZB
2. Generalized autoimmune diseases (SLE)
   NZB/WF$_1$, BXSB/Mp, Swan, SL/Ni
   MRL/Mp (*lpr*), Motheaten (*me*), Nude (*nu*)
3. Localized autoimmune diseases (Hashimoto disease)
   Experimental allergic diseases
   Postthymectomy autoimmune diseases[a]
   (Induced autoimmune diseases in nude mice)[a]

---

[a] A new approach proposed in this paper.

Introduction

Spontaneous development of autoimmunity has been reported in such strains of mice as NZB (22), NZB/WF$_1$ (22), BXSB/Mp (14), Swan (1), and SL/Ni (23), as well as in mice homozygous for the autosomal recessive mutations to lymphoproliferation (14), motheaten (20), and nude (12) (Table 21-1). In these mice, antinuclear, antierythrocyte or antithymocyte antibodies are the major type of autoantibodies found, and immune complex nephritis often develops. Roles of endogenous viruses have been suspected in some of them (22,23). Organ-localized autoimmune diseases, such as thyroiditis or gastritis, have not yet been observed in these strains or mutants. For the study of localized autoimmunity, a classical model, experimentally induced allergic disease, has been exclusively used in mice (Table 21-1).

Recently, Kojima et al. (9) reported on postthymectomy autoimmune thyroiditis in mice. This type of disease has been found in other organs, including ovary (15), and at least five different models of localized autoimmune diseases have so far been established. Some examples of diseases that were induced by neonatal thymectomy in mice at the Jackson Laboratory are presented in Table 21-2 (Kojima and Prehn, in press—Immunogenetics). Furthermore, this model has been strengthened by the successful induction of the same disease in nude mice following immunologic procedures (7,11). In this paper, we discuss the potentiality of our experimental system from the viewpoint of a useful approach to more critical understanding of localized autoimmunity.

Table 21-2. Examples of postthymectomy autoimmune diseases
(Kojima and Prehn, in press—Immunogenetics)

Oophoritis	A/J (90%)
Orchitis	SWR/J (35%)
Gastritis	BALB/cJ females (30%)
Thyroiditis	A/J females (5%)
Coagulating gland adenitis[a]	(C57BL/6J × A/J) F$_1$ (40%)

---

[a] We previously reported this lesion as prostatitis. In this contribution, however, we prefer to call it coagulating gland adenitis, because this gland, although phylogenically comparable with the anterior lobe of the prostate, is no longer called prostate after maturation.

Methods and Materials

Animals. Our colony of nude mice has a BALB/c background as previously described (11). They have been maintained under clean conditions as a closed colony and survive up to 6 months of age.

Homozygous (nu/nu) mice were produced by mating immunologically reconstituted nude males to heterozygous $(nu/+)$ females. Four-week-old specific pathogen-free BALB/c-nu/nu females derived from our colony of BALB/c-$nu/+$ mice were obtained from the Institute of Medical Science (Dr. K. Suzuki), University of Tokyo. The mice were housed in an air-conditioned box at about 26°C and each cage was covered with a filter cap. They were fed Oriental compressed pellet diet (CMF) plus cut carrot and processed cheese once a week and provided with tap water ad libitum.

Thymectomy and ovariectomy. Thymectomy was performed on day 0 (day of birth), 3, or 7 as previously described (9). Ovariectomy was performed on day 1.

Preparation of cell suspensions. Thymus, spleen, lymph node, and bone marrow cell suspensions were prepared as previously described (10). The cell suspensions were injected intraperitoneally (IP) into nude mice 4 weeks old. Peripheral blood was obtained from the axillary vessels of normal adult mice. Pooled blood (0.4 ml) was directly injected IP into the recipients.

Immunofluorescence test. Blood for immunologic tests was obtained from the axillary vessels of 3- or 15-month-old BALB/c-nu/nu mice; it was left at room temperature overnight and the serum separated and frozen at −20°C before use.

Indirect immunofluorescence tests were performed on unfixed frozen and paraffin sections of mouse ovaries and stomachs according to the method of Taguchi (21).

Histological examination. Nude mice given various treatments were killed, unless otherwise stated, at 3 months of age. The in situ thymic tissues were carefully examined histologically. Ovaries, stomach, and other organs were removed, fixed in formalin, embedded in paraffin, sectioned, and stained with hematoxylin and eosin.

Results

Difference in the Ability of Peripheral Lymphoid Cells from Normal Male and Female Mice to Induce Oophoritis

Oophoritis was induced in nude mice by injecting 10^5–10^6 normal male spleen cells (40%), but not by more (10^7) or fewer (10^4) cells. In contrast, normal female blood, spleen, and lymph node cells were generally ineffective in inducing the oophoritis (Table 21-3). Interestingly, spleen cells (10^6) from day-1 ovariectomized females were partially effective (30%).

Table 21-3. Difference in the ability of normal (+/?) male and female lymphoid cells to induce oophoritis in BALB/c-nu/nu mice[a]

Cell type	Cell dose	Induction of oophoritis (%)	
		By male cells	By female cells
Normal thymus cell	10^7	2/12 (17)	3/12 (25)
	10^6	2/12 (17)	2/10 (20)
Normal spleen cell	10^7	1/13 (8)	0/10 (0)
	10^6	5/12 (42)	0/21 (0)
	10^5	4/11 (36)	2/16 (13)
	10^4	0/13 (0)	1/11 (9)
Gx spleen cell[b]	10^6	—	7/23 (30)
Normal lymph node cell	10^7	1/15 (7)	—
	10^6	6/12 (50)	1/10 (10)
Normal peripheral blood	0.4 ml[c]	5/11 (45)	0/10 (0)

[a] The cells or peripheral blood were injected IP into female nude mice 4 weeks old and their ovaries were examined histologically two months later.
[b] Ovariectomy was performed on day 1.
[c] 0.4 ml blood contains about 10^6 lymphocytes.

Failure of Both Normal Male and Normal Female Peripheral Lymphoid Cells to Induce Gastritis

Both normal male and normal female peripheral lymphoid cells (10^6) were generally ineffective in inducing gastritis in nude mice (7). Spleen cells (10^6) from ovariectomized females were also ineffective (Table 21-4).

Table 21-4. Summary: Induction of oophoritis and gastritis in BALB/c-nu/nu mice[a]

Treatment	Oophoritis	Gastritis
Male		
Thymus cell	±	+
Spleen cell	+	—
Lymph node cell	+	—
Peripheral blood	+	—
Bone marrow cell	—	—
Female		
Thymus cell	±	+
Spleen cell	—	—
Lymph node cell	—	—
Peripheral blood	—	—
Gx female spleen cell	+	—
Tx male spleen cell	++	++
Tx female spleen cell	++	++

[a] Cells (10^6 from normal or Gx mice and 10^7 from Tx mice) or blood (0.4 ml) were injected IP into female nude mice 4 weeks old. Ovaries and stomach were examined histologically 2 months later. For detailed data concerning gastritis see ref. 7. Thymectomy (Tx) was performed on day 3 and ovariectomy (Gx) was on day 1. Incidences of the diseases were divided into four grades: —, 0–10%; ±, 10–30%; +, 30–50%; ++, over 50%.

Effectiveness of Thymus Cells in Inducing Both Diseases

Thymus cell injections usually resulted in some degree of oophoritis (20%) and gastritis (30%) (7). No sex or dose difference was noticed (Table 21-3). All the above data are summarized in Table 21-4.

Transfer of Oophoritis to Nude Mice

Spleen cells (10^7) from day-3 Tx mice are highly effective in inducing oophoritis and gastritis, as previously reported (7,11) (Table 21-4). Interestingly, spleen cells (10^7) from day-7 Tx mice, which were free from the disease, were ineffective (Table 21-5).

Using spleen cells (10^7, but not 10^6 or fewer) from disease-developing nTx female mice, transfer of oophoritis to nude females was always successful. However, normal mice injected with such cells failed to develop oophoritis (Table 21-5). In addition, the prior administration of normal adult spleen cells (10^7) at one week of age inhibited the later induction of the disease in those treated nude mice (data not presented).

Proliferative Changes in the Ovary and Stomach

In a pilot study we recorded the development of spontaneous tumors in three experimental groups of female nude mice observed for 15 months: Group 1 was given 10^7 normal male spleen cells at 4 weeks of age. This treatment does not induce autoimmune disease in nude mice. Group 2 was given 10^7 nTx female spleen cells, which induces both oophoritis and gastritis. Group 3 was given 10^7 normal female bone marrow cells, which does not induce these diseases. Granulosa cell tumors were found in 2 of 18 mice in group 2 and hyperplastic gastric lesions were found in 10 of 18 mice of the same group (Table 21-6). In contrast,

Table 21-5. Transfer of oophoritis to BALB/c-nu/nu and BALB/c-$+/?$ mice by the injection of spleen or lymph node cells from thymectomized mice[a]

Donor		Recipient	
Cell type	Oophoritis[b]	Mice	Oophoritis (%)
Day-3 Tx female spleen cell	±	4W-nu/nu-female	9/10 (90)
	+		10/10 (100)
	+	NB-nu/nu-female	6/6 (100)
	+	4W-$+/?$-female	0/10 (0)
	+	NB-$+/?$-female	0/4 (0)
Day-3 Tx female lymph node cell	±	4W-nu/nu-female	10/11 (91)
Day-7 Tx female spleen cell	−		0/10 (0)

[a] Spleen or lymph node cells (10^7) from the donor mice with or without oophoritis were injected i.p. into BALB/c-nu/nu or BALB/c-$+/?$ females of various ages, and their ovaries were examined at 3 months of age.

[b] Donor mice were divided into three groups: mice without oophoritis, −; mice with oophoritis, +; and a group of mice including both oophoritis-positive and -negative individuals, ±. Abbreviations are as follows: Tx, thymectomized; NB, newborn; 4W, four weeks of age.

Table 21-6. Ovarian and gastric lesions observed in nude mice after various treatments[a]

Treatment at 4 weeks of age	Granulosa cell tumors	Hyperplastic gastric lesions
Normal male spleen cells (10^7)	0/15 (0%)	0/15 (0%)
nTx female spleen cells (10^7)	2/18 (11%)	10/18 (56%)
Normal female bone marrow cells (10^7)	0/15 (0%)	0/15 (0%)

[a] Mice were observed for 15 months.

none of these lesions was found in mice of groups 1 (15 mice) and 3 (15 mice). Lymphoreticular tumors were found in mice of groups 2 and 3, but not in group 1, and lung adenomas were found in mice of all three groups.

Circulating antibodies (IgG) against ooplasm or zona pellucida were demonstrated by an indirect immunofluorescence test in some sera of disease-developing mice, but not in those of normal or untreated nude mice. Antiparietal cell antibodies (IgG) were always seen in the sera of gastritis-developing mice (7). Detailed studies on circulating antibodies are reported elsewhere (21).

Discussion

There are several possible mechanisms for postthymectomy autoimmunity. We have not yet ruled out the role of infectious agents in postthymectomy diseases. There is also the possibility that the diseases are induced in nude mice by graft versus host (GVH) reactions. We first devoted ourselves to describing the phenomenon of postthymectomy ovarian dysgenesis and interpreted it as indicating that some unknown thymic factor physiologically controls ovarian development (15–17). We also have limited understanding of the genetic factors regulating the susceptibility of these lesions. Confusion in interpreting our results might be attributed to our earlier failure to postulate a reasonable mechanism. We recently proposed an autoimmune mechanism to explain our data, although our understanding is still incomplete (7,9–11,21). At present, we think two other mechanisms, infectious and GVH etiologies, are less likely. There is a sex and dose effect in the ability of peripheral lymphoid cells to induce oophoritis in nude mice. Furthermore, the demonstration of specific antibodies is difficult to explain by those two mechanisms.

Studies on postthymectomy autoimmune disorders (and on induced diseases in nude mice) have led us to several points. First, we have previously presented a hypothesis that helper T cells (specific to thyroid antigens) already peripheralize in mice by 3 days of age while suppressor T cells do not (9,10). In other words, certain helper T cells appear earlier ontogenically than corresponding suppressor T cells. This may explain a clear-cut difference in the disease-inducing effect between day-3 and day-7 thymectomy. According to Cantor and Boyse (5), most, if not all, of the peripheral T cells of normal mice (without exogeneous stimuli) between birth and 2 weeks of age are composed of Ly123-positive cells. At about 2 weeks of age, two other T cell populations, $Ly1^+$ and $Ly23^+$

cells, start to appear. In adult mice, the distribution of peripheral Ly123+, Ly1+, and Ly23+ T cells is 50%, 30%, and 5–10%, respectively (5). Cantor also found that Ly123+ cells (a major population) and Ly1+ (a minor population) could be detected in the thymus early in life, but Ly23+ cells were present in very low numbers at that time (4). These data support our hypothesis. Although they speculated that the generation of functionally distinct T-cell subclasses is a differentiative process independent of antigen, it may be possible as well to speculate that at least some of spontaneously appearing Ly1+ and Ly23+ T cells are specific helper and suppressor T cells generated in response to newly differentiating (auto)antigens in several tissues after birth. Indeed, it has been shown by immunofluorescence that the specific oocyte or parietal cell antigens first appear between 3 and 7 days of age (Taguchi et al., unpublished). In contrast, several investigators, using their own experimental systems, have reported that the highest suppressor activity is observed in neonatal thymus or spleen cells (3,6,13). This indicates that Ly123+ T cells themselves or another category of cells function as suppressor cells, or some of neonatal T cells (Ly123+) give rise to mature suppressor T cells (Ly23+) in response to administered antigens.

Second, some relationships between the HLA type and a number of immunologic diseases have been reported in man. Experimentally, this type of association has been most clearly demonstrated in experimental allergic thyroiditis in mice, in which the disease is induced in association with certain H-2 haplotypes (19). However, if we examine other experimental systems, including ours, such a close relationship has not yet been confirmed (2,9,18). Our study on post-Tx autoimmunity, using many inbred strains of mice and their hybrids, has not yet confirmed the major role of the H-2 type in determining the type of disease (Kojima and Prehn, in press). For example, BALB/cJ (dd) females are susceptible to gastritis (30%), whereas DBA/2J (dd) females are not (0%). Furthermore, SWR/J (qq) mice are susceptible to orchitis (35%) whereas DBA/1J (qq) mice are not (0%). In order to assess a possible role of the H-2 locus in our system, studies are now underway using congenic strains of mice. A possible genetic linkage, if it exists between susceptibility gene(s) and other non-H-2 loci, is also being determined using recombinant inbred strains of mice.

Third, since the procedures employed in this study (neonatal thymectomy and intraperitoneal injections of lymphoid cells) have no apparent relevance to the ovary or stomach, it is unlikely that the present diseases are induced by altering the self-antigens. Therefore, autoimmune reactions must occur against native autoantigens. Furthermore, autoreactive cells in this system do not appear to be abnormal clones resistant to the thymic regulation but appear to be normally existing clones, because it is usually difficult to transfer the disease to normal adult mice by a sufficient dose of cells for induction in nude mice. It is also noteworthy that autoantibodies are usually directed to oocytes in oophoritis and to parietal cells in gastritis, although anti-steroid-producing-cell antibodies have also been demonstrated (21). Why are specific antibodies against other components of those organs rarely produced? We speculate that the localized autoimmunity may be directed to the most principal and specific cells of the organ. In other words, the present type of autoimmunity may develop in close relation to a postnatal differentiative process of organs.

Finally, our recent study clearly showed that a significantly increased tumor incidence in neonatally thymectomized mice was limited to endocrine and lymphoreticular tissues (8). This finding does not support any of the central immune surveillance hypotheses but suggests the importance of specific preneoplastic conditioning for de novo development of tumors. The finding also tells us that the prior induction of localized autoimmunity may favor the spontaneous development of solid tumors in nude mice. In this connection, it may be interesting that in a pilot experiment, two granulosa cell tumors, many hyperplastic gastric lesions reminiscent of Menetrier's disease, and some lymphoreticular tumors were observed in nude mice given spleen cells from nTx mice. Such a treatment is known to induce oophoritis and gastritis.

Development of localized autoimmune diseases after neonatal thymectomy is no longer an isolated phenomenon, since at least five different types of diseases have been confirmed following thymectomy in various strains of mice at the Jackson Laboratory. The induction of oophoritis and gastritis in nude mice was also confirmed in another colony of a BALB/c background (Kojima and Outzen, unpublished). Our system, in addition to other experimental models, will provide a new approach to understanding the localized type of autoimmunity.

Acknowledgments

The skillful technical assistance of Miss K. Tsuchimoto and Mrs. M. Izawa is greatly acknowledged. We also thank Drs. H.C. Outzen and L.D. Shultz for their critical reading of the manuscript.

This work was supported in part by a Grant-in-Aid for Cancer Research (53-5) from the Ministry of Health and Welfare, Japan, and also by Contract NO1-CP-55650 National Cancer Institute, USA. A part of the work reported in this paper was undertaken by A.K. under the sponsorship of Dr. R.T. Prehn (U.S. National Institutes of Health Research Grant CA 20920) during the tenure of an American Cancer Society-Eleanor Roosevelt-International Cancer Fellowship awarded by the International Union Against Cancer.

References

1. Bach, J.F., M.A. Bach, C. Carnaud, M. Dardenne, and J.C. Monier. Thymic hormones and autoimmunity, pp. 223–226. *In* N. Talal (ed.), Autoimmunity (genetic, immunologic, virologic and clinical aspects). New York: Academic Press, 1977.
2. Bigazzi, P.E. Autoimmune responses to spermatozoa in vasectomized rats and mice of different inbred strains, pp. 455–462. *In* N.R. Rose, P.E. Bigazzi, and N.L. Warner (eds.), Genetic control of autoimmune disease. New York: Elsevier/North-Holland, 1978.
3. Calkins, C.E., and O. Stutman. Changes in suppressor mechanisms during postnatal development in mice. J. Exptl. Med. 147:87–92, 1978.
4. Cantor, H. Lymphocyte communication and autoimmunity, p.. 161. *In* N.R. Rose, P.E. Bigazzi, and N.L. Warner (eds.), Genetic control of autoimmune disease. New York: Elsevier/North-Holland, 1978.
5. Cantor, H., and E.A. Boyse. Functional subclasses of T lymphocytes bearing different Ly antigens. I. The generation of functionally distinct T cell subclasses is a differentiative process independent of antigen. J. Exptl. Med. 141:1376–1389, 1975.

6. Hardy, B., and E. Mozes. Expression of T cell suppressor activity in the immune response of newborn mice to a T-independent synthetic polypeptide. Immunology 35:757–762, 1978.
7. Kojima, A., O. Taguchi, and Y. Nishizuka. Experimental production of possible autoimmune gastritis followed by macrocytic anemia in athymic nude mice. Lab. Invest. 42:387–395, 1980.
8. Kojima, A., O. Taguchi, T. Sakakura, and Y. Nishizuka. Prevalent types of tumors developing in neonatally thymectomized mice. Gann 70:839–843, 1979.
9. Kojima, A., Y. Tanaka-Kojima, T. Sakakura, and Y. Nishizuka. Spontaneous development of autoimmune thyroiditis in neonatally thymectomized mice. Lab. Invest. 34:550–557, 1976.
10. Kojima, A., Y. Tanaka-Kojima, T. Sakakura, and Y. Nishizuka. Prevention of postthymectomy autoimmune thyroiditis in mice. Lab. Invest. 34:601–605, 1976.
11. Kojima, A., Y. Tanaka-Kojima, and Y. Nishizuka. Experimental induction of autoimmunity in nude mice, pp. 127–137. In T. Nomura, N. Osawa, N. Tamaoki, and K. Fujiwara (eds.), Proceedings of the second international workshop on nude mice. Tokyo: University of Tokyo Press, 1977.
11a. MacKay, I.R., and F.M. Burnet. Autoimmune Diseases: Pathogenesis, Chemistry and Therapy. Springfield, Thomas. 1963.
12. Morse, H.C., A.D. Steinberg, P.H. Schur, and N.D. Reed. Spontaneous "autoimmune disease" in nude mice. J. Immunol. 113:688–697, 1974.
13. Mosier, D.E., and B. M. Johnson. Ontogeny of mouse lymphocyte function II. Development of the ability to produce antibody is modulated by T lymphocytes. J. Exptl. Med. 141:216–226, 1975.
14. Murphy, E.D., and J.B. Roths. Autoimmunity and lymphoproliferation: Induction by mutant gene lpr, and acceleration by a male-associated factor in strain BXSB mice, pp. 207–221. In N.R. Rose, P.E. Bigazzi, and N.L. Warner (eds.), Genetic control of autoimmune disease. New York: Elsevier/North-Holland, 1978.
15. Nishizuka, Y., and T. Sakakura. Thymus and reproduction: Sex-linked dysgenesia of the gonad after neonatal thymectomy in mice. Science 166:753–755, 1969.
16. Nishizuka, Y., and T. Sakakura. Ovarian dysgenesis induced by neonatal thymectomy in the mouse. Endocrinology 89:886–893, 1971.
17. Nishizuka, Y., and T. Sakakura. Effect of combined removal of thymus and pituitary on post-natal ovarian follicular development in the mouse. Endocrinology 89:902–903, 1971.
18. Penhale, W.J., A. Farmer, and W.J. Irvine. Thyroiditis in T cell-depleted rats: Influence of strain, radiation, dose, adjuvants and antilymphocyte serum. Clin. Exptl. Immunol. 21:362–375, 1975.
19. Rose, N.R., L.D. Bacon, R.S. Sundick, Y.M. Kong, P. Esquivel, and P.E. Bigazzi. Genetic regulation in autoimmune thyroiditis, pp. 63–87. In N. Talal (ed.), Autoimmunity (genetic, immunologic, virologic and clinical aspects). New York: Academic Press, 1977.
20. Shultz, L.D., and R.B. Zurier. "Motheaten": A single gene model for stem cell dysfunction and early onset autoimmunity, pp. 229–240. In N.R. Rose, P.E. Bigazzi, and N.L. Warner (eds.), genetic control of autoimmune disease. New York: Elsevier/North-Holland, 1978.
21. Taguchi, O., Y. Nishizuka, T. Sakakura, and A. Kojima. Autoimmune oophoritis in thymectomized mice: Detection of circulating antibodies against oocytes. Clin. Exptl. Immunol. 40:540–553, 1980.
22. Warner, N.L. Genetic aspects of autoimmune disease in animals, pp. 33–62. In N. Talal (ed.), Autoimmunity (Genetic, immunologic, virologic and clinical aspects). New York: Academic Press, 1977.
23. Yoshiki, T., T. Hayasaka, R. Fukatsu, T. Shirai, T. Itoh, H. Ikeda, and M. Katagiri. The structural proteins of murine leukemia virus and the pathogenesis of necrotizing arteritis and glomerulonephritis in SL/Ni mice. J. Immunol. 122:1812–1820, 1979.

General Discussion

KRUEGER: This is, seemingly, a very specific phenomenon—relative to time. This suggests some types of emergence of a specific cell at a specific time. Have you looked using specific antiserums for the emergence of such cells, e.g., suppressor cells?

KOJIMA: This point you make is important. At present we don't have good evidence about the T cells involved in our phenomenon; however, analysis of T-cell subpopulations using anti-Ly serum is in progress in Japan.

MITCHELL: You have emphasized autoimmunity rather than infectious disease throughout your presentation. Do you have evidence one way or the other for or against either of these mechanisms?

KOJIMA: In addition to autoimmunity, two other mechanisms are possible—an infectious etiology and a GVH mechanism. The evidence that nude mice and day-0-thymectomized mice do not develop the oophoritis may be against an infectious etiology. A large number of normal spleen cells failed to induce the diseases in nude mice and this argues against a GVH mechanism. In addition, the sex difference in the ability of peripheral lymphoid cells to induce oophoritis and the demonstration of specific antibodies are difficult to explain by those mechanisms. Therefore, at present, we believe an autoimmune mechanism is most likely.

J.D. SMALL: The possibility of one or more infectious agents as the cause of the gastritis should be considered. D.G. Brownstein et al. (Vet. Pathol. 14:606–617, 1977) reported the presence of a *Cryptosporidium* sp. in association with hypertrophic gastritis in snakes. From my own experience with snakes from this small collection, *Citrobacter freundii* was present in the gastric tissue and regression of the hyperplasia occurred when treated with oral chloramphenicol. At that time the presence of *Cryptosporidium* sp. was not recognized. *Citrobacter freundii* is the causative agent of transmissible murine colonic hyperplasia (S.W. Barthold, et al. Lab. Anim. Sci. 26:889–894, 1976). The gastric lesions illustrated in Dr. Kojima's presentation are strikingly similar to those reported by Brownstein.

22

The Nude Mouse in Autoimmune Disease

Karsten Buschard,* Sten Madsbad,† Erik Dabelsteen,‡ and Jørgen Rygaard

Pathological-Anatomical Institute, Kommunehospitalet DK-1399 Copenhagen K, Denmark, Hvidøre Hospital,† Klampenborg, Denmark, and Department of Oral Diagnosis,‡ The Royal Dental College, Copenhagen, Denmark.

Abstract

A new field of application of the nude mouse is suggested: as an in vivo model for studying autoimmune diseases. This is illustrated by two examples, concerning diseases with either a cellular or a humoral immunologic pathogenesis. Either the nude mouse's own organs or transplanted human tissue acts as target tissue. Several investigations of patients with insulin-dependent diabetes mellitus (IDDM) and animal studies suggest that lymphocytes reactive against islet components may be involved in the pathogenesis of IDDM. Both viral- and chemically-induced diabetes can be passively transferred from mouse to mouse with spleen cells. We attempted passive transfer of IDDM from man to nude mouse using 6×10^6 human peripheral blood lymphocytes. Control mice received lymphocytes from healthy persons. The recipient mice for 12 out of 23 patients showed significantly increased blood glucose values. This diabetic state in the recipient mice was transient but of some weeks' duration. In pemphigus, antibodies against cell membranes of the epidermis may play a pathogenetic role. Normal human oral mucosa was transplanted to nude mice; after 7 days these nude mice were injected with serum from a patient with pemphigus. After 24 h immunofluoresence microscopy showed ample deposits of antibodies in the cell membranes of the human transplanted epidermis, but only minimal amounts in the nude mouse oral mucosa. The examples show that the nude mouse may be of value in pathogenetic studies of diseases in which cellular or humoral immunologic mechanisms are suspected.

---

* To whom correspondence should be addressed.
© 1982 Gustav Fischer New York, Inc.
Proceedings of the Third International Workshop on Nude Mice.

Introduction

This study concerns a possible new application field of the nude mouse; namely, the use of the nude mouse in pathogenetic studies of autoimmune diseases. After transfer of cells and/or sera to the nude mice the *in vivo* reactions on the organs of the mice or transplanted human tissues are studied.

Two examples illustrate this in vivo model for studying autoimmune diseases: (a) insulin-dependent diabetes mellitus (IDDM), in which cellular immune reactions may be involved in destruction of the insulin-producing cells, and (b) pemphigus, in which antibodies against cells of the epidermis and oral mucosal membranes may play a pathogenetic role. Our experiments are presented in two parts, diabetes mellitus and pemphigus, with a common discussion.

Diabetes Mellitus

Several studies have indicated that cellular immune mechanisms may be involved in the pathogenesis of IDDM. Some of the most important are mentioned briefly here. Histologic studies have shown lymphocyte infiltration in the islets of Langerhans in patients with newly diagnosed IDDM (8); patients with IDDM display positive leukocyte migration tests against fetal calf or human pancreas homogenate (13,15); and lymphocytes from diabetic patients have been found to adhere to human insulinoma cells (10). The following investigations concerning experimental diabetes in animals have demonstrated the importance of the thymus-dependent immune system for diabetogenesis. Nude mice do not develop diabetes after infection with the diabetogenic encephalomyocarditis virus, in contrast to normal, immunologically intact mice (5,11). Furthermore, nude mice show significantly lower blood glucose values than normal mice after treatment with multiple dosages of streptozotocin (4).

On this background, and being encouraged by successful passive transfer in different diabetes–animal models [immunologic (12); encephalomyocarditis virus (1); streptozotocin (3)] the following experiments were carried out.

Materials and Methods

The first study of passive transfer of diabetes from man to mouse comprised six consecutive patients with newly diagnosed insulin-dependent diabetes mellitus (IDDM) referred to Hvidøre Hospital, Copenhagen (2). The patients were three men and three women aged 12–25 years; patient number 4, however, was 49 years old. None of the patients had a family history of diabetes mellitus and all were without other endocrine diseases. Three of the patients had had symptoms for less than 4 weeks. Four had weight loss up to 12 kg. All had ketonuria, and all had glycosuria of more than 5%. The patients' blood glucose was between 14 and 22 mmol/liter. Insulin treatment was started on the day of diagnosis, and all patients continued to take insulin thereafter.

Fifty to 60 ml of blood was collected from a cubital vein from both patients and control donors. The blood samples were taken from the patients within the first 5 days after diagnosis.

The lymphocyte fraction was separated by centrifuging the whole blood for 30 min on a Hypaque-Ficoll gradient. Each mouse was injected intraperitoneally with 0.5 ml Balanced Salt Solution (BSS) containing 6 million lymphocytes. Each person was donor to five or six nude mice, and the numbers of experimental mice were matched in the control group.

The recipient mice were 8-week-old nude BALB/c mice (eleventh backcross of a gene transfer to BALB/c/BOM). The animals were purchased from Gl. Bomholtgaard, Laboratory Animals Breeding and Research Center, DK-8680 Ry, Denmark, where they were reared under specific pathogen-free (SPF) conditions. Over the experimental period, they were kept two to six together in Makrolon cages at the Pathological-anatomical Institute, Kommunehospitalet, Copenhagen, as formerly described (17). Autoclaved feed pellets (Gl. Bomholtgaard) and sterile drinking water were supplied ad libitum.

Blood glucose for determinations was collected in 25-μl pipets from the paraorbital venous plexus of nonfasting animals every 3–5 days.

Results

The results are shown in Figure 22-1. Blood glucose values in the experimental mice were generally higher than in the control mice. The differences were statistically significant (according to the Mann-Whitney U test) at one or more sampling times.

Mice receiving lymphocytes from patient 1 displayed the highest blood sugars, maximal on the nineteenth day with 260 mg/100 ml—compared with 109 mg/100 ml in the four control mice. A slight increase in blood glucose values was observed in control mice after transfer of lymphocytes, the highest mean value being 129 mg/100 ml. A similar slight increase was also seen in mouse to mouse transfers and can probably be explained by the stress of cell transfer and blood sampling.

To date, 23 newly diagnosed diabetic patients have been investigated and the results appear in Table 22-1. As can be seen the passive transfer was positive for only 12 out of the 23 patients.

However, the primary finding of the study is that transfer of lymphocytes from patients with newly diagnosed insulin-dependent diabetes mellitus can induce a diabetic state in nude mice.

Table 22-1. Results of passive transfer of lymphocytes from diabetic man to nude mice

Mean blood glucose values of the recipient mice	Number of patients
>200 mg/100 ml	4
Between 150 and 200 mg/100 ml	5
<150 mg/100 ml but significantly higher than the control mice	3
Not statistically different from the control mice	11

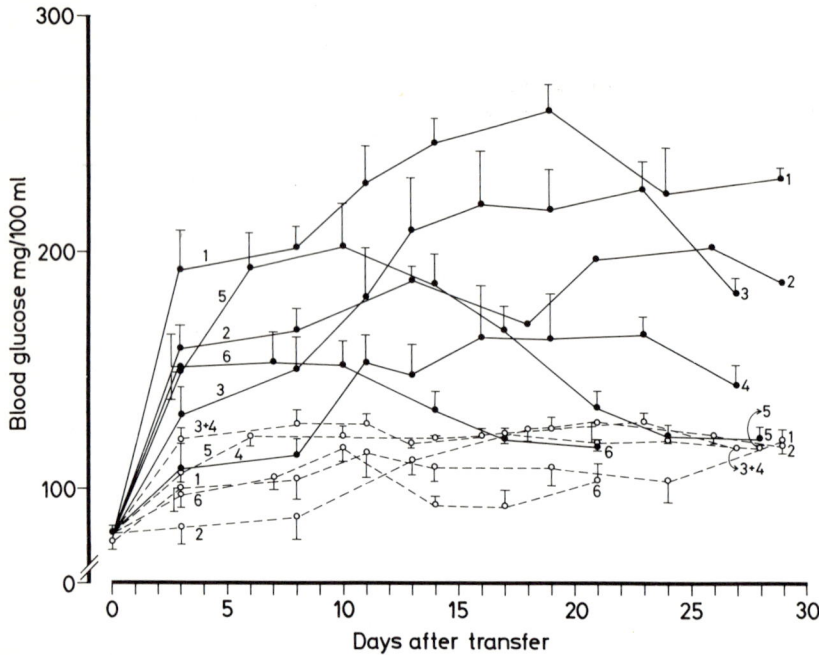

Figure 22-1. Blood glucose values (mean ± 1 SEM) in experimental (full line) and in control (broken line) groups. Numbers (1–6) adjacent to lines identify the human lymphocyte donor.

Pemphigus

The other experiment using nude mice in pathogenetic study of an autoimmune disorder concerns pemphigus, a serious blister-forming disease in skin and mucous membranes.

In patients with pemphigus, antibodies to the epidermal cell membranes can be demonstrated. A number of studies, both in vitro and in vivo, suggest that pemphigus antibodies are of pathogenetic significance, although this has never been proved (16). Among the most important in vitro findings is that addition of pemphigus serum to a tissue culture of normal human skin can induce acantholysis (18). The most suggestive in vivo observation is of bullous eruptions, which persist for some days, in newborn children of mothers with active pemphigus (19).

In this experiment, oral mucosal tissue from healthy donors was transplanted to the athymic nude mice, which, a week later, were injected with serum from pemphigus patients. Twenty-four hours after the injection, the epithelial transplants were removed and preparations were studied by immunofluorescence microscopy. Thus, a very important difference between this experiment and the diabetic transfer is that the target organ is transplanted human tissue.

Materials and Methods

The mucosa donors were seven 3- to 6-year-old children with no sign of pemphigus or other bullous skin lesions who were undergoing tonsillectomy. Punch mucosal biopsies, 3 mm in diameter, were taken from the tonsillectomy incision edge. No complications occurred.

Thirteen recipient mice, similar to the mice of the diabetic transfer experiment, were used. The mice were anesthetized with Epontol and an incision was made through the skin and panniculus carnosus in the flank region, and an area of deep fascia was exposed by blunt dissection. Prepared mucosal specimens were laid on the fascia and protected by hat-shaped capsules of 1 mm thick polyethylene sheet. These capsules were vacuum formed over a machined brass template, washed repeatedly, air dried to remove impurities, and finally gas sterilized with ethylene oxide before use. The skin was sutured over the capsules (14). To be sure of good vascularization to the transplanted mucosa, the serum was not injected until a week later. The serum was given intraperitoneally, 0.5 ml from patient or control to each mice. Seven mice were injected with pemphigus serum and six mice with control serum.

Sera from two patients with pemphigus vulgaris were investigated. Patient 1 was a 65-year-old man with clinically very active pemphigus. Patient 2 was a 44-year-old woman treated with Imuran (Wellcome Research Triangle Park, North Carolina), having a moderately active disease. In both patients the initial lesions appeared in the mouth rapidly followed by the debut of skin lesions. Both patients had a pemphigus history of 13 months. Three healthy individuals were donors of control serum.

Twenty-four hours after administration of serum, the mice were sacrificed and the transplants were removed. Immunofluorescence microscopy study was made as described previously (7,14). In brief, the tissue was rapidly frozen in dry ice and cryostate sections were cut. The sections were direct immunofluorescent stained using heavy chain-specific FITC conjugate rabbit immunoglobulins (DAKO, Westbury, New York), against human IgG, IgA, IgM, and complement, respectively.

Results

The results were the same for all the seven experimental mice. Figure 22-2 shows a section from one of the oral mucosa transplants. A bright fluorescence on the basal and parabasal cells of the epithelium can be seen. Only IgG deposits were detected. By contrast, no fluorescence was seen in the transplants from the six mice which had received serum from the control donors.

An important question is whether the antibodies responsible for the observed fluorescence are blood group antibodies. This seems to be very unlikely because of the distribution of the fluorescence—for blood group antibodies do not bind to the basal layer of the epithelium (6).

Using the pemphigus serum, however, there were no signs, in our study, of acantholysis or other pathological epidermal change. It may be that the 24-h study period was insufficient, and future longer studies, supplemented with irradiation provocation, will show whether the skin lesions can be reproduced in

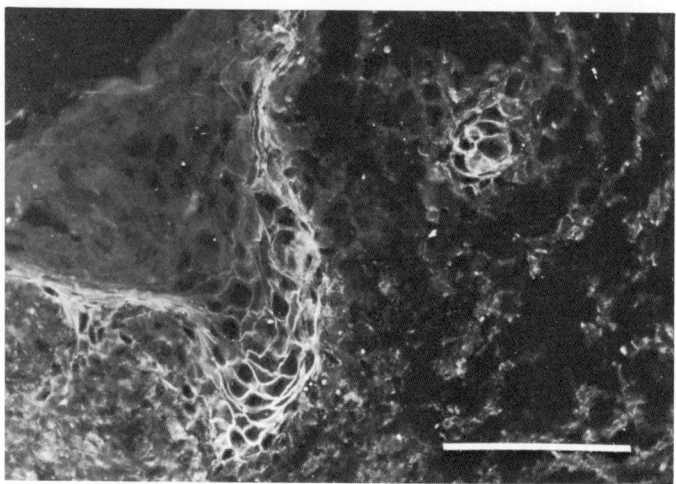

Figure 22-2. Direct immunofluorescence staining of human oral mucosa transplant to nude mice 24 h after administration of pemphigus human serum to the nude mice. The bright fluorescence on basal and parabasal cells indicates binding of human IgG. (Scale = 100 μm.)

this model. The effect of addition of complement should also be studied (9). However, this experiment has shown passive transfer of pemphigus antibodies to relevant human target tissue.

Discussion

This study has demonstrated by means of two examples—diabetes mellitus and pemphigus—that nude mice are useful in pathogenetic studies of autoimmune diseases, where cellular or humoral immunologic mechanisms are suspected.

Experimental transfer of autoimmune diseases from man to mouse has also been used in two other investigations. Volpé (21) found elevated serum thyroxin values in recipient mice after passive transfer to mice of blood lymphocytes from patients with Graves' disease. Toyka et al. (20) made passive transfers to normal mice of serum fractions from patients with myasthenia gravis. Immunoglobulin G from 15 of 16 patients significantly affected the recipient mice; among other things the mice showed the typically myasthenic features of reduction in acetylcholine receptors at neuromuscular junctions. Some of the mice showed weakness or decremental responses to repetitive nerve stimulation.

In this myasthenia gravis experiment normal mice were used. There are, however, two significant advantages obtained by using nude mice as recipients: (a) The passively transferred material itself and its effect are not influenced by a thymus-dependent immune system in the recipient. (b) Human transplanted tissue can function as target tissue.

The precise mechanisms by which the recipient mice or transplanted human tissue are affected are not clarified in any of the mentioned examples. However, the experimental model is new and far from fully explored. It seems reasonable to assume that an understanding of mechanisms behind the passive transfer disease in mice will give very important information about the pathogenesis of the corresponding disease in humans.

The described applications of the nude mouse may open new perspectives for in vivo study of the action in autoimmune diseases of lymphocytes or other cell types, serum, or combinations of cells and serum.

Acknowledgments

We wish to thank Dorte Fugman and Jytte Kræmmer for skillful technical assistance.

K.B. is a Research Fellow of the Michaelsen Foundation. The study was supported in part by the Danish Medical Research Council, Grant Nos. 512-8191, 512-10299, and 512-10808, and by the Michaelsen Foundation.

References

1. Buschard, K. Passive transfer of virus induced diabetes mellitus with spleen cells. Acta Pathol. Microbiol. Scand. (C) 86:29–32, 1978.
2. Buschard, K., S. Madsbad, and J. Rygaard. Passive transfer of diabetes mellitus from man to mouse. Lancet i:908–910, 1978.
3. Buschard, K., and J. Rygaard. Passive transfer of streptozotocin induced diabetes mellitus with spleen cells. Studies of synogeneic and allogeneic transfer to normal and athymic nude mice. Acta Pathol. Microbiol. Scand. (C) 85:469–472, 1977.
4. Buschard, K., and J. Rygaard. Is the diabetogenic effect of streptozotocin in part thymus-dependent? Acta Pathol. Microbiol. Scand. (C) 86:23–27, 1978.
5. Buschard, K., J. Rygaard, and E. Lund. The inability of a diabetogenic virus to induce diabetes mellitus in athymic (nude) mice. Acta Pathol. Microbiol. Scand. (C) 84:299–303, 1976.
6. Dabelsteen, E. Quantitative determination of blood group substances A of oral epithelial cells by immunofluorescence and immunoperoxidase methods. Acta Pathol. Microbiol. Scand. (A) 80:847–853, 1972.
7. Dabelsteen, E., S. Ullman, K. Thomsen, and J. Rygaard. Demonstration of basement membrane autoantibodies in patients with benign mucous membrane pemphogoid. Acta Dermatol. 54:189–192, 1974.
8. Gepts, W. Pathologic anatomy of the pancreas in juvenile diabetes mellitus. Diabetes 14:619–633, 1965.
9. Hashimoto, T., T. Nishikawa, S. Kurihara, H. Hatano. Complement-fixing pemphigus antibodies. Arch. Dermatol. 114:1191–1192, 1978.
10. Huang, S.-W., and N.K. Maclaren. Insulin-dependent diabetes: A disease of autoaggression. Science 192:64–66, 1976.
11. Jansen, F.K., O. Thurneyssen, and H. Müntefering. Virus induced diabetes and the immune system II—Evidence for an immune pathogenesis of the acute phase of diabetes. Biomedicine 31:1–2, 1979.
12. Korčáková, L., M. Titlbach, K. Novza. Adaptive transfer of immunodiabetes in guinea pig. Acta Diabetol. Lat. 11:112–135, 1974.
13. MacCuish, A.C., J. Jordan, C.J. Campbell, L.J.P. Duncan, and W.J. Irvine. Cell-mediated immunity to human pancreas in diabetes mellitus. Diabetes 23:693–697, 1974.
14. Mackenzie, I.C., E. Dabelsteen, and B. Roed-Petersen. A method for studying epithelial-mesenchymal interactions in human oral mucosal lesions. Scand. J. Dent. Res. 87:234–243, 1979.

15. Nerup, J., O.O. Andersen, G. Bendixen, J. Egeberg, R. Gunnarsson, H. Kromann, and J. Poulsen. Cell-mediated immunity in diabetes mellitus. Proc. R. Soc. Med. 67:506–513, 1974.
16. Ruocco, V., A. Rossi, G. Argenziano, C. Astarita, L. Aluiggi, B. Farzati, and G. Papaleo. Pathogenicity of the intercellular antibodies of pemphigus and their periodic removal from the circulation by plasmaphoresis. Br. J. Dermatol. 98:237–241, 1978.
17. Rygaard, J. Thymus and self: Immunobiology of the mouse mutant nude. Copenhagen: FADL; New York: J. Wiley & Sons, 1973.
18. Schiltz, J.R. and B. Michel. Production of epidermal acantholysis in normal skin in vitro by the IgG fraction from pemphigue serum. J. Investment. Dermatol. 67:254–260, 1976.
19. Terpstra, H., M.C.J.M. de Jong, and A.H. Klokke. In vivo bound pemphigus antibodies in a stillborn infant. Passive intrauterine transfer of pemphigus vulgaris? Arch. Dermatol. 115:316–319, 1979.
20. Toyka, K.V., D.B. Drachman, D.E. Griffin, A. Pestrouk, J.A. Winkelstein, K.H. Fischbeck, and I. Kao. Myasthenia gravis. Study of humoral immune mechanisms by passive transfer to mice. New Engl. J. Med. 296:125–131, 1977.
21. Volpé, R. The pathogenesis of Graves' disease: An overview. Clin. Endocrinol. Metab. 7: 3–29, 1978.

General Discussion

BEATTIE: Our laboratory has been unable to achieve passive transfer of diabetes; our study was published in *Lancet* in June, 1979.

BUSCHARD: I have no certain explanation of the discrepancy in our results. However, there are some points I want to mention: the different patients, the different methods of isolation, and the different mice.

In Denmark there is a seasonal variation in the incidence of insulin-dependent diabetes mellitus; briefly, there is a peak of new cases in autumn and winter and the transfer is positive among these patients. You may not have the same variation in California. In our method of isolation of the lymphocytes we may lose T-suppressor cells, which may be of importance. Finally, we use different mice, although of the same strain BALB/c.

I know that there have been some difficulties in reproducing the passive transfer of diabetes, but Doctor Paik, Albert Einstein College of Medicine, New York, has shown me a positive transfer to nude mice from one out of three patients.

Finally it should be mentioned that passive transfer of streptozotocin-induced diabetes from mice to mice has been performed by both Doctor Kiesel (U. Kiesel, et al., Streptozotocininduced diabetes: A transmissible disease. Diabetologia 14: 245, 1978) and Doctor Paik.

KRUEGER: The addition of complement probably won't lead to lesions—high-titered pemphigus sera, when injected into monkeys, fixes at the intercellular area of the epidermis but does not cause lesions (i.e., the blisters of pemphigus).

BUSCHARD: In the literature there are discussion about the role of complement in the pathogenesis of the pemphigus lesion. So far our studies are in accordance with the previous studies on monkeys, where no acantholysis was produced even when complement was present.

23

The Immunopathology of Congenitally Athymic (Nude) New Zealand Mice

M. Eric Gershwin* and Yoshiyuki Ohsugi

Section of Rheumatology-Clinical Immunology, Department of Internal Medicine, University of California, Davis, California 95616.

Abstract

Congenitally athymic (nude) mice on a New Zealand black (NZB), New Zealand white (NZW), and BALB/c backgrounds were produced by repetitive selective backcrossing, with parental selection based upon MLR and H-2 testing. $F'12$ generation nude mice of these three strains were compared to their littermates $nu/+$ controls with respect to survival, histology, blood counts, splenic surface markers, response to mitogens, spontaneous plaque-forming cells, appearance of naturally occurring thymocytotoxic antibodies (NTA), and number of splenic B cell clones. Under pathogen-free conditions, NZB, but not NZW or BALB/c, nude mice survive less than 3 weeks. A contributing factor to this premature death is a dramatic absence of T cell progenitor populations in the NZB nude group. Further, NZB nude mice have a significantly earlier appearance of NTA than $nu/+$ littermates and likewise appear to have heightened spontaneous polyclonal B-cell responses against the haptenes dansyl, NIP, TNP, 2,4-DNP, and sulfonate. NZB nude mice, in contrast, have significant and comparable elevations of splenic B-cell clones as their $nu/+$ littermates. It is suggested that NZB mice have several independent immunologic defects, including abnormalities of thymic epithelial cells, T-cell differentiation pathways, and polyclonally activated B-cell populations. Furthermore, the absence of detectable splenic Thy-1.2-bearing cells in NZB nude mice ($<1\%$), compared to values of 6–12% in other nude strains, suggests that NZB nude mice are an excellent model for study of T-cell differentiation and the origin of Thy-1.2-bearing cells in the nude mouse.

Introduction

Many of our current concepts of autoimmunity in New Zealand mice, as well as their genetic, viral, and lymphocyte subpopulation interactions, depend heavily on the use of surgically manipulated mice. However, the numerous age-dependent

---

* To whom correspondence should be addressed.
© 1982 Gustav Fischer New York, Inc.
Proceedings of the Third International Workshop on Nude Mice.

changes, the influence of sex hormones, and the variability and failure of neonatal thymectomy (even with irradiation and antitheta-treated bone marrow reconstitution) and splenectomy to eliminate the role of the thymus and spleen during gestation have led to a continued reevaluation of the data (1,9). Indeed, use of such models has led to the alternative suggestions that disease in New Zealand mice results from T-cell, B-cell, monocyte, thymic epithelial, and/or stem cell defects (1,5,6,9,10). These differences and the availability of both hereditary asplenic ($Dh/+$) and congenitally athymic (nude) mice in our laboratory prompted us to examine the underlying immune defect in NZB mice by selectively inbreeding colonies of both $Dh/+$ and nude NZB mice. We report herein that nude NZB mice, under specific pathogen-free (SPF) conditions and unlike comparably inbred nude NZW or BALB/c mice, have a marked deficiency of T-cell progenitor populations and a heightened polyclonal B-cell activation, suggesting independent defects in both T- and B-cell populations of NZB mice. Similarly NZB mice have an increase in B-lymphocyte clones in fetal liver, spleen, lymph node, and bone marrow. Moreover, this increase is comparably present in colonies of NZB nude and germ-free animals, including male and female mice, and is unrelated to thymic suppressor function.

Materials and Methods

Mice. Colonies of NZB (H-2d), NZW (H-2z), and BALB/c (H-2d) mice were maintained under SPF conditions at the Animal Resources Branch of the University of California School of Veterinary Medicine. Congenitally athymic (nude) NZB, NZW, and BALB/c mice were produced by initial mating of nude mice on an N:NIH (S) background to generate an F'1 generation that was heterozygous for the nu gene ($nu/+$). These offspring were thence backcrossed to parental NZB stock to produce litters that were roughly 50% heterozygous ($nu/+$) and 50% homozygous ($+/+$). These mice underwent test crossing and after the F'6 generation, only $nu/+$ mice found homozygous for H-2d were used in the inbreeding protocol (4,5). Such identified heterozygous $nu/+$ mice were backcrossed to parental stock; this has continued for 12 generations. Although inbreeding continues, all nude mice bred in this study were generated by the mating of F'12 $nu/+$ males × F'12 $nu/+$ females. Further, all animals studied herein are homozygous for H-2d (NZB; BALB/c) or H-2z (NZW), and the data presented are from mice of both sexes from 1 to 2 weeks old.

Functional studies. Complete blood counts, assay for natural thymocytotoxic antibody, splenic surface marker [Thy-1.2, rabbit antimouse brain (RAB) and surface Ig] quantitation, and background plaques to NIP = 4-OH-5-I-3-nitrophenyl, dansyl = 5-dimethylaminonapthalene-sulfonyl (NIP), trinitrophenyl (TNP), 2, 4 dinitrophenyl (2, 4-DNP), and sulfonate were performed as described earlier (2,3,7,8).

Colony formation. Lipopolysaccharide-induced and spontaneous B-lymphocyte clone formation in semisolid culture was studied in NZB, C57BL/6, BALB/c, DBA/2n, and NZW mice. Briefly, mice were sacrificed by cervical dislocation and their spleen resected. Single-cell suspension were prepared in RPMI 1640 medium containing 10% fetal calf serum and washed three times. Subsequently, the cells were resuspended at 37°C, in a final concentration of 2.5×10^4 cells/ml in McCoy's modified medium containing 15% fetal calf serum, 2 mM L-glutamine, 16 μg/ml L-asparagine, 8 μg/ml L-serine, 5×10^{-5} M 2-mercaptoethanol (2ME), and 0.3% Bacto-agar. This mixture was plated in sterile 35-mm tissue culture dishes and incubated at 37°C for 7 days in 10% CO_2 humidified air. All culture dishes were performed in duplicate and colonies (>20 cells) enumerated in a dissecting microscope. Culture of lower (10^4) and higher cell concentrations (10^5/ml) gave similar results for all strains; all data herein are derived from final cell concentrations of 2.5×10^4/ml. Finally, all experiments were performed by incubating the above mixtures either with 25 μg lipopolysaccharide (LPS) L-3129, *Escherichia coli* serotype 0127:B8, Sigma Chemical Co., St. Louis, Missouri) or without LPS (background colonies).

Results

Survival

Nude (F'12) NZB mice, even under SPF conditions, do not survive beyond 17–20 days of life. Although they appear healthy until 24 h before death and have only approximately 20–30% less body weight than their $nu/+$ littermates, they develop a rapidly progressive runting illness with diarrhea. The cause of death has generally been septicemia with normally noninvasive nonpathogenic organisms (i.e., *E. coli*); viral studies have not been performed but there is no evidence at autopsy of hepatitis. Furthermore, culling of nude offspring, for foster nursing, does not improve this picture. Because of the influence of morbidity on the parameters herein, all mice were studied between 7 and 14 days of life. Nevertheless, in contrast to the dramatic wasting of nude NZB mice, nude NZW and BALB/c mice are currently being maintained for more than 6 months under comparable conditions. Finally, it should be noted that nude NZB mice had no evidence of antibodies to erythrocytes or nucleic acid antigens.

Blood and Splenic Counts

The mean hematocrit of all mice studies ranged from 34 to 40, with no significant differences apparent. However, all groups of nude mice had a profound leukopenia compared to their $nu/+$ littermates ($P < 0.05$) (Table 23 1). Moreover, this leukopenia, when blood smears were examined, was primarily a lymphopenia. Similarly, the number of spleen cells recovered from NZB nude mice was profoundly reduced (6×10^6/spleen) compared to NZB $nu/+$ mice (24×10^6/spleen, $P < 0.05$). Although the number of spleen cells from both NZW and BALB/c nude groups was reduced relative to their $nu/+$ littermates, the differences were most profound in the NZB nude group (Table 23-2). Indeed, the hypocellularity and reduction of spleen cells in NZB nude mice were statistically lower than in NZW or BALB/c nude mice ($P < 0.05$). This reduction in spleen size was far greater than the gross difference (25%) in body weight between nude and $nu/+$ mice, including comparable counts of neonatal animals. Moreover, histologically the NZB nude spleen has very poorly developed white

Table 23-1. Hematocrit and leukocyte counts of nude NZB, NZW and BALB/c mice

Group	Hematocrit[a]	WBC (10^3 cells/mm^3)[a]	Differential
NZB nu/nu	35 ± 7	5.3 ± 2.5[b]	85% PMN, 15% lymph
NZB nu/+	38 ± 4	15.2 ± 4.3	22% PMN, 78% lymph
NZW nu/nu	40 ± 6	6.2 ± 2.1[b]	81% PMN, 19% lymph
NZW nu/+	37 ± 7	13.8 ± 3.4	14% PMN, 86% lymph
BALB/c nu/nu	34 ± 8	4.8 ± 2.9[b]	71% PMN, 29% lymph
BALB/c nu/+	38 ± 5	11.7 ± 3.6	18% PMN, 72% lymph

[a] Mean ± SEM; 1 week of age.
[b] $P < 0.05$, Student's t test.

pulp and a greatly reduced number of lymphocytes. NZB nude mice also have more erythrocytes and fewer mononuclear cells in red pulp than nu/+ littermates; these changes are also noted in NZW and BALB/c nude groups. Although such features are difficult to quantitate objectively, they appear more marked in the NZB nude group; this, however, may be secondary to the reduction in spleen cell number noted above (Table 23-2).

Splenic Surface Markers

NZB nude mice have a significant absence of T-cell markers using both a RAB and an AKR anti-Thy-1.2 serum (Table 23-3). Indeed, in both cases fewer than 1% of cells are detected as positive. This compares to levels in NZB nu/+ mice of 39% with RAB and 34% with anti-Thy-1.2 sera ($P < 0.001$). Moreover, there are readily detectable levels of T-cell markers with RAB and anti-Thy-1.2 sera in the spleen of NZW and BALB/c nude mice. The values are slightly but not significantly higher for RAB (13–16%) than for anti-Thy-1.2 sera (8–11%). However, both of these frequencies are significantly higher than NZB nude mice ($P < 0.001$) (Table 23-3). Indeed, the vast majority of NZB nude spleen cells appear to be Ig+, 91%, compared to 73% in nu/+ controls. This increase in Ig+ cells was likewise noted in the spleen of NZW and BALB/c nude mice

Table 23-2. Number of spleen cells recovered

Group	No. cells/spleen ($\times 10^6$)[a,b]
NZB nu/nu	6 ± 2[c,d]
NZB nu/+	24 ± 8
NZW nu/nu	14 ± 5[c]
NZW nu/+	28 ± 6
BALB/c nu/nu	19 ± 5[c]
BALB/c nu/+	32 ± 7

[a] Mean ± SEM, 10 days of age.
[b] Viability > 95%.
[c] $P < 0.05$, Student's t test, compared to nu/+ controls.
[d] $P < 0.05$, compared to NZW and BALB/c nu/nu mice.

Table 23-3. Splenic RAB⁺ and Thy 1.2-bearing cells in nude NZB, NZW, and BALB/c mice

Group	% Splenic RAB+[a]	% Splenic Thy 1.2[a]
NZB nu/nu	<1[b,c]	<1[b,c]
NZB $nu/+$	39 ± 8	34 ± 4
NZW nu/nu	13 ± 5[b]	8 ± 3[b]
NZW $nu/+$	37 ± 6	31 ± 5
BALB/c nu/nu	16 ± 6[b]	11 ± 3[b]
BALB/c $nu/+$	30 ± 3	32 ± 4

[a] Mean ± SEM, 10–12 days of age.
[b] $P < 0.01$, Student's t test, compared to $nu/+$ littermates.
[c] $P < 0.001$, Student's t test, compared to NZW and BALB/c nude mice.

relative to their $nu/+$ littermates. However, the Ig+ levels in NZB nude mice are elevated relative to NZW nude (64%) or BALB/c (66%) mice (Table 23-4). It must be emphasized that these changes in percentage of Ig+ cells must take into consideration the profound lymphopenia of NZB nude spleen (Table 23-2).

B-Cell Clones

There was a dramatic increase in B colony formation in both background and LPS-stimulated splenic cells from NZB mice in the age span 1 day to 3 months, with a peak of 860 per 2.5×10^4 seeded cells at 2 months. After 3 months of age the number of clones obtained declined sharply and reached significantly lower values at 10 months of age (fewer than 150 per 2.5×10^4 cells). In contrast, although there was some variation between the other strains studied, NZW, C57BL/6, C3H/HE, BALB/c, and DBA/2N, there were only minor differences noted within a strain with age. Of interest, however, was the significant elevation of NZW mice at 1 month (Table 23-5). It is important to note that there were no statistical differences between male and female mice for any strain and the data for both sexes are combined (for example at 6 weeks of age male NZB spleen colonies were 839 ± 112 compared to 683 ± 115 per 2.5×10^4 in female

Table 23-4. Frequency of immunoglobulin-bearing cells in spleens of nude NZB, NZW, and BALB/c mice

Group	% Ig+[a]
NZB nu/nu	91 ± 11[b]
NZB $nu/+$	73 ± 9
NZW nu/nu	64 ± 8
NZW $nu/+$	57 ± 10
BALB/c nu/nu	66 ± 7
BALB/c $nu/+$	51 ± 12

[a] Percentage of total spleen cells; mean ± SEM.
[b] $P < 0.05$, compared to NZW and BALB/c nude mice.

Table 23-5. Number of B cell colonies/2.5×10^4 spleen cells

Group	Age	Number[a]
NZB	1 day	140 ± 51[b]
NZB nude	1 day	200 ± 123
BALB/c	1 day	<30
NZB	1 week	350 ± 40[b]
BALB/c	1 week	106 ± 20
DBA/2n	1 week	13 ± 3
C57BL/6	1 week	85 ± 16
NZB	2 months	858 ± 109[b]
BALB/c	2 months	201 ± 20
DBA/2n	2 months	183 ± 29
C57BL/6	2 months	353 ± 38

[a] Mean ± SEM.
[b] $P < 0.01$ compared to age-matched BALB/c, DBA/2n, or C57BL/6 controls.

mice). Finally, in addition to increments in number of LPS-stimulated clones, NZB mice likewise had major increases in the number of background colonies. Indeed, the age tendency of background colonies revealed an early increase followed by a decline. For example, the number of spontaneous clones was significantly higher than control C57BL/6 mice at 1 week, 1, 2, 5, and 8 months of age ($P < 0.01$). Nonetheless, despite this increase the rate of colony formation and the time for peak colony formation (days 5 and 6) were similar in NZB and other strains. Indeed, this increase in number of B-lymphocyte clones was observed in both conventionally housed and germ-free NZB mice. Finally and significantly, this increment of NZB colony formation was thymic independent and was comparably seen at 1 day of age in both nude and $nu/+$ NZB mice.

Background Plaques

There was a significant increase in splenic plaque-forming cells of NZB mice to the haptens dansyl, NIP, TNP, 2,4-DNP, and sulfonate, compared to BALB/c controls ($P < 0.01$). However, NZB nude and $+/+$ littermate mice gave similar results.

Discussion

There have been multiple thymic abnormalities described or proposed in NZB mice, including loss of thymic suppressor, helper, and recirculating cells (9,10). Such features, although possibly contributing factors to disease expression, do not appear to be major primary etiologies. Indeed, some lymphocyte subpopulation alternations may be secondary to the appearance of NTA. For example, hereditary asplenic ($Dh/+$) mice have normal suppressor activity but still develop multiple autoantibodies (5). Nonetheless, it is relevant that NZB thymus undergoes premature involution. Indeed, histologically, this involution is preceded by vacuolization and degeneration of thymic epithelial cells (6). Moreover, recent observations from this laboratory have suggested a similar major defect in the functional capacity of NZB thymic epithelial cells. This relative deficiency may explain the reduced number of T-cell precursors in NZB nude mice and, if true, suggests a contributing role of thymic epithelial cells in the generation of T-cell markers in the nude mouse and of autoimmunity in New Zealand mice.

There are other characteristics of NZB nude mice that suggest a concurrent B-cell defect. First, there are greater numbers of Ig+-bearing spleen cells in NZB nude mice than in other groups. Nonetheless, the relative contribution of T-cell deficiency to these features becomes important to consider. For example, NTA appears very early in NZB nude but not NZB $nu/+$ mice, suggesting heightened clonal expansion of B cells in the absence of a thymus. Additionally, however, the number of spontaneous splenic antibody-forming cells to a variety of haptens is elevated in both NZB nude and $nu/+$ mice relative to all other groups. Although the responses are higher for NZB nude compared to NZB $nu/+$ mice, some correction must be made for the relatively higher value of NZB nude versus $nu/+$ splenic Ig+-bearing cells. Indeed, under such circumstances NZB mice appear to have heightened B-cell activity, independent of thymic status. Such data point significantly in the direction of a major underlying B-cell anomaly. Finally, although nude mice on other backgrounds have been reported to have both NTA and spontaneous plaque-forming responses, such observations are most noteworthy at older ages (i.e., 3–6 months) and are lower than we report herein (Table 23-6).

From the data herein, we cannot categorically state the extent of T- and B-cell defects in NZB mice. The data do strongly imply a major T-cell deficiency and polyclonal B-cell activation. The importance of B-cell population defect characterization in New Zealand mice is more than academic. The major thrust of therapeutic studies in the past has depended on pharmacologic and immunologic manipulation of thymic populations, including intensive study of several synthetic and chemically isolated thymic extracts. Although some such agents have been demonstrated to have beneficial effects on select autoantibody production, they have only minor influence on disease mortality. In contrast, murine and human systemic lupus erythematosus (SLE) studies have demonstrated that the most efficacious therapeutic agents inhibit B-cell function. Although such drugs may have multiple actions it is nonetheless of significance that they appear more

Table 23-6. Naturally occurring thymocytotoxic antibody in nude NZB, NZW, and BALB/c mice

Group	No. positive total	Titer[a]
NZB nu/nu	7/9[b]	3.3[b]
NZB $nu/+$	0/8	—
NZW nu/nu	0/5	—
NZW $nu/+$	0/8	—
BALB/c nu/nu	0/7	—
BALB/c $nu/+$	0/7	—

[a] Log_2 sera dilution; positive is $>50\%$ cytotoxicity.
[b] At 1 week of age.

beneficial than many of the so-called thymic enhancing factors. Because the thrust of research in New Zealand mice has as its primary and immediate goal application the development of clinically promising trials in humans, this study as well as other observations of B-cell function must be strongly considered. Although extrapolation of data from mouse to human must be done with caution, it seems reasonable to propose that therapeutic programs must take into account the observation that New Zealand mice appear to have several independent defects of both T- and B-cell populations. Finally, we must emphasize that the defect in B-cell clones of NZB mice was independent of sex. Thus, although hormonal factors may influence expression of disease, they do not appear to govern this specific thymic-independent B-cell anomaly.

Acknowledgment

Work reported here was supported in part by the U.S. National Cancer Institute, Grant RO 20816. M.E.G. is the recipient of Research Career Development Award AI 00193.

References

1. East, J., M.A. de Sousa, D.M.V. Parrott, and H. Jaquet. Consequences of neonatal thymectomy in New Zealand Black mice. Clin. Exptl. Immunol. 2:203–215, 1967.
2. Garner, G.M., M.E. Gershwin, and A.D. Steinberg. Properties of fractioned spleen cells from NZB/W mice. Cell. Immunol. 15:129–142, 1975.
3. Gelfand, M.C., G.J. Elfenbein, M.M. Frank, and W.F. Paul. Ontogeny of B lymphocytes. II. Relative rates of appearance of lymphocytes bearing surface immunogloubulin and complement receptors. J. Exptl. Med. 139:1125–1141, 1974.
4. Gershwin, M.E., J.J. Castles, K. Erickson, and A. Ahmed. Studies of congenitally immunologic mutant New Zealand mice. II. Absence of T cell progenitor populations and B cell defects of congenitally athymic (nude) New Zealand black (NZB) mice. J. Immunol. 122:2020–2025, 1979.
5. Gershwin, M.E., J.J. Castles, R.M. Ikeda, K. Erickson, and J. Montero. Studies of congenitally immunologic mutant New Zealand mice. I. Autoimmune features of hereditarily asplenic ($Dh/+$) NZB mice: reduction of naturally occurring thymocytotoxic (NTA) antibody and normal suppressor function. J. Immunol. 122:710–717, 1979.
6. Gershwin, M.E., R.M. Ikeda, W.L. Kruse, F. Wilson, M. Shifrine, and W. Spangler. Age-dependent loss in New Zealand mice of morphological and functional characteristics of thymic epithelial cells. J. Immunol. 120:971–979, 1978.
7. Gershwin, M.E., B. Merchant, and A.D. Steinberg. The effects of synthetic polymeric agents on immune responses of nude mice. Immunology 32: 327–336, 1977.
8. Milich, D.R., and M.E. Gershwin. T cell differentiation and the congenitally athymic (nude) mouse. Dev. Comp. Immunol. 1:289–298, 1977.
9. Morton, J. I., and B. V. Siegel. Transplantation of autoimmune potential. III. Immunological hyper-responsiveness and elevated endogenous spleen colony formation in lethally irradiated recipients of NZB bone marrow cells. Immunology 34:863–868, 1978.
10. Moutsopoulos, H.M., M. Boehm-Truitt, S.S. Kassan, and T.M. Chused. Demonstration of activation of B lymphocytes in New Zeland mice at birth by an immunoradiometric assay for murine IgM. J. Immunol. 119:1639–1644, 1977.

24

Athymic Mice: An Experimental Animal for the Isolation of "Crohn's Disease Agent"*

Kiron M. Das,† Isabel Valenzuela, and Rachel Morecki

Departments of Medicine,† Pathology, and Microbiology, Albert Einstein College of Medicine, 1300 Morris Park Avenue, Bronx, New York 10461.

Crohn's disease is a chronic inflammatory disease of unknown etiology which may affect any portion of the gastrointestinal tract. Recently, several "transmissible agents" have been implicated in the pathogenesis of this disease. Therefore, we studied the effects of tissue extracts from patients with Crohn's disease in homozygous nude mice (nu/nu). Mesenteric lymph nodes were obtained at surgery from four patients with active Crohn's disease, two suffering from ulcerative colitis, and two subjects with cholecystitis. Cervical lymph nodes from a patient with sarcoidosis were also obtained. The nodes were homogenized and filtered (0.45 μm), and approximately 0.2–0.4 mg of total protein was injected intraperitoneally into 10- to 16-week-old nu/nu mice. Thirty-four mice were injected with homogenates of lymph nodes from the patients with Crohn's disease, and 20 nu/nu mice received homogenates of lymph nodes from the other patients.

Four nu/nu injected with homogenates from four different Crohn's disease patients developed generalized lymphadenopathy and splenomegaly. Histological examination showed this to be lymphocytic lymphoma. Two lymphomas were homogenized, filtered, and injected intraperitoneally into five nu/nu mice; two of them developed lymphoma similar to the parent tumors.

Nu/nu injected with the lymph node homogenates obtained from the other three groups did not develop any lesion. Two of the four lymphomas had B-cell

---

* For details please refer to: Das, K.M. et al., Proc. Natl. Acad. Sci. USA 77:588–592, 1980.
† To whom correspondence should be addressed.
© 1982 Gustav Fischer New York, Inc.
Proceedings of the Third International Workshop on Nude Mice.

surface markers. They reacted with sera from other patients with symptomatic Crohn's disease but not with any sera from patients with active ulcerative colitis or normal subjects, indicating that the tumors contain an "antigen (s)" recognized by the sera of patients with Crohn's disease.

General Discussion

GAUTSCH: Did you examine the nude mouse lymphomas which were induced by Crohn's disease filtrates for the presence of C-type virus particles?

DAS: Yes, we did. In three of the four lymphomas viruslike particles in close association with rough endoplasmic reticulum were seen. These particles had close morphological similarities to type B and C murine particles; size varied from 80 to 120 nm.

HOLLAND: Because under germ-free conditions athymic nude mice show a 12- to 15-fold increased risk for spontaneous lymphoreticular neoplasms (Holland et al., J. Natl. Cancer Inst. (USA) 61:1257, 1978), have you considered the possibility that the lymphomas induced by Crohn's disease lymph node extracts might occur as the result of an indirect effect, such as adjuvant stimulation, peritonitis, or some other nonspecific accelerating phenomenon?

DAS: Yes, we did. Certainly several factors may activate oncogenic murine viruses to produce lymphoma. However, the interesting phenomenon is the recognition of the lymphomas by the sera from other patients with active Crohn's disease and not the sera from control patients. This suggests the presence of an antibody in the sera of patients with Crohn's disease directed to an antigen present in the lymphoma. Obviously the exact mechanism of production of lymphoma is still unclear and further studies are needed to answer these issues.

KRUEGER: Will serum from most patients with Crohn's disease react with these nude mouse lymphomas? Can this be used to diagnose Crohn's disease?

DAS: I hope so. So far we studied 10 patients with Crohn's disease. Recently we initiated a blinded study with about 80 sera from patients with inflammatory bowel disease and other gastrointestinal disorders. On completion of this study the diagnostic value of the "immunofluorescence test" will be more clear.

KOROS: Have you examined by electron microscopy any of the primary tissues from patients whose lymph node homogenates produce lymphomas in nude mice?

In addition to doing electron microscopic studies of the lymphomas which develop in nude mice injected with Crohn's disease patient's lymph node extract, do you plan to study other tissues of the tumor-bearing mice?

If any virus particles are found to be associated with a tumor, I believe it is very important to determine from where the virus particles came. For example, my colleagues (Drs. Olson and Acevedo, Allegheny General Hospital) and I have found by electron microscopy that one particular tumor carried in long-term passage in nude mice (No. 293 human oat cell carcinoma from Dr. Lola Reid) has virus particles which have also been seen by Drs. A. Knowles and J. Gautsch in an independent study. My colleagues and I have looked at other tissues of the nude mouse and found no evidence of such particles in liver, lung, or spleen of the tumor-bearing nude mice. The origin of these particles is not known. I believe it is important in such studies to examine by electron microscopy the primary material as well as the nude mouse recipients in order to elucidate the origin of any viral components found in subsequent tumors developing in the nude mice.

DAS: We have not yet examined the primary tissues from patients with Crohn's disease or non-tumor-bearing organs of the mice. We intend to perform the electron microscopic examination of mouse tissues in the near future.

GERSHWIN: May I ask you to expand upon the original purpose of your study?

DAS: Current literature on the etiology of Crohn's disease suggests that a transmissible agent(s) may be related to this disease. In conventional animals following injection of Crohn's disease tissue filtrates, granulomatous changes developed both at the injection site and in the ileum. However, a minimum of 6–8 months are needed before such tissue changes can be seen. The role of nude mice in the studies of different infectious agents which cannot be grown in the conventional animals is well known. If indeed a transmissible agent is involved in the etiology of Crohn's disease, we thought, the "agent" might grow or express well in the nude mice. However, production of lymphoma was rather unexpected.

25

Antibody Response to Allogeneic and Xenogeneic Skin Grafts in Nude Mice

Peter J.A. Capel,* Simon P.M. Lems, and Robert A.P. Koene

Department of Medicine, Division of Nephrology, Sint Radboudziekenhuis, University of Nijmegen, Nijmegen, The Netherlands.

Abstract

Grafting of allogeneic or xenogeneic skin on nude mice results in a primary antibody response, and during this response the antibody activity switches from IgM toward IgG. We studied the nature of this antibody response in different nude strains (C3H *nu/nu*, C57B1/6 *nu/nu*, B10.LP *nu/nu*, and BALB/c *nu/nu*). At day 21 after transplantation, when antibody titers were falling, the transplantation of a second graft of similar donor type resulted in a secondary antibody response only if the first graft had been removed before the appearance of IgG antibodies. Removal of the first graft after the appearance of IgG antibodies resulted in a nonresponsiveness of the recipient to the second graft. This suggested a suppressive role of the IgG antibodies. This was supported by the finding that passive transfer of nude sera containing specific antibody activity of IgG class suppressed the primary response to a skin graft in the nude mice, whereas sera containing only IgM activity did not induce suppression. We conclude that, in nude mice, an allograft or xenograft induces a primary antibody response that switches from IgM toward IgG, and that the concomitant presence of antigen and IgG antibodies results in the induction of specific unresponsiveness.

Introduction

Congenitally athymic nude mice accept allografts and xenografts permanently (12,13). The absence of graft rejection results from the lack of cytotoxic T cells and not from the absence of a humoral response, because nude mice generate cytotoxic antibodies against allogeneic and xenogeneic skin grafts (4,7,13). Although the antibody is formed in concentrations sufficient to cause rejection

---

* To whom correspondence should be addressed.
© 1982 Gustav Fischer New York, Inc.
Proceedings of the Third International Workshop on Nude Mice.

after administration of rabbit complement (4,7), such rejection does not occur spontaneously in the nude mice because of the inefficiency of the mouse's own complement system in this type of rejection (6).

Earlier we found that in nude mice the antibody activity against allogeneic and xenogeneic skin grafts switched from IgM toward IgG during the primary response (4). In this study we have analyzed this antibody response further, giving special attention to the genetic background of the recipients and the characteristics of the secondary response.

Materials and Methods

Animals. Inbred B10.D2/new Sn mice were originally obtained from the Jackson Laboratory (Bar Harbor, Maine) and inbred PVG/c rats from the Institute of Psychiatry, Bethlem Royal Hospital (Beckingham, Kent, U.K.). In our laboratory these strains were kept by continuous brother–sister matings. B10.LP *nu/nu* were obtained from the Radiobiological Institute T.N.O. (Rijswijk, The Netherlands) and C3H *nu/nu*, C57BL/6 *nu/nu*, and BALB/c *nu/nu* were obtained from Gl. Bomholdgard Ltd. (Ry, Denmark).

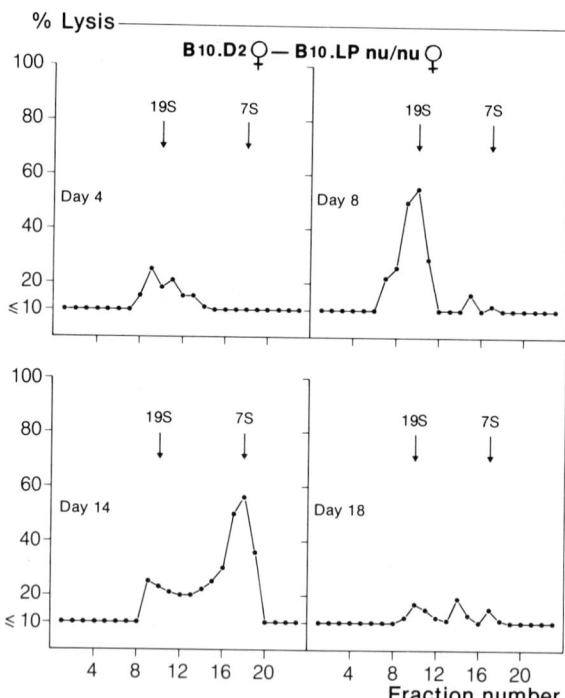

Figure 25-1. Cytotoxic activity of B10.LP *nu/nu* serum after ultracentrifugation on sucrose gradients. Sera were taken at days 4, 8, 14, and 18 after grafting of B10.D2 skin.

Skin grafts. Female tail skin was grafted onto the right dorsal flank of female *nu/nu* recipients by a modified fitted graft technique (2). Donor and recipient mice were between 6 and 8 weeks old. Second grafts were placed onto the left dorsal flank.

Serology. On different days after transplantation nude mice were bled and their sera stored at $-90°C$. The lymphocytotoxic activity was measured in a trypan blue exclusion test using rabbit serum as a complement source as previously described (3). IgM and IgG antibodies were determined after separation of the sera on isokinetic sucrose gradients (4).

Results

Primary Antibody Response

In different strains of nude mice the primary antibody response was measured by a lymphocytotoxicity assay at various days after grafting of allogeneic B10.D2 or xenogeneic PVG/c skin. As recipients of the xenografts, B10.LP *nu/nu*, C57BL/6 *nu/nu*, C3H *nu/nu*, and BALB/c *nu/nu* were used, whereas in the allogeneic models B10.D2 skin was grafted onto B10.LP *nu/nu* or C57BL/6 *nu/nu* recipients. Analysis of the sera on sucrose gradients showed cytotoxic activity in the 19S as well as in the 7S peak. The 19S activity reached a maximum at day 8 after grafting, whereas the 7S activity was maximal around day 14.

The response of a B10.LP *nu/nu* recipient of a B10.D2 skin graft given in Figure 25-1 is similar to the responses obtained with other donor–recipient combinations of which the results are summarized in Table 25-1.

Secondary Antibody Response

The secondary antibody response to xenogeneic PVG/c skin was studied in B10.LP *nu/nu*, C57BL/6 *nu/nu*, C3H *nu/nu*, and BALB/c *nu/nu* recipients by transplantation of a second PVG/c graft on day 21 after transplantation. The antibody activity was measured on days 4, 8, and 14 after regrafting. Similarly, the secondary response to allogeneic B10.D2 skin was studied in B10.LP *nu/nu*

Table 25-1. Antibody response in nude mice

Donor	Recipient	Maximal titer cytotoxic	Cytotoxic activity at:[a]							
			Day 4		Day 8		Day 14		Day 21	
			19s	7s	19s	7s	19s	7s	19s	7s
PVG/c	B10.LP *nu/nu*	1/256	+	±	++	++	+	++	+	++[b]
PVG/c	BALB/c *nu/nu*	1/64	+	−	++	+	+	++	+	+
PVG/c	C57BL/6 *nu/nu*	1/512	+	−	++	+	+	++	±	+
PVG/c	C3H *nu/nu*	1/64	+	−	++	+	+	++	−	+
B10.D2	B10.LP *nu/nu*	1/256	+	−	+	±	+	+	±	±
B10.D2	C57BL/6 *nu/nu*	1/128			Not done					

[a] Activity in 19s or 7s fraction after separation on sucrose gradients.
[b] Percentage lysis: $< 10\%$, −; 10–25%, ±; 25–90%, +; $>90\%$, ++.

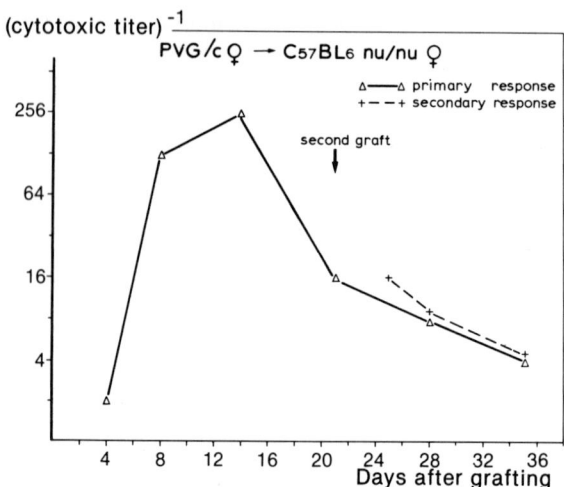

Figure 25-2. Primary and secondary antibody response of C57BL/6 *nu/nu* mice after transplantation of a PVG/c skin and regrafting of a second skin at day 21.

and C57BL/6 *nu/nu* recipients. In none of the cases did antibody activity exceed the residual activity still present after the first transplantation. Regrafting with a third-party graft resulted in a normal response. An example of the absence of the secondary response is given in Figure 25-2.

The absence of a secondary response might be caused by either a generalized immunologic incompetence of nude mice or the induction of suppression by the presence of antigen from the first graft during the production of immunosuppressive antibodies. We therefore studied how graft removal during the primary response influenced the secondary response. A B10.D2 skin graft was removed on days 8, 16, or 21 after grafting. Twenty days after removal of the graft the C57BL/6 *nu/nu* recipients were regrafted with B10.D2 skin. Eight days after regrafting, the animals were bled and the antibody activity was measured. The data given in Table 25-2 show that removal of the graft at day 8, i.e., before the appearance of IgG antibodies, resulted in a strong secondary response. Graft removal on days 16 or 21, i.e., after IgG antibodies had appeared, lead to suppression of this response.

Table 25-2. Abrogation of secondary response by the presence of antigen (B10.D2 → C57BL/6 *nu/nu*)

Removal of first graft on:	Regrafting on	Secondary response at day 8 after regrafting
Day 8	Day 28	1/512[a] (21)[b]
Day 16	Day 36	0 (3)
		1/64 (3)
Day 21	Day 41	0 (8)

[a] Cytotoxic titer.
[b] Number of animals tested.

Induction of Suppression by Immune Serum of Nude Mice

We have determined the suppressive capacity of immune sera of nude mice to find out whether the absence of the secondary response was the result of the induction of unresponsiveness by antibodies. C57BL/6 nu/nu mice were grafted with B10.D2 or PVG/c skin. On days 8, 16, and 28 the mice were bled and 0.25 ml of this serum was administered intraperitoneally (IP) to C57BL/6 nu/nu recipients of B10.D2 or PVG/c skin graft immediately after grafting. The primary response was measured on day 8 after grafting. The results obtained after administration of normal C57BL/6 nu/nu serum or immune nude serum are given in Table 25-3. These results show that suppression of the primary response occurred with nude serum obtained 16 days after grafting. This suppression is most likely caused by IgG antibodies and not by IgM antibodies because antiserum obtained at day 8 after grafting, which had a similar or even higher cytotoxic activity but lacked IgG activity, did not alter the alloantibody response. In the xenogeneic combination low amounts of IgG activity were detectable at day 8, and only a slight suppression was observed with this serum.

Discussion

Allogeneic and xenogeneic skin grafts induce a primary antibody response in congenitally athymic nude mice. This is in agreement with the findings of Rygaard (12,13), who found an antibody response to xenografts. The absence of alloantibodies in his study is not in contrast with our findings because he

Table 25-3. Suppression of antibody response by immune nude serum

Treatment[a]	Response at day 8, cytotoxic titer[b]
A. PVG/c → C57BL/6 nu/nu	
None	1/512 (5)[b]
Normal C57BL/6 nu/nu serum	1/512 (4)
C57BL/6 nu/nu anti-PVG/c serum:[c]	
Day 8 (1/512)[d]	1/128 (3)
Day 16 (1/256)	1/16 (4)
Day 28 (1/16)	1/512 (4)
B. B10.D2 → C57BL/6 nu/nu	
None	1/64 (6)[b]
Normal C57BL/6 nu/nu serum	1/64 (6)
C57BL/6 nu/nu anti-B10.D2 serum:[c]	
Day 8 (1/64)[d]	1/64 (4)
Day 16 (1/64)	0 (3)
	1/8 (3)
Day 28 (0)	1/64 (4)

[a] Intraperitoneal administration of 0.25 ml serum on day of transplantation.
[b] Number of animals tested in parenthesis.
[c] Serum obtained at day 8, 16, or 28 after grafting.
[d] Cytotoxic titer of the serum.

only searched for antibodies at day 30 after grafting. We found that the alloantibody response has been extinguished by that time.

During the primary antibody response the activity switched from IgM toward IgG. To exclude that this response was dependent on the genetic background of a particular nude strain, nude mice with different haplotypes were used as recipients and in all cases IgM and IgG antibodies were formed.

That the formation of antibodies does not result in graft rejection by nude mice is because of the inefficiency of the mouse complement system, as we have shown earlier (4,7). The finding that nude mice can evoke a primary antibody response to allografts and xenografts leads to the conclusion that major histocompatibility (MHC) antigens are T-cell independent or, alternatively, that MHC antigens need T-cell help, which is provided by a residual T-cell function present in nude mice. This T-cell help should then be located in the very low number of theta-positive cells which are present in nude mice (9,10). Although residual T-cell function cannot be completely excluded it seems more likely that MHC antigens can induce a T-cell-independent antibody response. In keeping with this hypothesis are the findings that the switch from IgM toward IgG in the response to T-cell-independent antigens does not depend on a T-cell function (1,8,11,14).

The findings by Klein et al., (5), who found H-2 antigens being T-cell dependent, are in contradiction with our results. However, in their study they used the absence of 2-mercaptoethanol (2-ME)-resistant antibodies as a criterion for the absence of IgG antibodies. This assumption is probably not completely valid since we have found that IgG is inactivated by 2-ME, especially at low Ig concentrations (3).

In all donor–recipient combinations a secondary response was absent, although the mice could respond to an unrelated second graft. The inability to reject the graft and the occurrence of a primary antibody response results in the simultaneous presence of antigen and antibody, a situation that in normal mice can induce antibody-mediated suppression of the immune response. Two conditions must be fulfilled before one can decide that the absence of a secondary response is the result of an active suppression mechanism and is not a general property of nude mice: (a) The antibodies produced by nude mice must be able to induce suppression after passive transfer. (b) The removal of antigen before the appearance of suppressive antibodies must result in the occurrence of a secondary response. Treatment of nude recipients with immune serum obtained from nude mice after grafting indeed induced suppression of the primary antibody response. It is likely that this suppression is only induced by IgG antibodies and not by IgM antibodies because serum obtained at day 8 after grafting, which had no detectable IgG activity when directed against an allograft or very low IgG activity when directed against a xenograft, induced no or only slight suppression, respectively. This was found in spite of the fact that these sera had the same or even higher cytotoxic titers than the suppressive IgG-containing serum obtained at day 16 after grafting.

The occurrence of a secondary response after removal of the graft before the appearance of IgG antibodies shows that nude mice are able to mount a secondary response and that the absence of this response is a result of antibody-mediated suppression.

We conclude that allografts and xenografts evoke a primary response in nude mice. This response, including the switch toward IgG, is most likely T-cell independent, although we cannot exclude a possible residual T-cell function being responsible. When concomitantly present with antigen, IgG antibodies induce specific unresponsiveness by which the response is abrogated. In contrast to IgG, IgM antibodies are unable to induce this suppression.

Acknowledgments

We thank J.F.H.M. Hagemann and P. Daamen for excellent technical assistance. We are indebted to the staff of the animal laboratory.
This study was supported by the Netherlands Kidney Foundation and the Netherlands Foundation for Medical Research (FUNGO).

References

1. Barthold, D.R., B. Prescott, P.W. Stashak, D.F. Amsbaugh, and P.J. Baker. Regulation of the antibody response to type III pneumococcal polysaccharide. J. Immunol. 112: 1042–1050, 1974.
2. Berden, J.H.M., P.J.A. Capel, and R.A.P. Koene. The role of complement factors in acute antibody mediated rejection of mouse skin allografts. Eu. J. Immunol. 8:158–162, 1978.
3. Capel et al. J. Immunol. Methods 36:77–80, 1980.
4. Gerlag, P.G.G., P.J.A. Capel, J.H.M. Berden, and R.A.P. Koene. Antibody response and skin graft rejection in the nude mouse. Transplant. Proc. 9:1179–1182, 1977.
5. Klein, J., S. Livnat, V. Hauptfeld, L. Jerabek, and I. Weissman. Production of anti H-2 antibodies in thymectomized mice. Eu. J. Immunol. 4:41–44, 1974.
6. Koene, R.A.P., P.G.G. Gerlag, J.F.H.M. Hagemann, U.J.G. van Haelst, and P.G.A.B. Wijdeveld. Hyperacute rejection of skin allografts in the mouse by the administration of alloantibody and complement. J. Immunol. 111:520–526, 1973.
7. Koene, R.A.P., P.G.G. Gerlag, J.L.J. Jansen, J.F.H.M. Hagemann, and P.G.A.B. Wijdeveld. Rejection of skin grafts in the nude mouse. Nature (London) 251:69–70, 1974.
8. Marchalonis, J.J. Antibodies and surface immunoglobulins of immunized congenitally athymic (nu/nu) mice. Aust. J. Exptl. Biol. Med. Sci. 52:535, 1974.
9. Raff, M.C. Theta-bearing lymphocytes in nude mice. Nature (London) 246:350–351, 1973.
10. Roelants, G.E., K.S. Mayor, L.B. Hägg, and F. Loor. Immature T lineage lymphocytes in athymic mice. Presence of TL, lifespan and homeostatic regulation. Eu. J. Immunol. 6:75–81, 1976.
11. Rüde, E., J. Wrede, and M.L. Gundelach. Production of IgG antibodies and enhanced response of nude mice to DNP-AE-Dextran. J. Immunol. 116:527–533, 1976.
12. Rygaard, J. Skin grafts in nude mice. 1. Allografts in nude mice of three genetic backgrounds (Balb/c, C3H, C57Bl) Acta Pathol. Microbiol. Scand. (A) 82:80–92, 1974.
13. Rygaard, J. Skin grafts in nude mice. 2. Rat skin grafts in nude mice of three genetic backgrounds (Balb/c, C3H, C57Bl). The effects after preparation by Thymus grafts. Acta Pathol. Microbiol. Scand. (A) 82:93–104, 1974.
14. Sharon, R., P.R.B. McMaster, A.M. Kask, J.D. Owens, and W.E. Paul. DNP-Lys-Ficoll: A T independent antigen which elicits both IgG and IgM anti-DNP antibody secreting cells. J. Immunol. 114:1585–1589, 1975.

26

Preliminary Studies of Normal Untreated and/or Carcinogen-Treated Adult Human Breast, Prostate, and Esophagus as Xenografts in Nude Mice

Marion G. Valerio,* Elliot L. Fineman, Ronald L. Bowman, Curtis C. Harris,† Benjamin F. Trump,‡ Elizabeth A. Hillman,‡ and Barry M. Heatfield‡

Litton Bionetics, Inc., 5516 Nicholson Lane, Kensington, Maryland 20795; Division of Cancer Cause and Prevention,† National Cancer Institute, Bethesda, Maryland, and Department of Pathology,‡ School of Medicine, University of Maryland, Baltimore, Maryland.

Abstract

The major goal in the xenotransplantation segment of these collaborative studies has been to develop methods for the long-term survival of normal adult human epithelial tissues as xenografts in the nude mouse. If successful, this xenograft system will be used as a model for the in vivo study of both function and differentiation of normal human epithelial cells and of neoplastic progression of carcinogen-exposed human tissues. Breast, prostate, and esophagus are now in the early stages of investigation. They have been transplanted into nude mice after 2.5 days to 10 weeks as explants in organ culture. Prostatic tissue has been maintained at least 22 weeks with viability, esophageal mucosa at least 16 weeks, and breast tissue up to 1 year in the nude mouse.

Introduction

The transplantation of normal adult human tissues into the nude mouse and their subsequent use in evaluating carcinogenic agents and carcinogenesis has been limited. Normal, hyperplastic, preneoplastic, and neoplastic human mam-

---

* To whom correspondence should be addressed.
© 1982 Gustav Fischer New York, Inc.
Proceedings of the Third International Workshop on Nude Mice.

mary tissue have been successfully xenotransplanted into the nude mouse (6,9, 15). The nude mouse accepts human skin grafts (8,10) and adult adipose tissue (1) and bone marrow (12) have also been successfully xenotransplanted.

The nude mouse xenograft system to be discussed here is one segment of a collaborative program studying carcinogenesis in various human epithelial tissues. Carcinogenesis data in humans are still obtained largely from retrospective epidemiological studies and by extrapolation from experimental animals. An alternative approach was to develop a model system using human tissues (2). The nude mouse offers the potential to study human tissues in an in vivo environment. Our findings with adult human bronchus, pancreatic duct, and colon have been reported (14). These tissues have been maintained for 715, 145, and 89 days, respectively, as evidenced by a viable-appearing epithelium with normal histology and the incorporation of tritiated thymidine into epithelial cells of the grafts.

In this contribution we describe the preliminary results of studies with normal untreated and/or carcinogen-treated adult human breast, prostate, and eosphagus transplanted into nude mice.

Materials and Methods

Animals. We used athymic nude mice of NIH-Swiss background. The original breeding nucleus was received from Dr. Carl Hansen of the Veterinary Resources Branch, National Institutes of Health, USA, in 1975. The colony was maintained by breeding nu/nu males with $nu/+$ females. They were bred, reared, and maintained in a modified barrier facility and housed in transparent plastic cages covered with nonwoven, spun polyester filter covers and the cages were kept in laminar air-flow rack units (Lab Products, Inc., Rochelle Park, New Jersey) within the barrier facility. Room temperature was 27–30°C; humidity varied with ambient conditions. The mice were fed NIH 31 autoclavable rat and mouse ration (Zeigler Bros., Gardners, Pennsylvania) and given acidified water (pH 2.5) ad libitum. The mice have lifespans of 18–24 months under these conditions. Maximum age reached in the colony was 1311 days.

Human tissue specimens. Tissues were obtained at the time of "immediate autopsy" (13) or surgery for either malignant or nonmalignant disease. Breast was also obtained at the time of reductive mammoplasty. Specimens were collected under clean conditions and transferred in L-15 culture medium at 4°C to the laboratory for processing. In most cases prior to xenotransplantation, the tissues were cultured as explants for various time periods in CMRL-1066 medium with 5% heat-inactivated or charcoal-absorbed fetal calf serum. Esophagus was trimmed of external muscle layers and cut into 0.5–1 cm^2 fragments (3). Epithelial tissue was dissected out of the breast tissue and cut into approximately 1-cm^2 pieces (4). At the time of xenotransplantation, representative sections of the fresh tissue were fixed in gluteraldehyde and/or formalin to compare with xenografts harvested after scheduled periods of time. Tissues were implanted subcutaneously in the flank of the nude mouse except for breast tissue, which was implanted in the cleared and uncleared mammary fat pad, the shoulder fat pad, and subcutaneously.

Normal and Carcinogen-Treated Human Xenografts in Nude Mice

Carcinogen and hormone treatment of tissues. The explant culture and carcinogen and hormone treatment of the explants was done in the Department of Pathology at the University of Maryland School of Medicine (4,11).

Explants of prostate were cultured at 37°C in CMRL-1066 supplemented with fetal calf serum and antibiotics for several days prior to carcinogen treatment. N-methyl-N'-nitro-N-nitrosoguanidine (MNNG) was obtained from the Carcinogenesis Standard Reference Compound Bank, through the Information and Resources Segment, Carcinogenesis, DCCP, NCI, NIH, Bethesda, Maryland. MNNG was dissolved in dimethyl sulfoxide (DMSO, Fisher Scientific Co., Springfield, New Jersey) at final concentrations of 5.0 μg/ml and 10.0 μg/ml and added to the culture medium once a week for 4 weeks using a micropipet (Schwarz/Mann, Rockville, Maryland). DMSO alone was added to control cultures at concentrations identical to experimental cultures. The medium was changed the day following treatment. The explants were xenotransplanted after 4 weeks.

Explants of breast tissue were cultured in control breast culture medium for 2 weeks prior to treatment with either MNNG or dimethylbenzanthracene (DMBA). Control breast culture medium consisted of CMRL-1066 medium supplemented with 0.1 μg/ml hydrocortisone, 1 μg/ml bovine recrystallized insulin, 5% heat-inactivated fetal calf serum, 2 mM L-glutamine, and antibiotics. DMBA and MNNG were dissolved in DMSO at final concentrations of 0.1 μg/ml and 2.5 μg/ml, respectively. They were added to the culture medium three times at 2-day intervals. DMSO alone was added to control cultures at concentrations identical to the carcinogen-treated cultures.

Explants of breast tissue were cultured in control breast culture medium for 2 weeks and then transferred to medium with 50 ng/ml or 500 ng/ml insulin, 5% charcoal-absorbed, heat-inactivated fetal calf serum and varying concentrations of hormones (progesterone, aldosterone, estradiol, and prolactin) at a constant level or in a 28-day cycling regimen. Further details of the hormone treatment are presented elsewhere (4).

Retrieval of grafts. Grafted tissues were removed at predetermined intervals after grafting or at the time of a spontaneous death of the host. Two hours prior to scheduled retrieval of grafts each mouse was injected intraperitoneally with 200 μCi of tritiated thymidine (20 Ci/mmole, New England Nuclear, Boston, Massachusetts).

Histological examination. Tissues for routine light microscopy were fixed in 10% neutral buffered formalin, paraffin embedded, sectioned at 4–6 μm, and stained with hematoxylin and eosin. Tissues processed for high-resolution light microscopy were fixed in mixed aldehydes (4F:1G), epon embedded, sectioned at 1 μm, and stained with toluidine blue. Light microscopic autoradiography was performed as previously described (7).

Results

Xenotransplantation of Human Prostate

Forty-eight normal adult human prostate specimens from three immediate autopsy patients have been transplanted into nude mice and recovered 1–22 weeks later (Table 26-1). Histologically, the normal prostate is made up of

Table 26-1. Xenotransplantation of adult human tissue[a]

Experiment	No. grafts/ No. mice	In vitro treatment[b]	Time of assessment (weeks)
HP 79-1	8/8	10 days CMRL-1066	1, 2, 4, 7
HP 79-2	30/8	4 weeks CMRL-1066 and 5 or 10 μg/ml MNNG once/week	2, 4, 6, 8, 22
HP 79-3	10/5	10 days CMRL-1066	1, 2, 3, 4, 5
HE 77-1	10/5	2.5 days CMRL-1066	1, 2, 3, 4, 6
HE 78-2	12/6	2.5 days CMRL-1066	Mice died at 12 days[c]
HE 78-3	8/4	2.5 days CMRL-1066	2 4, 6, 8
HE 78-4	6/6	2.5 days CMRL-1066	8, 16
HE 79-5	30/16	14 days medium B without serum	1, 2, 4, 8, 16, hold
HM 78-1	18/9	4 weeks, 2 weeks post-carcinogen treatment	4, hold
HM 78-2	48/12	10 days CMRL-1066	1, 2, 3, 4, 10
HM 78-3	18/9	8 weeks, 6 weeks post-carcinogen treatment	4, hold
HM 78-4	18/9	14 weeks, 12 weeks post-carcinogen treatment	4, hold
HM 78-5	36/9	10 days CMRL-1066	4, hold
HM 78-6	16/16	10 weeks, hormone treatment	24, hold
HM 78-7	7/7	14 weeks, hormone treatment	9, hold

[a] Abbreviations used are: HP, human prostate; HE, human esophagus; HM, human breast; MNNG, N-methyl-N'-nitro-N-nitrosoguanidine.

[b] Tissues were maintained in culture as explants for varying times and in varying media prior to xenotransplantation. Prostate was cultured in CMRL-1066 supplemented with fetal calf serum and antibiotics. Esophagus was cultured in CMRL-1066 supplemented with insulin, hydrocortisone, and fetal calf serum or medium B without serum. Breast was cultured in CMRL-1066 supplemented with hydrocortisone, insulin, heat-inactivated fetal calf serum, glutamine, and antibiotics.

[c] All mice died 12 days after xenografting; the grafts had evidence of a viral infection.

tubular-alveolar glands and ducts surrounded by dense fibromuscular stroma. The glands are lined by secretory columnar cells and nonsecretory basal cells, which are situated among the basal portions of the columnar cells. In culture the explant becomes covered by a sheet of cells continuous with the epithelium lining the ducts or glands. After 10 days in culture the columnar secretory cells become necrotic and slough, leaving basal cells (Figure 26-1). One week after xenotransplantation the surface is covered by one to two layers of flattened to cuboidal epithelial cells and the glands are filled with degenerating cells. One week later the surface is covered with epithelial cells and the glands are barely discernible (Figure 26-2). By 4 weeks the graft is encapsulated by mouse tissue. The epithelium that covered the surface of the graft is still present in many areas and there are epithelial growths into the prostatic connective tissue with basal cells repopulating the glands. At 7 weeks duct or gland formation was evident (Figure 26-3). They were lined by polygonal cells, which are probably basal cells. Ultrastructurally at 7 weeks differentiation of the basal cells into

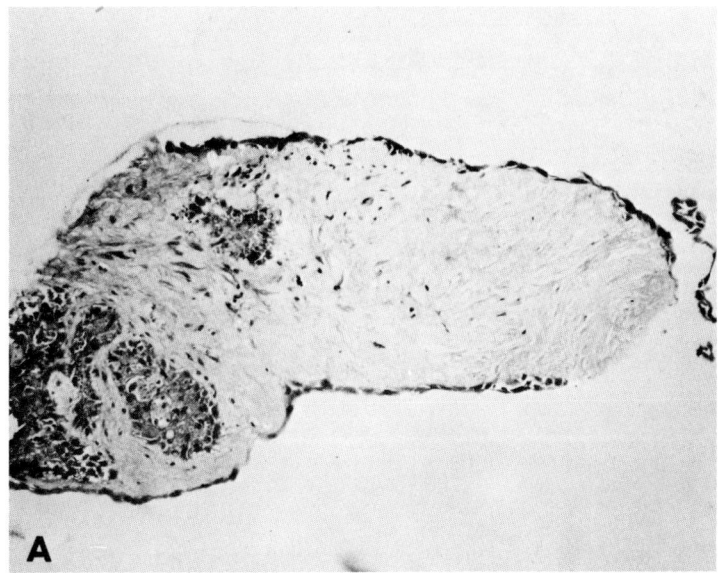

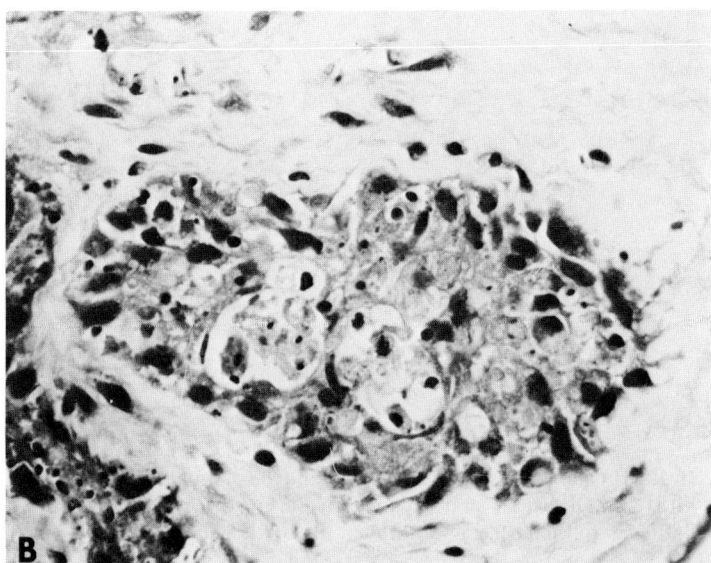

Figure 26-1. Explant of prostatic tissue after 10 days in culture and prior to xenotransplantation. (A) The surface of the explant is covered with epithelial cells and glands are seen in the stroma. ×45. (B) Higher magnification of gland with necrotic debris filling the lumen. ×448.

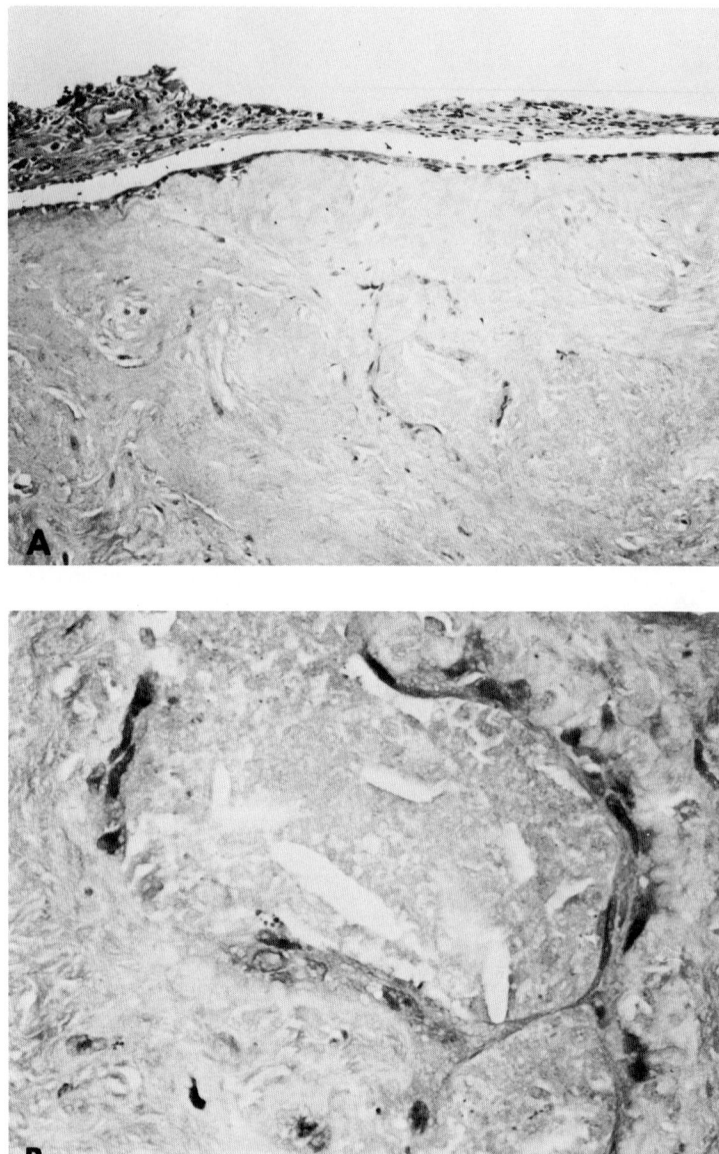

Figure 26-2. Prostate xenograft 1 week after transplantation. (A) The surface is covered with epithelial cells and glands are barely discernible. ×112. (B) Gland lined by flattened epithelial cells and lumen filled with debris. ×448.

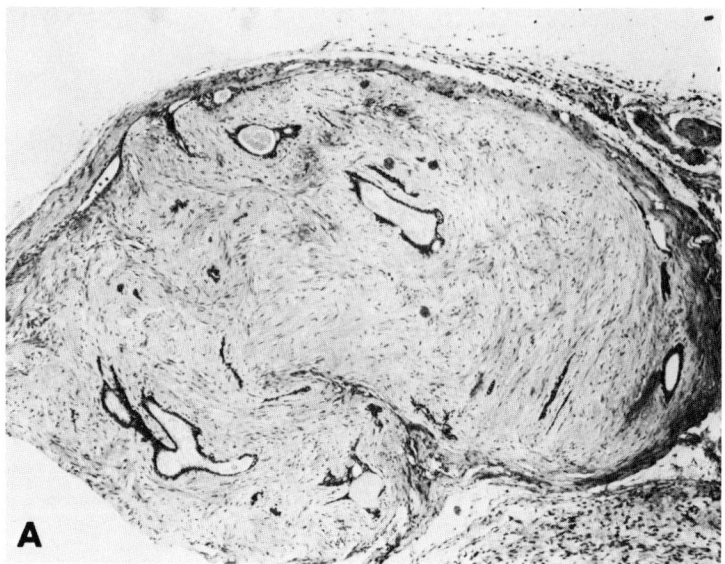

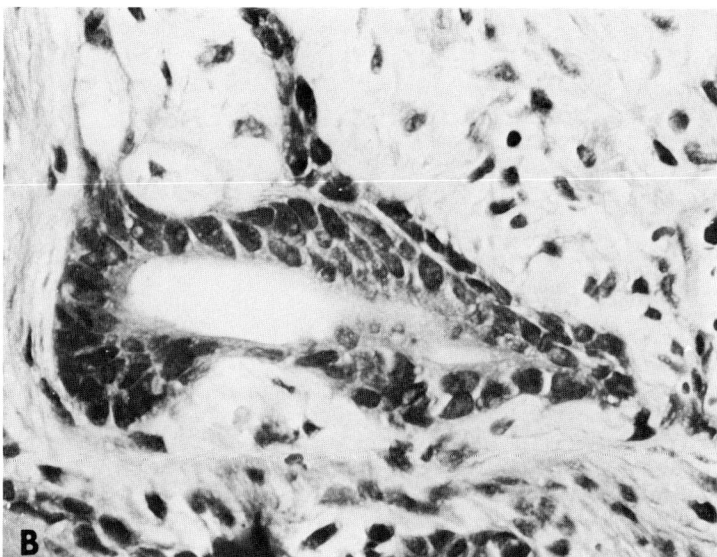

Figure 26-3. Prostate xenograft 7 weeks after transplantation. (A) Graft is encapsulated by mouse tissue but surface epithelium is evident in several areas along with ductal or glandular structures in the stroma. ×45. (B) Higher magnification showing ductal or gland formation by cuboidal to polygonal epithelial cells. ×448.

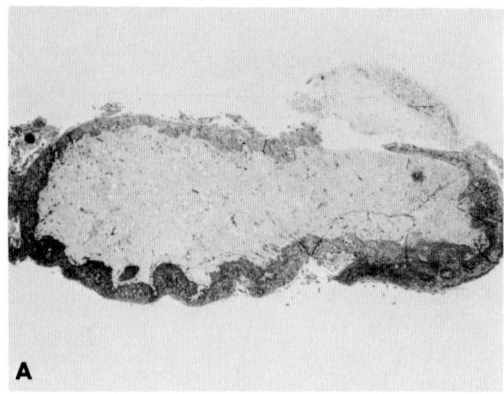

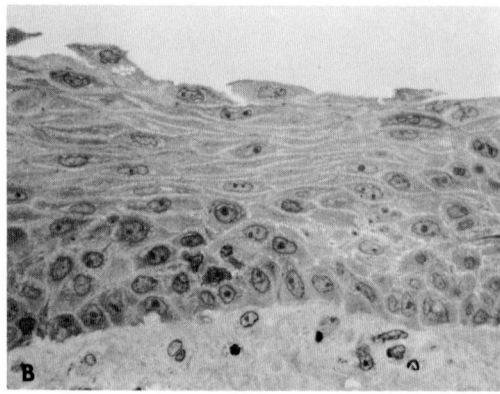

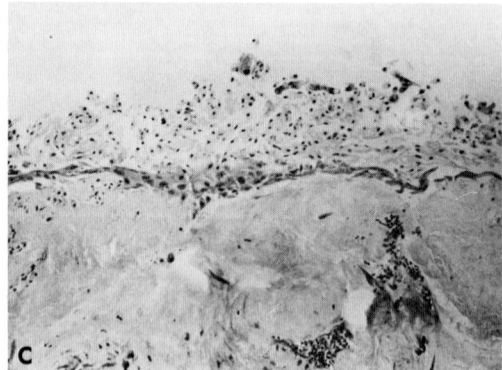

Figure 26-4. Esophageal explant prior to xenotransplantation. (A) Esophageal explant after 2.5 days in culture. ×19. (B) Higher magnification of (A) showing basal layers and moderately thick stratified squamous epithelium. ×308. (C) Mucosal epithelium explant one to two cell layers thick, after 10 days in culture. ×77.

secretory cells was not evident. One untreated graft was retrieved at 5 months. Nests of epithelial cells with ductal or acinar patterns were present.

In explant culture morphological evidence of transformation in prostatic epithelial cells occurred with MNNG (3). These explants were transplanted into nude mice but 22 weeks later there was no gross indication of any tumor induction. Microscopic evaluation of these tissues has not been completed at this time.

Xenotransplantation of Human Esophagus

Sixty esophageal fragments from five immediate autopsy cases were transplanted. Tissues from the first four cases were cultured as explants in CMRL-1066 medium for 2.5 days, whereas those from the fifth case were cultured as explants for 2 weeks prior to xenotransplantation (Table 26-1). These experiments were to assess survival and growth of normal human esophagus as xenografts in the nude mouse. The esophagus has not been exposed to carcinogens.

After 2.5 days in explant culture, normal, nonkeratinizing, stratified squamous epithelium is present. However, with prolonged culture there is a desquamation of the superficial epithelial layers and after 2 weeks in serum-free medium the squamous epithelium is decreased in thickness (Figure 26-4).

Grafts were retrieved 1–16 weeks after transplantation. Those that had been cultured as explants for 2.5 days generally formed a large cyst that was filled with desquamated squamous cells (Figure 26-5). The cyst was lined by stratified squamous epithelium showing normal differentiation from basal to superficial layers with desquamated cells in the lumen of the cyst. Mitotic figures in the basal layers were not infrequent. Cells labeled with tritiated thymidine were prominent in the basal layers up to 4 months after xenotransplantation. In contrast, those grafts from explants that had been cultured for 2 weeks prior to xenotransplantation generally did not form a single large lumen but several smaller lumina that were lined by stratified squamous epithelium which in most cases was thinner than those grafts that formed a single large cyst.

Xenotransplantation of Human Breast

One hundred sixty-one breast specimens from four patients have been transplanted into 71 nude mice (Table 26-1). Those specimens exposed to carcinogens were from one patient (HM 78-1,3,4) and those specimens subjected to hormone treatment were from another patient (HM 78-6,7). The other two experiments each used specimens from a different patient. The treatment regimens these specimens received in explant culture are presented elsewhere (4). Three different groups of studies have been done in the nude mice: (a) experiments to assess the survival of breast tissue grafted in different sites and in male and female mice (HM 78-2,5), (b) exposing explants to the carcinogens MNNG and DMBA in culture prior to xenotransplantation (HM 78-1,3,4), and (c) treating the explants in culture with various cycling or constant hormone regimens (HM 78-6,7).

In our studies, the site of transplantation (subcutaneous, shoulder fat pad, cleared or noncleared mammary fat pad) and the sex of the host mice appeared to play little role in the survival of the graft. In contrast to what occurs in

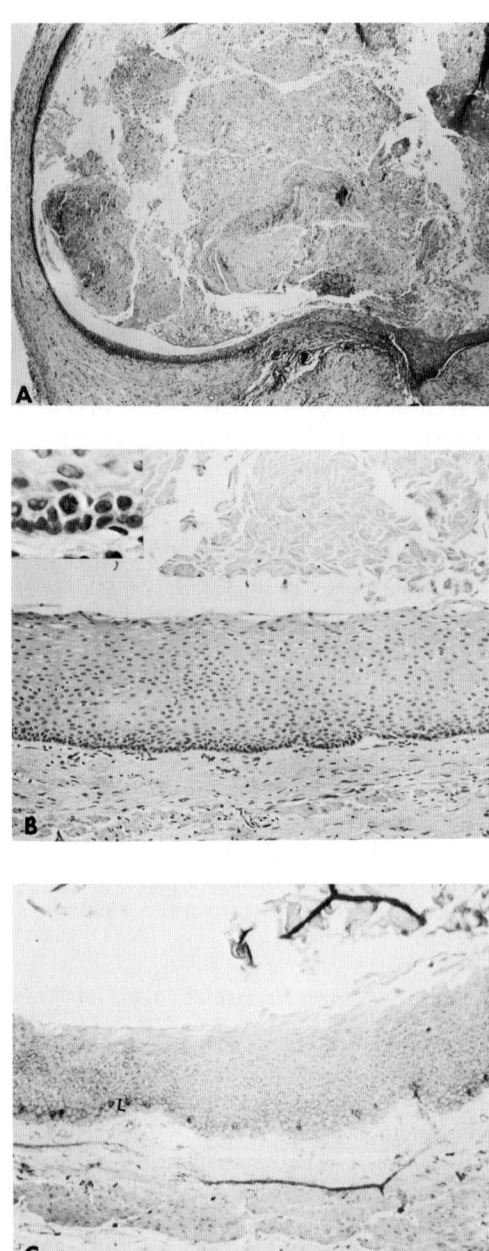

Figure 26-5. Esophageal xenograft 16 weeks after transplantation. (A) Graft has formed a cyst lined by stratified squamous epithelium and the lumen is filled with desquamated squamous cells. ×19. (B) Thick stratified squamous epithelium of graft. ×77. Inset shows basal cell in mitosis. ×308. (C) Autoradiogram with numerous basal cells incorporating tritiated thymidine. ×77.

prolonged explant culture of breast tissue and in xenotransplantation of prostatic tissue, the epithelium covering the surface of the explant disappeared after grafting and epithelium that remained was within the stroma of the graft. In the first week after transplantation much of the epithelium degenerated and sloughed; however, after 2–4 weeks it regenerated and ducts or ductules lined by one to two layers of epithelium were present. Occasional epithelial cells in mitoses were seen and tritiated thymidine was incorporated into the epithelial cells. In general there were fewer lobules present in xenografts than in the cultured tissue prior to xenotransplantation or in fresh breast tissue from surgery or autopsy specimens.

Grafts exposed to the carcinogens MNNG and DMBA in culture and then transplanted were not morphologically distinguishable from untreated grafts after 4 weeks. The remaining grafts from these experiments are currently being held in the nude mice for long-term evaluation.

Grafts treated in culture with breast trophic hormones had increased lobular tissue at least up to 4 months after xenotransplantation when compared with untreated grafts (Figure 26-6). The normal structure of the ducts, ductules, and supporting stroma was well preserved and periodic acid Schiff (PAS)-positive secretory material was present within the lumens.

Discussion

An interesting phenomenon associated with the xenotransplantation of certain tissues (esophagus, bronchus, pancreatic duct) in the nude mouse is that the epithelial cells proliferate and grow out from the periphery of the explant to eventually form a cyst. Once the cyst is formed and reaches a certain size the stimulus for proliferation is apparently lost. The cells are then maintained in this state for a long period of time. This is at least apparent in the bronchus and pancreatic duct. The esophagus may not react the same, as there is still active cell proliferation after 4 months in the nude mouse, and it is not known whether this will continue for prolonged periods. The fact that the epithelium of the bronchus and pancreatic duct becomes quiescent is a factor that must be taken into account when planning carcinogen exposure studies.

Prostate and breast are hormonally controlled and as such may have to be manipulated in the nude mouse to stimulate continued growth and differentiation. The use of androgens is currently being examined in the prostate (3). Administration of appropriate hormonal stimulus to the nude mouse has been reported to result in growth and differentiation of xenografted bovine mammary tissue (16).

The lack of any carcinogen-induced morphological change in the xenografts to date and the fact that some epithelial abnormalities present at the time of transplantation tend to revert to normal in the nude mouse poses a number of questions. Why do these epithelial abnormalities observed in the cultured prostatic epithelium revert to normal or at least fail to progress? Could the residual immune response of the nude mouse be selecting out these cells and destroying them before they have a chance to proliferate or will these grafts require constant or continuing stimulus by a carcinogen or other factor? These are questions that can only be answered after more experimentation with varied carcinogen treatments.

These preliminary observations demonstrate that adult human prostate, esophagus, and breast tissues can be maintained as xenografts in the nude mouse with retention of normal morphology for relatively long periods of time, as we have previously demonstrated with adult human bronchus, pancreatic duct, and colon. Thus, the xenograft system is potentially useful for the study of human tissues.

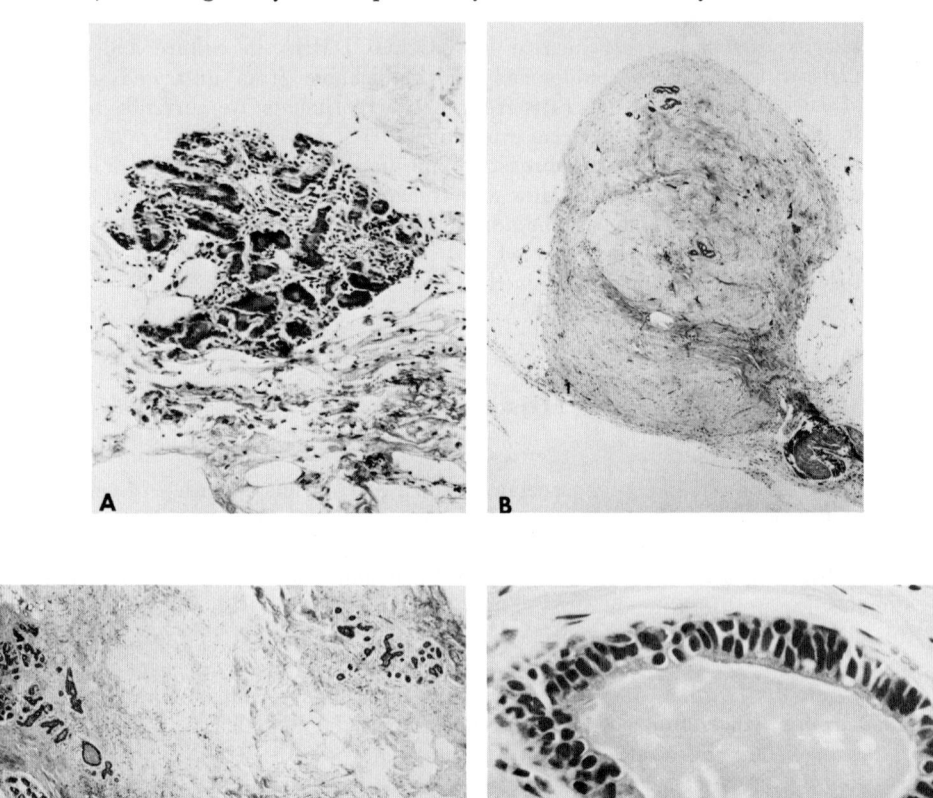

Figure 26-6. Human breast tissue in explant culture and after xenotransplantation (experiment HM 78-6). (A) Explant prior to transplantation. Explant had been exposed to cycling hormone regimen for 10 weeks in culture. ×77. (B) Breast xenograft 30 weeks after transplantation. This graft had been exposed to a constant hormone level in culture; a few ductules are present. ×19. (C) Breast xenograft 30 weeks after transplantation. This graft had been exposed to a cycling hormone regimen in culture; several well-preserved lobules are evident. ×19. (D) Higher magnification of (C) showing a duct filled with secretory material. ×308.

Acknowledgments

The authors wish to acknowledge the animal care and technical assistance of Mr. James Mason, Mr. Douglas Hevener, and Ms. Barbara Sheatz.
This work was supported in part by U.S. Public Health Service Contracts NO1 CP 43274, NO1 CP 95640, and NO1 CP 7-5909 and U.S. National Cancer Institute Grant CA 15798.

References

1. Bach-Mortensen, N., P. Romert, and S. Ballegaard. Transplantation of human adipose tissue to nude mice. Acta Pathol. Microbiol. Scand. (C) 84:283–289, 1976.
2. Harris, C. Chemical carcinogenesis and experimental models using human tissues. Beitr. Pathol. 158:389–404, 1976.
3. Heatfield, B.M., H. Sanefuji, and B.F. Trump. Studies on carcinogenesis of human prostate. IV. Comparison of normal and neoplastic prostate during long-term explant culture. Scanning Electron Microscopy (SEM)/1979/Vol. III, SEM Inc. 645–655.
4. Hillman, E.A., M.J. Vocci, J.W. Combs, H. Sanefuji, T. Robbins, D.L. Janss, C.C. Harris, and B.F. Trump. Human breast organ culture studies, pp. 79–106. In C.C. Harris, B.F. Trump, and G.D. Stoner (eds.), Methods in cell biology, Vol. 21B. New York: Academic Press, 1980.
5. Hillman, E.A., M.J. Vocci, W. Schürch, C.C. Harris, and B.F. Trump. Human esophagus organ culture studies, pp. 331–348. In C.C. Harris, B.F. Trump, and G.D. Stoner (eds.), Methods in cell biology, Vol. 21B. New York: Academic Press, 1980.
6. Jensen, H.M., and S.R. Wellings. Preneoplastic lesions of the human mammary gland transplanted into the nude athymic mouse. Cancer Res. 36:2605–2610, 1976.
7. Jones, R.T., L.A. Barrett, C. van Haaften, C.C. Harris and B.F. Trump. Carcinogenesis in the pancreas. 1. Long-term explant culture of human and bovine pancreatic ducts. J. Natl. Cancer Inst. (USA) 58:557–565, 1977.
8. Manning, D.D., N.D. Reed, and C.F. Shaffer. Maintenance of skin xenografts of widely divergent phylogenetic origin on congenitally athymic (nude) mice. J. Exptl. Med. 138:488–494, 1973.
9. Outzen, H.C., and R.P. Custer. Brief communication: Growth of human normal and neoplastic mammary tissues in the cleared mammary fat pad of the nude mouse. J. Natl. Cancer Inst. (USA) 55:1461–1466, 1975.
10. Rygaard, J. Skin grafts in nude mice. 3) Fate of grafts from man and donors of other taxonomic classes. Acta Pathol. Microbiol. Scand. (A) 82:105–112, 1974.
11. Sanefuji, H., B.M. Heatfield, and B.F. Trump. Studies on carcinogenesis of human prostate. V. Effects of the carcinogen N-methyl-N'-nitro-N-nitrosoguanidine (MNNG), on normal prostate during long-term explant culture. Scanning Electron Microscopy (SEM)/1979/Vol. III, SEM, Inc. 657–663.
12. Thorling, E.B., and B. Pedersen. Proliferation of human bone marrow in NMRI nu/nu Bom mice. Acta Pathol. Microbiol. Scand. (A) 82:345–347, 1974.
13. Trump, B.F., J.M. Valigorsky, J.H. Dees, B.A. Wolfgang, J. Mergner, K.M. Kim, R.T. Jones, R.E. Pendergrass, J. Garbus, and R.A. Cowley. Cellular change in human disease. A new method of pathological analysis. Hum. Pathol. 4:89–109, 1973.
14. Valerio, M.G., E.L. Fineman, R.L. Bowman, C.C. Harris, G.D. Stoner, H. Autrup, B.F. Trump, E.M. McDowell, and R.T. Jones. Long-term survival of normal adult human tissues as xenografts in congenitally athymic nude mice. J. Natl. Cancer Inst. (USA) 66:849–858, 1981.

15. Welsch, C.W., B.J. Mann, and M.M. Pienkowski. Successful transplantation of hyperplastic and carcinomatous human breast tissue to the athymic "nude" mouse, p. 58. *In* V.H. James (ed.) Fifth International Congress of Endocrinology, Hamburg, Germany, July 18–24, 1976 Elsevier.
16. Welsch, C.W., M. McManus, V. DeHoog, T. Goodman, and H. Tucker. Hormone-induced growth and lactogenesis of grafts of bovine mammary gland maintained in the athymic "nude" mouse. Cancer Res. 39:2046–2050, 1979.

27

Heterotransplantation of Embryonic Human Organs into Athymic (Nude) Mice

Hideo Nishimura,* K. Arishima, C. Uwabe, and K. Shiota†

Central Institute for Experimental Animals, 1430 Nogawa, Takatsu, Kawasaki 213, Japan, and Department of Anatomy,† Faculty of Medicine, Kyoto University, Kyoto, Japan.

Abstract

The ability of athymic nude mice to accept organs taken from normal human embryos of conceptional age 32–54 days has been studied by both subcutaneous and intratesticular grafting with the following results:

1. Among 10 embryonic organs examined, three organs, the limb bone, lung, and gastrointestinal tract, grafted subcutaneously were accepted in almost 100% of cases and maintained long term up to 18 weeks.
2. Intratesticular grafts of embryonic eyeballs showed a high acceptability.
3. Differentiation of the readily accepted organs generally occurred in a normal manner.
4. Grafts of the embryonic livers occasionally showed development of cartilage or bile duct, but no hepatocytes. Thus, this technique offers advantages in enabling experiments to be conducted in comparative physiology, biochemistry, pharmacology, and teratology as well as in allowing microbiological studies that require human tissues.

Introduction

Although a number of studies on grafting of human adult organs and malignant tumors into nude mice have been reported, only a few have been concerned with xenografting of human fetal organs. The pioneering studies of Povlsen et al. (5) and Skakkebaek et al. (6) demonstrated that such human organs as

---

* To whom correspondence should be addressed.
© 1982 Gustav Fischer New York, Inc.
Proceedings of the Third International Workshop on Nude Mice.

thymus, lung, pancreas, adrenal, kidney, testis, and ovary taken from 14- to 22-week-old fetuses could be successfully transplanted subcutaneously into nude mice and that the appearance of such grafts more or less resembled their in vivo counterparts. The similar study (2) was also done using human fetal pancreas in which growth of the implants was indicated by a branched system of small ductules and cells containing insulin and glucagon. The ages most successfully transplanted were around 7–12 weeks of gestation. In addition, successful transplantation of human fetal pituitaries at 7–22 weeks gestational age into nude mice has been reported (1). These implants showed development of typical adenohypophyseal cells but not neurohypophyseal cells and resulted in production of human growth hormone in the blood of recipient mice.

Because induced abortuses are more easily available in the first trimester, we attempted similar research on grafting using human embryos of conceptional age 4.5–8 weeks at the organogenic stage to examine whether such an approach could be used in a more applied way. The present study, which deals with the acceptability, maintenance, and histological differentiation of several grafted embryonic organs, differs from earlier studies in that much earlier conceptuses have been used.

Materials and Methods

Human donor material. Human embryos at Carnegie stages 14–22 (standardized conceptional age: 35–50 days) (4) were legally obtained from healthy women, whose pregnancy was interrupted by means of dilatation and curettage for socioeconomic reasons. A total of 131 undamaged or partially damaged specimens without external malformations and any sign of intrauterine death were used. The embryos were stored in sterilized petri dishes at 4°C and transferred to the laboratory as quickly as possible. Four to 24 h after recovery (4–7 hours in the majority of cases), the specimens were dissected under stereomicroscope and some of the organs of limb bones, lung, gastrointestinal tract (stomach and/or intestine), liver, heart, kidney, gonad and its duct system, adrenal, brain, and eyeball were removed. Each of the organs was minced aseptically into small pieces of approximately 0.5 mm size.

Recipient mice. Eight- to 12-week old female (for subcutaneous grafting) or male (for intratesticular grafting) BALB/cA nude mice (nu/nu) were bred and maintained in flexible vinyl film isolators under pathogen-free conditions at the Central Institute for Experimental Animals, Kawasaki, Japan. In addition, BALB/cA littermates ($nu/+$) were used as control in order to confirm the difference in acceptability of human embryonic organs.

Grafting technique. Grafting was done subcutaneously and intratesticularly. In the former 10–60 mg of embryonic tissue fragments were grafted into the subcutaneous tissue of the back of the mice using a trocar (inside diameter: 2–3 mm), and in the latter the testis of the nude mouse was pulled out through a small incision in the wall of the lower part of the abdomen under anesthesia with chloroform. Using a handmade glass pipet with a slender end portion (inside diameter: about 0.6 mm), 3–8 mg of the embryonic tissue fragments were taken and then blown into the testis, which was then returned to the original position and the wound was closed with stitches.

Examination of the grafted tissues. Transplantation sites were observed daily for gross outward appearance. Five and a half to 18 weeks after subcutaneous grafting and 1.5–7 weeks after intratesticular grafting, the animal was sacrificed and the site of subcutaneous grafting or the testis were examined grossly. Accepted transplants in the back, which were defined easily, and grafted testes were fixed in 10% neutral formalin, embedded in paraplast, sectioned serially, and stained with hematoxylin and eosin. Histological findings were compared with those of the corresponding sections taken from the whole-body histological specimens of stage-matched normal human embryos which are stored in the Congenital Anomaly Research Center, Faculty of Medicine, Kyoto University, Kyoto, Japan.

Results

Subcutaneous Grafting

Acceptability. The results of subcutaneous grafting of several human embryonic organs into nude or control mice are shown in Table 27-1. It is noteworthy that grafting of three organs, limb bones, lung, and gastrointestinal tract, was successful in almost all cases, and that these grafts were maintained for up to 18 weeks. Since the two rejected limb bone primordia were taken from embryos at Carnegie stages 14 or 16 and the two rejected lung grafts were from embryos at Carnegie stage 18, it is apparent that embryos older than stage 19

Table 27-1. Acceptability of human embryonic organs grafted subcutaneously into nude and control mice

Grafted embryonic organ	Carnegie stage of embryo [standard conceptional age in days (4)]	Time interval between grafting and sacrifice (days)	Proportion of grafts accepted, accepted/total No. (%)	
			BALB/cA-nu/nu	BALB/cA-nu/+
Limb bones	14–22 (35–50)	37–116	9/11 (82)	0/5
Lung	17–21 (39–48)	44–127	8/10 (80)	0/1
Gastrointestinal tract (stomach and/or intestine)	17–22 (39–50)	56–127	10/10 (100)	0/4
Liver	17–21 (39–48)	44–127	0/15[a]	—
Heart	14–21 (35–48)	56–127	1/12 (8)	—
Kidney	19–22 (45–50)	58–92	1/5 (20)	—
Gonad and its duct system	17–22 (39–50)	56–127	0/5	—
Adrenal	19–22 (45–50)	61–98	0/4	—
Brain	17–21 (39–48)	56–127	0/10	—
Eyeball	17–20 (39–47)	56–98	0/4	—

[a] Presence of the tissues other than the parenchyma are ignored.

are required for successful transplantation. However, the smaller amounts of tissues obtained from the younger embryos could explain the lack of acceptability of these grafts.

Table 27-1 shows that seven organs other than the three mentioned resulted in low or no acceptance. The heart and kidney grafts that were accepted were from embryos older than stage 21.

Differentiation of the accepted organs. Only the successfully developing grafts will be described.

1. *Limb bones.* The bone primordia grafts, consisting of condensed mesenchymal cells, precartilage, and cartilage tissues (Figure 27-1A), grew and developed into an apparently cartilaginous mass of irregular shape (Figure 27-1B). It was noteworthy that 6 of 11 grafts in total showed ossified regions (Figure 27-1C) and that the shortest time required for ossification was 37 days. No sign of degenerative or necrotic changes was noticed (see Figure 27-1C).
2. *Lung.* The original grafts of bronchopulmonary buds had an internal lining of simple ciliated columnar epithelium (Figure 27-2A). After grafting, these developed into dilated multilocular structures with the cavities lined with ciliated columnar, nonciliated cuboidal, or squamous epithelia

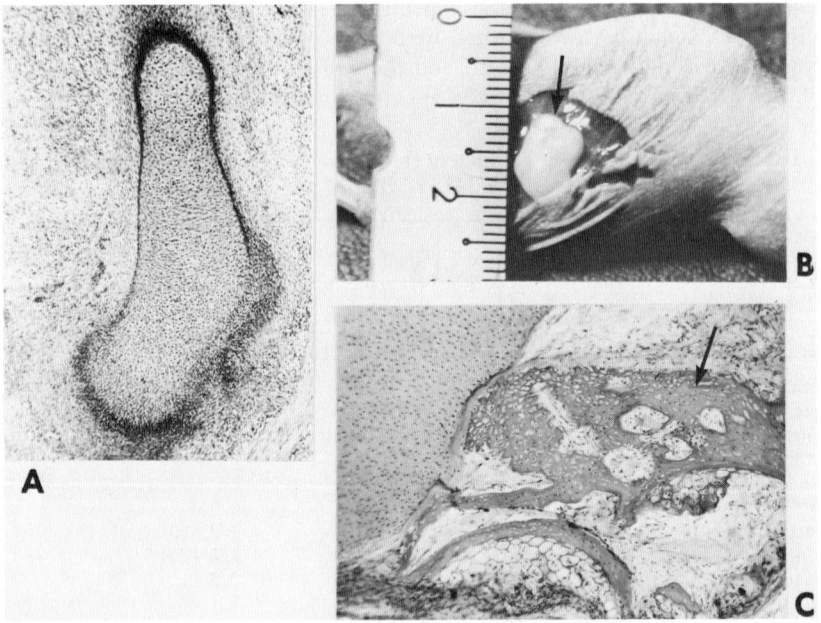

Figure 27-1. (A) Upper limb bone primordia (Carnegie stage 20) showing the stage of the original graft. ×94. (B) Gross appearance of the graft as a white mass (arrow) shown at 116 days after subcutaneous grafting of (A) staged upper limb bone. (C) Cartilage mingled with ossified regions (arrow) at 82 days after subcutaneous grafting of (A) staged upper limb bone. ×94.

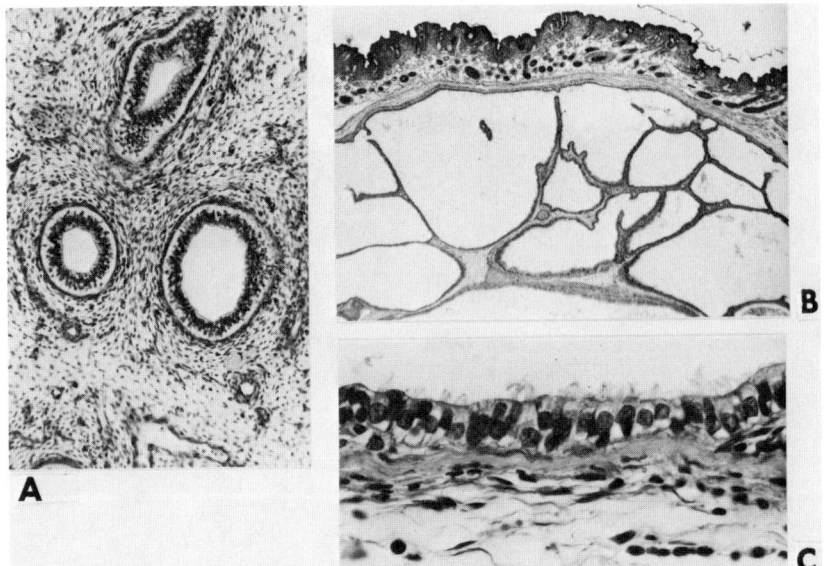

Figure 27-2. (A) Bronchopulmonary bud (Carnegie stage 20) showing the stage of the original graft. ×188. (B) Dilated multilocular structures at 65 days after subcutaneous grafting of (A) staged bronchopulmonary bud. ×38. (C) A part of inner wall of the cavity in (B) showing ciliated columnar intermingled with nonciliated cuboidal epithelia. ×375.

(Figure 27-2, B and C). Also, the differentiation was marked by the appearance of bronchial cartilages in some places. No degenerative or necrotic alterations were shown (see Figures 27-2, B and C).

3. *Gastrointestinal tract.*
   a. *Stomach.* The original graft showed an entodermal lining of simple or pseudostratified columnar epithelium as well as the layers of submucosa, muscularis, and serosa, but no gland buds were seen (Figure 27-3A). After grafting, a mass consisting of multiple cysts without degenerative or necrotic changes was formed (Figure 27-3B). As the tissue differentiation, gastric pits and glandular epithelial cells and the well-developed muscular layers were seen in the wall of the cysts (Figure 27-3C).
   b. *Small intestine.* In the original grafts, four layers of the wall lined with the multilayered columnar epithelium were seen (Figure 27-4A). After grafting, a mass of multiple cysts quite similar to the grafted stomach developed. The wall of the cysts showed the typical structure of the fetal intestine, such as presence of the villi and simple columnar epithelium, appearance of goblet cells and other glands, and clear-cut existence of nerve fibers (Figure 27-4B).

An interesting observation in the case of the grafted livers (see Figure 27-5A) was that in 4 of 15 grafts a mass of tissues grew, although no signs of hepato-

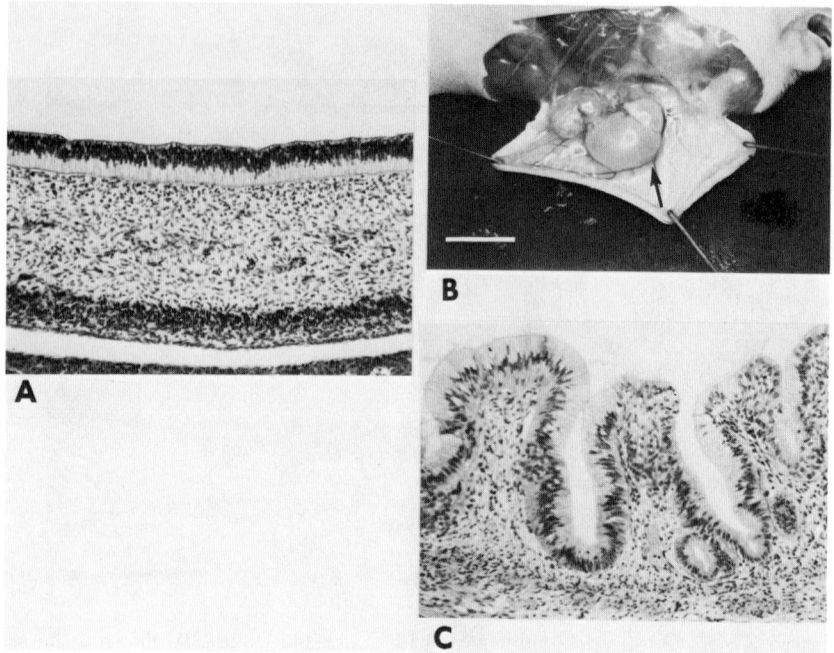

Figure 27-3. (A) Wall of stomach lined with simple columnar epithelium (Carnegie stage 20) showing the state of the original graft. ×175. (B) Gross appearance of multiple cysts (arrow) developed at 74 days after subcutaneous grafting of (A) staged stomach. (C) Wall of cysts showing gastric pits and glandular epithelial cells at 58 days after subcutaneous grafting of (A) staged stomach. ×175.

cytes were seen; in two of these grafts bile ducts surrounded by fibrous tissues were noticed, whereas in the two other cases either round cartilaginous bodies or a mass of fibrous tissues were observed.

Intratesticular Grafting

The intratesticular grafting studies differed from the earlier subcutaneous studies in that smaller numbers of animals were used and observation was made with a shorter time interval after the grafting.

Acceptability. The results of such experiments are shown in Table 27-2. Similar to the results obtained with subcutaneous grafting, three organs, the limb bones, lung, and gastrointestinal tract, were accepted in the majority of cases. The detailed data also revealed a tendency for the organs from older embryos to be more successfully grafted. Unlike the subcutaneous grafting, eyeballs transplanted into the testes were accepted in three of four cases. The other organs showed poor or no acceptability.

Differentiation of the accepted organs. Generally, the results after grafting of the limb bones, lung, and gastrointestinal tract were similar to those

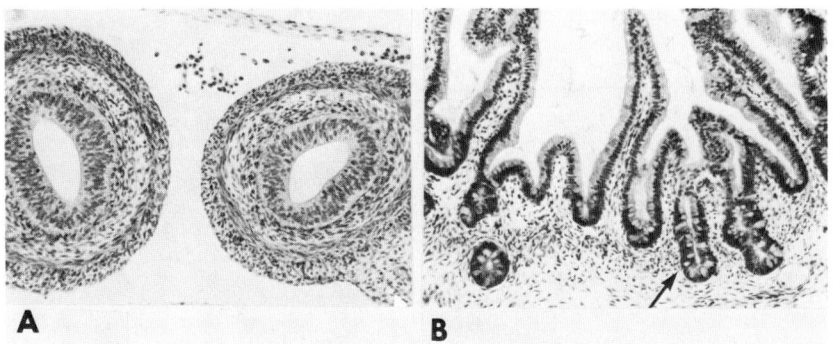

Figure 27-4. (A) Small intestine (Carnegie stage 20) showing the stage of the original graft. ×175. (B) Wall of cyst showing villi lined with simple columnar epithelium, goblet cells, and other glanduar epithelium (arrow) at 58 days after subcutaneous grafting of (A) staged small intestine. ×175.

obtained using subcutaneous grafts except that the limb bone primordia had grown cartilage but no ossified region, the stomach showed no gastric pits, and one successful intestinal graft had neither villi nor goblet cells. Such differences between intratesticular and subcutaneous grafts may result from an earlier stage of sacrifice in the former.

Grafting of the embryonic livers at the stage shown in Figure 27-5A yielded the results similar to these shown using subcutaneous grafts; all 6 livers after grafting showed no hepatocytes, but one of those indicated formation of the

Table 27-2. Acceptability of human embryonic organs grafted into the testis of nude mice (BALB/cA-nu/nu)

Grafted embryonic organ	Carnegie stage of embryo [standard conceptional age in days (4)]	Time interval between grafting and sacrifice (days)	Proportion of grafts accepted, accepted/total No. (%)
Limb bones	14–20 (35–47)	10–44	4/7 (57)
Lung	16–21 (38–48)	14–44	3/4 (75)
Gastrointestinal tract (stomach and/or intestine)	17–21 (39–48)	12–30	4/6 (67)
Liver	15–18 (37–41)	10–44	0/6[a]
Heart	14–21 (35–48)	10–21	0/5
Kidney	19–21 (45–48)	12–43	2/5 (40)
Gonad and its duct system	18–21 (41–48)	37–51	0/3
Adrenal	19–21 (45–48)	12–24	1/3 (33)
Brain	18–20 (41–47)	14–31	1/2 (50)
Eyeball	16–20 (38–47)	31–51	3/4 (75)

[a] Presence of the tissues other than the parenchyma are ignored.

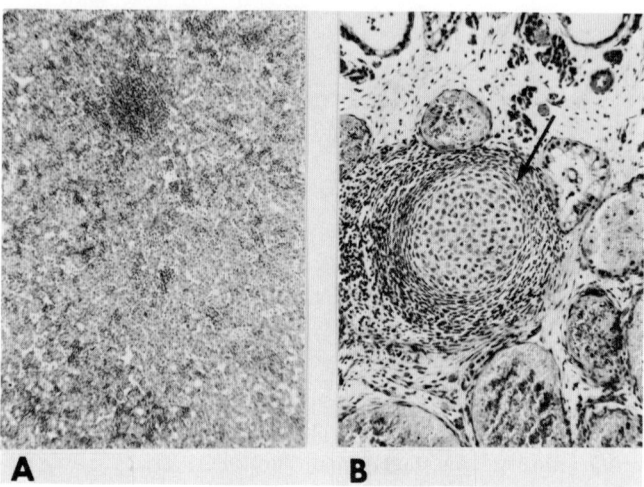

Figure 27-5. (A) Liver (Carnegie stage 20) showing the stage of the original graft. ×94. (B) Development of cartilage (arrow) at 10 days after intratesticular grafting of (A) staged liver. ×188.

round cartilaginous bodies (Figure 27-5B) and another liver developed the bile ducts.

The eyeballs, especially their neural ectoderm, showed an interesting finding of differentiation. The outer layer of the original graft, as shown in Figure 27-6A, developed into an irregularly arranged net, and pigmentation in the net was greatly increased. The arrangement of the retinal layers became irregular and showed rosette formation following grafting, and there appeared to be an increase in the number of cell layers (Figure 27-6, A and B). In two cases lenses were seen showing some growth and disappearance of their cavity.

Discussion

The important results can be outlined as follows.

Three embryonic organs, limb bones, lung, and gastrointestinal tract, grafted subcutaneously into nude mice were accepted in almost 100% of cases and maintained for up to 18 weeks. Interestingly, intratesticular grafting resulted in a high acceptability of eyeballs. Differentiation of the accepted organs occurred in principle in a normal manner with some irregularity. Grafting of the embryonic livers occasionally resulted in development of cartilage or of bile duct.

Our findings with grafted human limb bones, gastrointestinal tract, liver, and eyeballs are the first to be reported. The characteristic development of multiple cysts of the grafted gastrointestinal tract is in accord with the report that the fetal mouse digestive tracts transplanted into syngeneic mice showed organlike growth. In the case of lung, our results are similar to those previously published

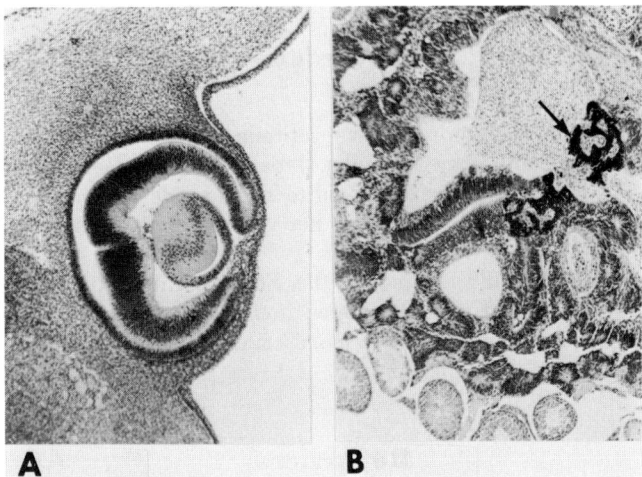

Figure 27-6. (A) Eyeball (Carnegie stage 18) showing the stage of the original graft. ×94. (B) Differentiated neuroectoderm of eye showing netlike arrangement of the outer layer with markedly increased pigments (arrow) and irregularly arranged and increased retinal layers with rosette formation at 44 days after intratesticular grafting of (A) staged eyeball. ×94.

(5). It should be noted that occasional success of the subcutaneous grafts of gonad and adrenal obtained by earlier workers (5) was not seen in the present studies, possibly because donor embryos were much younger.

Our results as a whole suggest that the present approach of human embryonic organ grafting is realistic for several selected organs. There is a prospect that other organs could be more easily transplanted if other sites of grafting, such as anterior eye chamber (3) are adopted. It should be added that to acquire human embryonic abortuses suitable for this approach the procedures of dilatation and curettage, hysterotomy, or hysterectomy are necessary. In Japan at least, use of the method of dilatation and curettage is likely to be continued, enabling further research to be done.

It should be emphasized that the value of this approach is to enable investigators to conduct experiments with developing human organs. Compared to the method of organ culture, the approach of grafting seems to have three advantages: First, the tissues grow under more physiological conditions, second, it enables observation of long-term differentiation, and, third, the effect of chemical substances can more easily be studied, since they can be administered systemically even in an insoluble form.

In conclusion, application of this technique is expected to be useful for studying the physiology, metabolism, and teratology in developing human organs. In addition, this method may provide a clue to comparison of various developmental phenomena between human and laboratory animals. Also, important information regarding infections with microbes which affect preferentially human organs, such as influenza virus B in respiratory organs and Rota viruses in intestines, can be obtained.

Acknowledgments

The authors wish to express their gratitude to Mr. Minoru Matsuura in Shionogi Research Laboratory in Osaka, Japan, for his generous assistance in preparation of histological specimens and to Dr. Yvonne Rosenberg of the U.S. National Institutes of Health (NIH) for her valuable help in the preparation of the manuscript.

The authors also greatly appreciate the Fogarty International Scholarship, U.S. National Institutes of Health, granted in 1979 to one of us (H.N.), which facilitated the work of completing the manuscript.

References

1. Bastert, G., P. Althoff, K.H. Usadel, H.P. Fortmeyer, and H. Schmidt-Matthiesen. Heterotransplantation of human fetal pituitaries in nude mice. Endocrinology 101:365–368, 1977.
2. Bowker, C.H., and P. Turmer. Human fetal pancreas transplants in nu/nu mice. Lancet February 12:365, 1977.
3. Gallie, B.L., D.M. Albert, J.J.Y. Wong, N. Buyukmihci, and C.A. Puliafito. Heterotransplantation of retinoblastoma into the athymic "nude" mouse. Invest. Ophthalmol. Vis. Sci. 16:256–259, 1977.
4. Nishimura, H., K. Takano, T. Tanimura, and M. Yasuda. Normal and abnormal development of human embryos: First report of the analysis of 1,213 intact embryos. Teratology 1:281–290, 1968.
5. Povlsen, C.O., N.E. Shakkebaek, J. Rygaard, and G. Jensen. Heterotransplantation of human foetal organs to the mouse mutant nude. Nature (London) 248:247–249, 1974.
6. Skakkebaek, N.E., G. Jensen, C.O. Povlsen, and J. Rygaard. Heterotransplantation of human foetal testicular and ovarian tissue to the mouse mutant nude. Acta Obstet. Gynecol. Scand. (Suppl.) 29:73–75, 1974.
7. Zinzar, S.N., G.J. Svet-Moldavsky, B.I. Leitina, and B.G. Tumyan. Enormous organ-like growth of transplants of fetal digestive tract. Transplantation 11:499–502, 1971.

General Discussion

PESCE: We have observed that several of our human bladder tumor cell lines form cysts rather than viable tumors. (1) Do certain organs form cysts? (2) Do you have any suggestions as to the biology of the cause of cyst formation?

NISHIMURA: (1) The accepted stomach and intestine developed into a mass consisting of multiple cysts regardless the routes of grafting, and their wall showed the typical fetal gastrointestinal structures. (2) The above-mentioned findings seem to suggest that minced small pieces of the wall of gastrointestinal tract maintain a potency of forming the structure with lumens.

OUTZEN: Have you observed any neoplastic transformation in your transplants similar to that reported by Svet-Moldovsky on transplantation of mouse embryonic tissues into syngeneic mice?

NISHIMURA: We have adopted the method of intratesticular grafting in addition to the subcutaneous route following the suggestion of Dr. Stevens in Bar Harbor and expected the occurrence of neoplastic transformation of some of the grafted organs. However, so far, we have never encountered such a case.

RYGAARD: We have been transplanting such human embryonic organs obtained from abortions performed by curettage or surgery [Povlsen et al., Nature (London) 248:247–249, 1974]. Now, however, abortion is induced with prostaglandins in all clinics. We tested the viability of such aborted fetuses immediately after spontaneous delivery. We found viable tissue in only 2 of 18 cases (Aagaard, Andersen, and Rygaard, unpublished data). I mention this as a word of warning—before starting a study of embryonic tissue be sure the method of abortion is appropriate and not going to be changed.

UNIDENTIFIED: A comment to Dr. Rygaard. If you wait long enough they will quit the prostaglandins—the side effects are too bad. I have done plaque-forming cells in such fetuses and indeed the viability is very low in spleen cells. A question for Dr. Nishimura. Have you done any electron microscopy on the tissues transplanted to the nude mice? Again, thinking about picking up virus particles.

NISHIMURA: No.

HELSON: Can you tell us if some of the brains took in the intratesticular site? You mentioned they didn't take subcutaneously.

NISHIMURA: Brain was accepted in one of two cases (Table 27-2).

28

Athymic Nude Mice: Ex Vivo In Vivo Models to Study the Development, Growth, and Differentiation of the Normal and Neoplastic Xenogeneic Mammary Gland

Clifford W. Welsch* and M. Jean McManus

Department of Anatomy, Michigan State University, East Lansing, Michigan 48824.

Abstract

Human breast tissues. Slices from 14 benign breast tumors were grafted subcutaneously (SC) to 20 adult female nude mice, 10 grafts per mouse. Five grafts were removed from each mouse at 30 days. Five mice were injected SC with human placental lactogen (HPL) (0.5 mg twice daily) for 18 days (days 42–60). The remaining grafts were taken from HPL-treated and control mice at 60 days. DNA synthesis of grafts was determined by [^{3}H]thymidine radioautographic analysis at 30 and 60 days. Mean change in [^{3}H]thymidine labeling index before and after treatment was +0.4 for controls and +20.4 for HPL-treated group ($P < 0.01$).

Bovine mammary tissues. Slices of bovine mammary gland from a midpregnant Jersey cow were grafted SC to 15 adult female mice, 10 grafts per mouse. After 30 days the mice were divided into three groups and received the following daily injections for 10 days: (1) controls, 0.9% NaCl; (2) bovine growth hormone (GH) (1 mg) + bovine prolactin (PRL) (1 mg) + 17β-estradiol (E) (1 μg) + progesterone (P) (1 mg); and (3) GH + PRL + E + P followed by ovariectomy and then GH + PRL + hydrocortisone (200 μg) for 7 days. Five grafts were removed before treatment and the remaining grafts were removed after treatment. Grafts of groups 1 and 2 were analyzed for DNA synthesis and grafts of group 3 were analyzed for α-lactalbumin by RIA. Mean change in [^{3}H]thymidine labeling index from pre- to posttreatment for controls was +2.6 and for hormone treatment was +30.0 ($P < 0.001$). Mean α-lactalbumin increased from 0.2 to 57.2 μg/mg tissue ($P < 0.001$). These results provide evidence that the nude mouse can accept and maintain mammae from foreign species and under suitable hormonal stimulus allow these tissues to grow and differentiate, simulating that which occurs in situ.

---

* To whom correspondence should be addressed.
© 1982 Gustav Fischer New York, Inc.
Proceedings of the Third International Workshop on Nude Mice.

Introduction

Normal and hyperplastic (benign) breast tissues, freshly exercised from human patients, have been successfully transplanted and maintained in the untreated athymic nude mouse (4). Although these transplants are readily accepted and retain their original morphology, they do not appear to grow; their ductal epithelium only exhibits a low maintenance rate of DNA synthesis (2).

During the past 2 years, our laboratory has initiated experiments designed to determine whether xenografts of mammary tissues maintained in the nude mouse have the capability of responding to endocrine manipulation of the host animal. In essence, are the grafted mammary tissues, derived from foreign species, capable of growth and differentiation in the presence of the appropriate hormonal milieu? Can these tissues be induced to function, qualitatively and quantitatively, as they did in the donor animals? In this communication, we provide evidence that not only can the nude mouse accept and maintain mammae from foreign species but, under the appropriate hormonal stimulus, these tissues can be induced to grow, simulating both quantitatively and qualitatively what is observed in situ, i.e., in the donor animal.

Materials and Methods

Human breast tissues. Fourteen human benign breast tumor biopsy specimens (HBT) (fibroadenoma or fibrocystic disease) were obtained from 14 women ranging in age from 19 to 66 years. The HBT were processed, according to the method described previously (1), into 20–40 slices (approximately $4.0 \times 4.0 \times 0.1$ mm) per specimen for transplantation into adult female nude mice (BALB/c background). The HBT slices were transplanted SC in the dorsal area of 20 mice, 10 slices per mouse.

Five grafts were removed from each mouse 30 days after transplantation. On day 42 posttransplantation, five mice, each grafted with one HBT (one mouse per biopsy, five HBT), received twice daily injections of 0.5 mg human placental lactogen (HPL) (Nutritional Biochemical Corp., Cleveland, Ohio). Four mice were injected for 18 days and one mouse for 14 days. For controls, 15 mice (9 HBT) received no treatment. At the termination of HPL treatment or at 60 days after HBT transplantation, the remaining grafts were removed from all mice. The 30- and 60-day grafts (pre- and posthormone treatment) were processed for [^{3}H]thymidine radioautographic and histological analyses as previously described (2). The differences in labeling indices (LI) [number of [^{3}H]thymidine labeled epithelial cells per unit area of epithelium (0.1 mm^2)] between 30 and 60 days were calculated and significance of mean changes between HPL-treated and untreated mice was determined by the Student's t test.

Bovine mammary tissues. A midpregnant Jersey cow was sacrificed and a portion of the mammary gland was processed, according to the method described previously (1), into $5.0 \times 5.0 \times 0.1$ mm slices for transplantation into

15 adult female "nude" mice (BALB/c background). The slices were transplanted SC into the dorsal area of each mouse, 10 slices per mouse.

Thirty days after grafting, the 15 graft-bearing mice were divided into three groups of five each. Mice of group 1 (controls) were given SC injections once daily for 10 days with 0.9% NaCl solution. Mice of group 2 were given SC injections once daily for 10 days with the following hormones: bovine growth hormone (GH) (NIH-B17, 1.0 mg/mouse/day); bovine prolactin (PRL) (NIH-B5, 1.0 mg/mouse/day); 17β-estradiol (E) (Nutritional Biochemical Corp., Cleveland, Ohio, 1.0 μg/mouse/day) and progesterone (P) (Nutritional Biochemical Corp., Cleveland, Ohio, 1.0 mg/mouse/day). Mice of group 3 were given SC injections once daily for 10 days with GH + PRL + E + P, after which they were ovariectomized and then given SC injections once daily for 7 days with GH + PRL + hydrocortisone (HC) (Nutritional Biochemical Corp., Cleveland, Ohio; 200 μg/mouse/day).

Five grafts were removed from each mouse before hormone treatment on day 30 posttransplantation and the remaining five grafts were removed immediately after hormone treatment. Grafts of groups 1 and 2 were processed for [^{3}H]thymidine radioautographic and histological analyses as previously described (2,6). Grafts of group 3 were processed for α-lactalbumen analyses by radioimmunoassay (RIA) as previously described (6). The data were analyzed by Student's t test for paired observations and analysis of variance with Dunnett's t test (radioautography) and one-way analysis of variance (RIA).

Results

Human Breast Tissues

Figure 28-1 illustrates the changes between 30 and 60 days in [^{3}H]thymidine labeling index (LI) of grafts from 14 transplanted HBT. Five mice (one HBT per mouse) received HPL injections from day 42 to day 60 (four mice) or from day 42 to day 56 (one mouse). Mean change in the LI of grafts from the HPL-treated mice was +20.4, in contrast to only +0.4 in controls ($P < 0.01$). HBT of HPL-treated mice were diagnosed as fibrocystic disease (HBT 7, 8, and 10) and fibroadenoma (HBT 9 and 11). HBT of controls included eight fibrocystic disease and one fibroadenoma. All 14 HBT were readily accepted by the mice; the original morphology was maintained in all transplants.

Bovine Mammary Tissue

Figure 28-2 illustrates the changes between 30 and 40 days in [^{3}H]thymidine labeling index (LI). The hormonal combination significantly ($P < 0.001$) increased the LI above control values. The LI of the original biopsy specimen (day 0) was 40.6 ± 6.7, a value not significantly different from the LI of the hormone-treated group after hormone treatment (37.9 ± 13.5). The LI of the control group was 11.9 ± 3.4. Morphologically, the grafts in the hormone-treated group showed more evidence of epithelial hypertrophy and secretion than

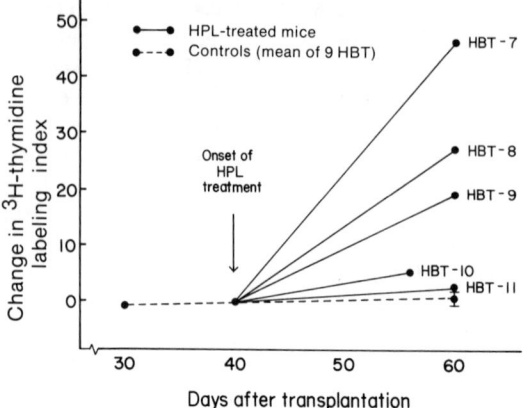

Figure 28-1. Changes in [³H]thymidine labeling indices (LI) of human benign breast tumor (HBT) grafts maintained in the athymic nude mouse after treatment with human placental lactogen (HPL). LI = number of [³H]thymidine labeled epithelial cells per unit area of epithelium. Day 30 LI: controls, 9.3 ± 1.3; HBT-7, 4.0; HBT-8, 6.7; HBT-9, 8.6; HBT-10, 3.3 and HBT-11, 7.1. HPL vs. controls, $P < 0.01$. [From ref. (2).]

did the grafts in the control group. Morphological viable duct and alveolar epithelium similar to that in the original biopsy specimen was present in all of the grafts removed from each mouse.

A highly significant ($P < 0.001$) increase in α-lactalbumen content of grafts

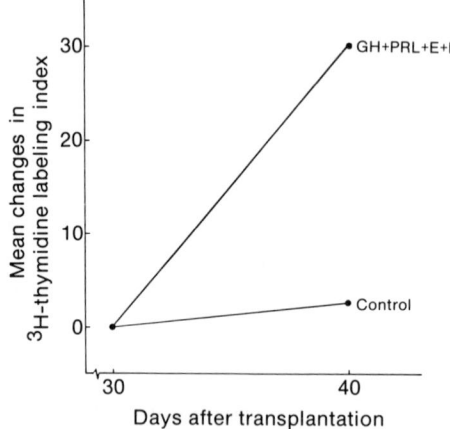

Figure 28-2. Effects of injections of bovine growth hormone (GH), bovine prolactin (PRL), 17β-estradiol (E), and progesterone (P) on [³H]thymidine labeling index of grafts of bovine mammary gland maintained in the athymic nude mouse. LI = number of [³H]thymidine labeled epithelial cells per unit area of epithelium. Day 30 LI: controls, 9.3 ± 2.0; hormone treatment, 7.9 ± 2.0. Day 40 LI: controls, 11.9 ± 3.4; hormone treatment, 37.9 ± 13.5. GH + PRL + E + P vs. controls, $P < 0.001$. [From ref. (6).]

Table 28-1. Effects of injections of 17β-estradiol, progesterone, bovine prolactin, and bovine growth hormone (days 30–39) followed by bovine prolactin, bovine growth hormone, and hydrocortisone injections (days 40–47) on α-lactalbumin synthesis in grafts of bovine mammary gland maintained in the athymic nude mouse[a]

Nude mouse number	Number of grafts Day 30	Day 47	α-Lactalbumin/mg tissue (μg) Onset of hormone treatment (day 30)	Termination of hormone treatment (day 47)
1	5	5	0.1	64.7
2	5	6	0.2	8.9
3	5	6	0.2	40.3
4	5	5	0.2	45.4
5	5	4	0.2	104.3
			$\bar{m} = 0.2$[b]	52.7[b]

[a] From ref. (6).
[b] Mean α-lactalbumin content of the original biopsy specimen (day 0) was 5.1 μg α-lactalbumin/mg tissue. $P < 0.001$.

obtained after hormonal treatment (day 47), when compared with those obtained at the onset of hormone treatment (day 30), was observed (Table 28-1). The grafts of all five nude mice responded positively to the hormonal treatment. More than 250 times more α-lactalbumin was observed in the grafts after hormone treatment, a quantity of α-lactalbumin over 10 times greater than that observed in the original biopsy specimen.

Discussion

The results reported in this study clearly show, for the first time, that xenografts of mammary tissue transplanted to the nude mouse, can respond with significant growth and differentiation in the presence of the appropriate hormonal milieu. The DNA-synthetic activity of the bovine mammary grafts prior to hormone treatment was quite low; the LI ranged from 7.9 to 9.3, values one-fifth to one-fourth of that which was observed in the original day 0 biopsy specimen (40.6). After 10 days of injections of bovine growth hormone, prolactin, 17β-estradiol, and progesterone, the LI increased to 37.9, a value statistically indistinguishable from that observed in the day 0 biopsy specimen. Bear in mind that the biopsy specimen was obtained from a cow in midpregnancy; thus the bovine mammary gland was already in a state of vigorous growth at the onset of grafting. That the LI returned to "normal" after hormone administration provides evidence indicating that the nude mouse is capable of supporting growth of the bovine mammary gland to a degree quantitatively comparable to that which occurs in situ. The hormonal induction of α-lactalbumin synthesis, a milk-specific protein, was also very striking. The α-lactalbumin content of the grafts was increased 10 times when compared with the original biopsy specimen and more than 250 times when compared with the prehormonal treatment grafts. It is clear, therefore that the hormonally treated athymic nude mouse is not only capable of maintaining growth of the xenogenic mammary gland but in addition is capable of supporting differentiation (lactation) as well.

We have reported previously that biopsy specimens of normal, hyperplastic (benign dysplasias) and carcinomatous human breast specimens can be successfully grafted to the nude mouse (1,2,4). The results of this study provide evidence that grafts derived from benign human breast biopsy specimens can be stimulated to grow in the nude mouse by the exogenous administration of HPL. In the results shown in Figure 28-2 the HBT sample with the greatest response to HPL (change in LI pre- to posttreatment = +46.2) (HBT-7) demonstrated a comparable increase in growth to that of bovine mammary tissue exposed to both steroid and pituitary hormones. Therefore, it appears that the human breast not only can be maintained in the nude mouse but in addition can respond to a hormonal growth stimulus. HPL has been reported to stimulate growth of normal and hyperplastic human breast tissues in organ cultures (3,5) and this hormone is chemically and physiologically similar to prolactin, a pituitary peptide important in murine mammary tumorigenesis (7). We have not yet tried to induce differentiation of human breast tissue grafts in nude mice because of a lack of sufficient quantities of purified human prolactin and human growth hormone, pituitary peptides probably crucial for complete growth and differentiation of the human breast. When these pituitary peptides become available in sufficient quantities for in vivo studies, then complete growth and differentiation of human breast tissue grafts in the nude mouse should be possible, as we have shown with bovine mammary gland grafts in this communication. Once we are able to demonstrate hormonally induced growth and differentiation of the human breast in the nude mouse, we will have for the first time an ex vivo in vivo model to study development, growth, differentiation, and ultimately cancerigenesis of the human breast.

Acknowledgment

Work reported here was supported by Research Grant BC-220C from the American Cancer Society.

References

1. McManus, M.J., S.E. Dombroske, M.M. Pienkowski, T.M. Anderson, L.C. Mann, J.S. Schuster, L.L. Vollwiler, and C.W. Welsch. Successful transplantation of human benign breast tumors in the athymic nude mouse and demonstration of enhanced DNA synthesis by human placental lactogen. Cancer Res. 38:2343–2349, 1978.
2. McManus, M.J., and C.W. Welsch. DNA synthesis of benign human breast tumors in the untreated athymic "nude" mouse: And in vivo model to study hormonal influences on growth of human breast tissues. Cancer 45:2160–2165, 1980.
3. Welsch, C.W., S.E. Dombroske, and M.J. McManus. Effects of insulin, human placental lactogen and human growth hormone on DNA synthesis in organ cultures of benign human breast tumours. Br. J. Cancer 38:258–262, 1978.

4. Welsch, C.W., B.J. Mann, and M.M. Pienkowski. Successful transplantation of hyperplastic and carcinomatous human breast tissues to the athymic "nude" mouse, p. 58. *In* Fifth International Congress of Endocrinology, Hamburg, Germany, July 18–24, 1976.
5. Welsch, C.W., and M.J. McManus. Stimulation of DNA synthesis by human placental lactogen or insulin in organ cultures of benign human breast tumors. Cancer Res. 37: 2257–2261, 1977.
6. Welsch, C.W., M.J. McManus, J.V. DeHoog, G.T. Goodman, and H.A. Tucker. Hormone-induced growth and lactogenesis of grafts of bovine mammary gland maintained in the athymic "nude" mouse. Cancer Res. 39:2046–2050, 1979.
7. Welsch, C.W., and H. Nagasawa. Prolactin and murine mammary tumorigenesis: A review. Cancer Res. 37:951–963, 1977.

General Discussion

HONG: You'll hate this question—but do the grafted mammary glands make secretory IgA? Are there lymphoid cells?

McMANUS: I do not know. We have done no immunological studies on the bovine or human grafts from the nude mouse. It would indeed be interesting to know whether IgA, if present in the lacting bovine grafts, is of mouse or bovine origin. There are considerable numbers of lymphoid cells in most of the bovine and some of the human grafts. We assume that these originate from the host. (Immunofluorescence studies indicate that the stroma, but not the epithelium, of grafted human breast tissue may be replaced by nude mouse stromal components.)

29

Endocrine Morphology and Reproductive Function in Athymic Nude Mice

Kowetha A. Davidson,* J. Michael Holland, Jerry W. Hall, and Lawrence C. Gipson

Biology Division, Oak Ridge National Laboratory, Oak Ridge, Tennessee 37830.

Abstract

Athymic BALB/c nude mice were reared and maintained in a specific pathogen-free (SPF) environment in order to assess the impact of genetic athymia on endocrine morphology and reproductive performance. Mice were killed at 4, 8, 12, 16, and 19 weeks of age. Kidneys, testes, ovaries, adrenals, thyroid, and submaxillary gland were processed for histology. Plasma content of estradiol (females) and testosterone (males) was determined. Plasma levels of estradiol and testosterone were comparable in nude and normal mice. The morphology of the ovaries of nude females was similar to normal females, although fewer nude females (42%) had corpora lutea at 8 weeks than normal females (84%). Degeneration of the adrenal x zone was slightly retarded in nude females. The thyroid gland was normal in nude mice; age-related changes were observed in both phenotypes. Sexual dimorphism was observed in the morphology of the kidney and submaxillary glands in nude males. The appearance of these dimorphic characteristics corresponded with the apearance of mature spermatozoa in the seminiferous tubules of the testes (8 weeks). Male and female nude mice were fertile; however, the survival of young to weaning was less in litters from nude females (28%) than from normal females (42%). These results suggest that abnormalities in endocrine morphology and reproductive performance are less severe in nude mice reared and maintained in SPF environment than those reported by other investigators.

Introduction

The athymic nude mouse is a pleiotropic mutant that is hairless and has reduced body growth and low fertility (4). Thymus development is arrested

---

* To whom correspondence should be addressed.
© 1982 Gustav Fischer New York, Inc.
Proceedings of the Third International Workshop on Nude Mice.

during embryogenesis, and the nude mouse does not have an effective thymus organ (10,12); consequently immune function is impaired (6,11). The impairment of normal immune function should establish this mutant as an excellent animal model for studying thymic–endocrine interactions. There is evidence which does suggest that the thymus modulates endocrine activity in the nude mouse. Abnormalities have been observed in the morphology of the adrenal cortex (17), thyroid gland (14), and ovaries (7). The absence of sexual dimorphism in submaxillary gland and kidneys has also been reported (17) as well as alterations in plasma levels of testosterone, estradiol, progesterone, corticosterone, and thyroxine (13).

Nude mice are highly susceptible to stresses of infectious diseases and wasting (5); therefore it is possible that endocrine as well as reproductive abnormalities previously observed may result from environmental stresses. In order to evaluate the effect of athymia on endocrine and reproductive function, nude mice should be reared and maintained in a carefully controlled pathogen-free environment so that environmental stresses are minimized. In this study we have assessed the effect of genetic athymia on morphology of endocrine organs and reproductive performance. Our observations suggest that abnormalities are less severe than those reported by other investigators.

Materials and Methods

Animal housing and experimental procedure. Athymic nude mice were produced by mating heterozygous normal females with heterozygous normal males. Breeding pairs had been backcrossed with inbred BALB/c mice (N10). These mice were bred and reared in the specific pathogen-free (SPF) barrier facilities at the Biology Division, Oak Ridge National Laboratory. The animals rooms were maintained at 26.7°C (80°F). Food and water were given ad libitum. Nude male, female, and normal littermates were killed at 4, 8, 12, 16, and 19 weeks of age. The mice were removed from the SPF facility the morning of sacrifice. They were weighed, anesthetized with methoxyflurane, and bled by cardiac puncture. Plasma was collected and stored at $-70°$ until assayed for hormone content. Adrenals, thyroid, submaxillary gland, and kidneys were taken for histological examination. In addition, the ovaries and reproductive tract were removed from females and testes from males.

Reproductive performance was determined by mating 9- to 10-week old females with males in the following combinations: $nu/nu \times +/?$, $+/? \times +/?$, $+/? \times nu/nu$. After pregnancy was observed the mating pairs were checked daily for birth of litters. The females were allowed to nurse their litters until weaning, at which time nude females were killed and examined for the presence of a thymus.

Radioimmunoassay. Plasma content of 17β-estradiol and testosterone (androgens) was determined by radioimmunoassay using rabbit antisera against 17β-estradiol-6-carboxymethyloxime:BSA (1:25,000) and testosterone-3-carboxymethyloxime:BSA (1:10,000), respectively (antigens purchased from Steraloid, Wilton, New Hampshire). The procedure for estradiol and testosterone radioimmunoassays was a modification of that reported by Cameron and Jones (2).

Table 29-1. Body weights of athymic nude mice and normal littermates

Age (weeks)	Females[a]		Males[a]	
	Nude	Normal	Nude	Normal
4	13.0 ± 0.28 (22)[b]	15.9 ± 0.31 (26)	12.3 ± 0.55 (13)	18.1 ± 0.79 (10)
8	18.6 ± 0.35 (10)	20.6 ± 0.29 (10)	23.1 ± 0.38 (10)	25.4 ± 0.58 (10)
12	20.4 ± 0.37 (10)	22.9 ± 0.41 (10)	25.2 ± 0.46 (7)	25.9 ± 0.83 (7)
16	22.6 ± 0.39 (10)	24.1 ± 0.61 (10)	24.9 ± 0.53 (10)	29.4 ± 0.53 (10)
19	21.4 ± 0.53 (5)	24.8 ± 0.49 (10)	26.2 ± 0.29 (7)	31.0 ± 0.80 (9)

[a] Mean weight in grams ± SE. $P \leq 0.01$ nude female vs. normal female except at 16 weeks of age (Student's t test). $P < 0.01$ nude male vs. normal male except at 12 weeks of age. $P \leq 0.05$ nude male vs. normal females except at 16 weeks of age.
[b] Number of animals.

Results

Growth and Reproduction

Body weights. The nude mice used in this study were healthy and showed no signs of wasting. When nude and normal mice were reared and maintained in a SPF environment, nudes were smaller than normal littermates of the same sex (Table 29-1). Mature nude males, 8 weeks and older, were consistently larger than both normal and nude females. Thus sexual dimorphism was observed in both nude and normal mice.

Steroid hormone levels. Mean plasma estradiol levels were low in both nude and normal females 4 weeks of age; nevertheless the levels were significantly higher in nudes than in normal females (Table 29-2). At 8 weeks of age mean estradiol levels were elevated twofold and threefold in nude and normal females, respectively. In both groups mean estradiol levels were lower in animals older than 8 weeks than in those 8 weeks of age. Estradiol levels in nude females were comparable to normal females and were not indicative of abnormalities in ovarian function.

Table 29-2. Plasma steroid hormone levels in athymic nude mice and normal littermates

Age (weeks)	Females (Estradiol)[a]		Males (Testosterone)[a]	
	Nude	Normal	Nude	Normal
4	30.3 ± 2.10[b] (12)[c]	23.2 ± 2.03 (13)	151 ± 12 (13)	253 ± 58 (10)
8	68.2 ± 7.54 (10)	75.5 ± 3.72 (10)	1157 ± 695 (10)	1829 ± 987 (10)
12	56.3 ± 4.26 (10)	52.4 ± 2.94 (10)	4844 ± 1663[b] (7)	497 ± 53 (7)
16	59.1 ± 2.26[b] (10)	46.3 ± 3.12 (10)	2029 ± 1146 (10)	2594 ± 1092 (10)
19	38.4 ± 2.60 (5)	40.7 ± 3.80 (10)	2004 ± 1501 (7)	4898 ± 1671 (9)

[a] Mean pg/ml ± SE.
[b] $P \leq 0.05$ nude vs. normal same sex (Student's t test).
[c] Number of assays.

Mean plasma testosterone levels are exceptionally low in both nude and normal males at 4 weeks of age (Table 29-2). By 8 weeks mean testosterone levels were elevated 7.7- and 7.2-fold in nude and normal males, respectively. Between 8 and 19 weeks plasma testosterone levels in nude males were not indicative of abnormal testicular function.

Reproductive performance (Table 29-3). These animals were observed from pregnancy to weaning of the first litter. Of the 10 nude and normal females paired with normal males nine females of each phenotype were observed to be pregnant and subsequently produced litters. Six of the 10 normal females paired with nude males became pregnant and produced litters. The average number of mice per litter was less in the nude females × normal male mating pairs than in the other two mating pairs, and postnatal deaths were exceptionally high. Only 28% of the young produced by nude females survived to weaning; 42% and 74% survived to weaning in the normal × normal and normal × nude mating pairs, respectively. Consequently the average litter size weaned from nude females was lower than in the other mating pairs. Five nude females nursed between one and three young to weaning. Six nude females were either pregnant for a second time or had produced a second litter when their first litters were weaned (21 days). Thus nude females were not only capable of supporting pregnancies from conception through parturition but they were also capable of suckling a limited number of young to weaning.

Morphology of Endocrine Organs

Nude and normal females. The ovaries in nude and normal females 4 weeks old were characterized by the absence of corpora lutea and the presence of numerous follicles, many of which were atretic. The ovaries were also compact and had very little interstitial tissue. At 8 weeks of age corpora lutea were observed in ovaries of 42% of the nude and 84% of the normal females. The ovaries were no longer compact and the amount of interstitial tissue was increased. There were fewer follicles than at 4 weeks of age. More atretic follicles were observed in ovaries of nude than in ovaries of normal females. Up to 19 weeks of age, changes in the ovaries of both phenotypes were indicative of changes in the estrous cycle as determined by vaginal and uterine histology.

Table 29-3. Reproductive performance of nude and normal littermates

	Nude ♀ × Normal ♂	Normal ♀ × Normal ♂	Normal ♀ × Nude ♂
No. of pairs	10	10	10
No. observed pregnant	9	9	6
No. of litters	9	9	6
No. live births	40	55	34
Litter size	4.4 ± 0.34[a]	6.1 ± 0.26[b]	5.7 ± 0.92
No. weaned	11	23	25
Litter size weaned	2.2 ± 0.49	5.75 ± 0.25[b]	5.0 ± 1.04

[a] Mean ± SE.
[b] $P \leq 0.05$ normal × normal compared with nude × normal.

Differences in adrenal gland morphology between nude and normal females were confined to the x zone. The x zone degenerated at a slower rate in nude females than in normal females. However, there were variations in the degree of degeneration of the x zone in both groups. Differences in zona glomerulosa, faciculata, and reticulata were not observed with the light microscope.

The morphology of the thyroid gland in nude and normal females was similar. The gland was composed of follicles of various sizes separated by thin layers of stroma. The cells of the follicles were simple cuboidal with centrally located nuclei. The inner portion of the follicles contained an eosinophilic colloid. As the animals aged some of the follicles enlarged and a larger portion of the follicular cells became more flattened in shape rather than cuboidal.

Nude and normal males. The seminiferous tubules in the testes of nude and normal males contained germ cells in various stages of spermatogenesis, although mature spermatozoa were not present at 4 weeks of age. Spermatids were observed in several sections of seminiferous tubules. Interstitial tissue was observed but was sparse. The testes in nude males were similar to normal males but fewer exhibited the presence of spermatids. At 8 weeks mature spermatozoa were observed in seminiferous tubules in both nude and normal males. The amount of interstitial tissue was increased relative to that observed at 4 weeks of age. Thus both nude and normal males were sexually mature by 8 weeks of age. Characteristic features observed in testes of both nude and normal males at 8 weeks did not change by 19 weeks of age.

The adrenal gland in nude and normal males was characterized by the absence of the x zone even at 4 weeks of age. The cortical zones were normal in appearance and no distinct changes were observed by 19 weeks of age.

The submaxillary gland in 4-week-old males of both phenotypes was similar to the gland in immature and mature females. The acini were composed of basophilic cells, which were pyramidal in shape. The terminal ducts were composed of eosinophilic cuboidal cells with basal striations and centrally located nuclei. The submaxillary gland in mature males was morphologically distinct from the gland in mature females and immature males. In both nude and normal males 8 weeks or older the terminal ducts were hypertrophied. The cells of the terminal ducts were columnar in shape rather than cuboidal. The nuclei of these cells were basally located and basal striations were absent. Thus the submaxillary gland in nude males was sexually dimorphic.

The kidney was similar in 4-week-old males and females of both phenotypes. At 8 weeks of age the dimorphic characteristic was observed. Bowman's capsule was partially or completely surrounded by a thickened parietal layer similar to the proximal convoluted tubules. This characteristic feature was observed in both nude and normal males between 8 and 19 weeks of age.

The morphology of the thyroid gland in nude and normal males was similar to the thyroid in females.

Discussion

Athymic nude mice reared and maintained in an environment that minimizes infectious disease and temperature stress exhibit less severe growth and endocrine abnormalities relative to those reported by other investigators. Flanagan (4) and Shire and Pantelouris (17) did not observe sex differences in body and kidney weights in nudes reared in conventional facilities. Sexual dimorphism is observed in nude mice reared in an SPF environment. The structure of the parietal layer of Bowman's capsule in mature normal and nude males is composed of a high cuboidal epithelium very similar to the proximal convoluted tubules, as described by Crabtree (3). Further confirmation of sexual dimorphism in nude males is found in the structure of the submaxillary gland, in which the terminal ducts were hypertrophied in mature nude males but not in immature males or mature females. Shire and Pantelouris (17), as well as Ruitenberg and Berkvens (16), did not observe sexual dimorphism in the submaxillary gland in nude males maintained in conventional facilities. It is possible that the lack of sex differences reported by other investigators may not result from athymia of the nude mouse but may be caused by adverse environmental factors which interfere with normal postnatal growth and development in athymic nude mice. However, when nude mice are reared and maintained in an SPF environment these abnormalities are significantly reduced.

It has been reported that the adrenal gland in the nude mouse is abnormal in structure (4,14,17). If the abnormalities of the adrenal x zone in nudes result indirectly from runting, as suggested by Shire and Pantelouris (17), then this may explain the absence of gross abnormalities in the adrenal gland of healthy nude mice used in this study.

Pierpaoli and Sorkin (14) observed that thyroxine levels are reduced in nudes and the thyroid gland appears structurally hypoactive in wasting nudes. In contrast, Ohsawa et al. (9) observed that thyroxine levels were normal in nude mice maintained under SPF conditions. More recently, Ruitenberg and Berkvens (16) also observed abnormalities in the morphology of thyroid glands in nudes 6 weeks and older. The morphology of the thyroid gland in nudes reared and maintained under SPF conditions is not different from that of normal mice. Age differences, however, were observed in both phenotypes.

In the present report steroid hormone levels are not indicative of abnormal gonadal function in males. Mean plasma testosterone levels are similar in both groups of males except at 12 weeks. The variability in testosterone levels is not unusual in mature males; Bartke et al. (1) observed a 30-fold difference in plasma testosterone levels in normal male mice housed and bled under identical conditions. The increase in plasma testosterone levels at 8 weeks of age coincides with the appearance of mature spermatozoa in the lumen of seminiferous tubules and the appearance of sexual dimorphic characteristics in kidneys and submaxillary glands. We also observed, as have other investigators (4,15), that nude males were fertile.

Lintern-Moore and Pantelouris (7,8) observed that the nude mouse ovary was smaller than that of the normal mouse, the amount of stroma was reduced, the follicles were contracted, and the number of growing follicles was reduced. They also observed that pregnant mare serum gonadotrophin (PMSG) stimulated follicle growth and inhibited follicular contraction. Weinstein (18), however, observed that serum follicle-stimulating hormone (FSH) and luteinizing hormone (LH) levels are similar in diestrus nude and normal females, but serum LH levels are not elevated in ovariectomized nude females. LH levels, however, were elevated in response to gonadotrophic hormone releasing hormone. These mice were maintained under conventional conditions.

Our results demonstrated that ovarian morphology and ovarian function as indicated by plasma estrogen levels are comparable in nude and normal females. Plasma estradiol levels are also significantly elevated in both phenotypes at 8 weeks of age. Although all nude females did not show definite signs of puberty, only half had at least one corpora luteum at 8 weeks of age. Thus it is possible that puberty was delayed in a portion of athymic nude mice.

The primary test for normal gonadal and reproductive function is the ability to support pregnancies from conception to parturition and support young during lactation. We observed that nude females reared and maintained under SPF conditions are capable of both. The number of mice per litter born and weaned, however, was reduced in nude females compared with normal females.

Using the various functional and morphological criteria described here, there is less evidence that there are endocrine and reproductive abnormalities in male and female athymic nude mice reared and maintained in an SPF environment. These data do not rule out more subtle abnormalities that are not demonstrated by the criteria used here or that require more time to be expressed. Nevertheless, the data indicate that abnormal endocrine and reproductive functions observed by other investigators may not result entirely from thymic aplasia; rather, they may result from environmental stress, to which these very sensitive mutants are susceptible.

Acknowledgments

This research was sponsored by the Office of Health and Environmental Research, U.S. Department of Energy, under contract W-7405-eng-26 with the Union Carbide Corporation.

References

1. Bartke, A., R.E. Steele, N. Musto, and B.V. Caldwell. Fluctuations in plasma testosterone levels in adult male rats and mice. Endocrinology 92:1223–1228, 1973.
2. Cameron, E.H.D., and D.A. Jones. Some observations on the measurement of estradiol-17β in human plasma by radioimmunoassay. Steroids 26:737–759, 1972.
3. Crabtree, C. Sex differences in the structure of Bowman's capsule in the mouse. Science 91:299, 1940.
4. Flanagan, S.P. 'Nude,' a new hairless gene with pleiotropic effects in the mouse. Genet. Res. 8:295–309, 1966.
5. Institute of Laboratory Animal Resources. A Report of the Committee on Care and Use of the "Nude" Mouse. Guide for the care and use of the nude (thymus-deficient) mouse in biomedical research. Inst. Lab. Anim. Resources News 19(2):3–20, 1976.
6. Kindred, B. Antibody response in genetically thymus-less nude mice injected with normal thymus cells. J. Immunol. 107:1291, 1971.
7. Lintern-Moore, S., and E.M. Pantelouris. Ovarian development in athymic nude mice. I. The size and composition of the follicle population. Mech. Ageing Dev. 4:385–390, 1975.

8. Lintern-Moore, S., and E.M. Pantelouris. Ovarian development in athymic nude mice. III. The effect of PMSG and oestradiol upon the size and composition of the ovarian follicle population. Mech. Ageing Dev. 5:33–38, 1976.
9. Ohsawa, N., F. Matsuzaki, K. Esaki, and T. Nomura. Endocrine functions of the nude mouse, pp. 221–226. In J. Rygaard, and C. O. Povlsen (eds.), Proceedings of the first international workshop on nude mice. Stuttgart: Gustav Fischer Verlag, 1974.
10. Pantelouris, E.M. Absence of thymus in a mutant mouse. Nature (London) 217:370–371, 1968.
11. Pantelouris, E.M. Observations on the immunobiology of 'nude' mice. Immunology 20: 247–252, 1971.
12. Pantelouris, E.M., and J. Hair. Thymus dysgenesis in nude (nu/nu) mice. J. Embryol. Exptl. Morphol. 24:615–623, 1970.
13. Pierpaoli, W., and H.O. Besedovsky. Role of the thymus in programming of neuroendocrine functions. Clin. Exptl. Immunol. 20:323–338, 1975.
14. Pierpaoli, W., and E. Sorkin. Alterations of adrenal cortex and thyroid in mice with congenital absence of thymus. Nature (London) 238:282–285, 1972.
15. Poiley, W.M., A.A. Ovejera, A.P. Otis, and C.R. Reeder. Reproductive behavior of athymic nude (nu/nu-BALB/c/A/BOM Cr) mice in a variety of environments, pp. 189–202. In J. Rygaard, and C.O. Povlsen (eds.), Proceedings of the first international workshop on nude mice. Stuttgart: Gustav Fischer Verlag, 1974.
16. Ruitenberg, E.J., and J.M. Berkvens. The morphology of the endocrine system in congenitally athymic (nude) mice. J. Pathol. 121:225–231, 1977.
17. Shire, J.G.M., and E.M. Pantelouris. Comparison of endocrine function in normal and genetically athymic mice. Comp. Biochem. Physiol. 47A:93–100, 1974.
18. Weinstein, Y. Impairment of the hypothalamo-pituitary-ovarian axis of the athymic "nude" mouse. Mech. Ageing Dev. 8:63–68, 1978.

General Discussion

SORDAT: Do you have any comments on the poor lactation of nudes?

DAVIDSON: We did find that the largest number of young that were suckled was three. I don't know if larger numbers would have been suckled if they had survived more than 48 h or not—but nude females were able to suckle at least three young.

OHSAWA: At the first workshop in Aarhus, we reported similar results—that the endocrine function of nude mice, including pituitary growth hormone, thyroid, adrenal, and gonads, are essentially normal when SPF nude mice are used.

DAVIDSON: I have read your manuscript. We carried out this study to confirm those results and also to stress the point, for other investigators, that if nude mice are reared and maintained in an SPF environment then endocrine morphology and function are normal.

ROMIJN: We have very recently also started the determinations of plasma levels of steroid hormones. I'd like to ask if you have carried out also the assay of other hormones? I can imagine, for example, that abnormal levels of prolactin in the nude mother are responsible for the poor lactation.

DAVIDSON: No. We have not carried out assays of other hormones.

OUTZEN: In a paper by Eaton et al. in 1975 (Lab. Anim. Sci. 25:309) they noted that prolactin levels were essentially the same in nudes and $nu/+$ mice. Also, germ-free, SPF, and conventional female nudes were observed to be fertile and could wean their pups.

INDEX

Adjuvants
  alum, 148, 150, 151, 153, 239–241, 243
  Bordetella pertussis extract, 147–149, 151
  concanavalin-A, 149, 151
  lipopolysaccharide, 147, 149, 151
Anaphylaxis
  passive cutaneous, 148, 149, 239–241, 243
  Schultz-Dale reaction, 149
Antibodies
  ablastic, 111, 113–120
  alloantibody, 279, 280
  homocytotropic, 147–150, 152, 153, 239, 243
  monoclonal, W3/13, 44
  specific for,
    Ascaris, 147, 149, 151
    asialo GM1, 413–415, 419, 420, 509, 518, 522
    Bordetella pertussis, 147–149, 151
    bovine γ-globulin, 148
    brain–associated Thy 1, 413, 415, 417, 419
    egg albumin, 147–151, 153, 239
    erythrocytes
      burro, 217
      chicken, 217
      sheep, 45, 103, 113, 217
    Ia antigens, 413, 415
    keyhole limpet hemocyanin, 147, 151
    lipopolysaccharides, 147, 149, 151
    Ly antigens, 505, 509, 516, 518, 519, 522
    Nippostrongylus brasiliensis, 147–150, 154
    Plasmodium berghei, 91–94, 97, 98
    pneumococcal capsular polysaccharide, 147, 149, 151
    tetanus toxoid, 243
    ThB 413, 417, 419
    Thy 1.2, 263, 264, 266, 413, 415, 417, 419, 505, 509, 518 519, 522
    Vi antigen, 147, 151
Antigens and haptens
  Ascaris extract, 147, 149, 151
  Bordetella pertussis, 147–149, 151
  bovine γ-globulin, 148
  carcinoembryonic antigen (CEA), 527, 536, 537

dansyl, 263, 264, 268
DNP, 263, 268
egg albumin, 147–151, 153, 239
erythrocytes
  burro, 217
  chicken, 217
  sheep, 45, 103, 113, 217
keyhole limpet hemocyanin, 147, 151
lipopolysaccharides, 147, 149, 151
NIP, 263, 264, 268
Nippostrongylus brasiliensis, 147–150, 154
pneumococcal capsular polysaccharide, 147, 149, 151
sulfonate, 263, 264, 268
tetanus toxoid, 243
TNP, 263, 264, 268
tumor
  T-dependent, 367, 371, 380
  T-independent, 98, 367, 371, 375, 380
Vi antigen, 147, 151
Antilymphocyte serum, 379, 381, 383, 385, 386
Antimacrophage serum, 637, 638
Antimicrobics
  clindamycin, 59, 60, 62–64
  diaminodiphenylsulfone, 59, 60, 63, 64
  minocycline, 59, 63, 64
  OK-432, 79–82, 87, 88
  pyrimethamine, 91
  rifampicin, 59–62, 64
Antispleen serum, 381, 383, 387
Antithymocyte serum, 246, 391
Autoimmunity
  adoptive transfer, 249, 250
  diabetes mellitus, 255, 259, 260
  gastritis, 245, 246, 248, 249, 252
  oophoritis, 245–249, 252, 255
  pemphigus, 255, 258–260

Bacteria
  Corynebacterium kutscheri, 19
  Corynebacterium parvum, 165, 217–223
  Escherichia coli, 19
  Listeria monocytogenes, 73
  Mycobacterium leprae, 59–64, 189–191, 193–195
  Mycobacterium tuberculosis, 157–164

325

326   Index

Mycoplasma pulmonis, 19
Neisseria gonorrhoeae, 67, 73
Pasteurella pneumotropica, 19
Pseudomonas aeruginosa, 19
Salmonella spp., 19
Staphylococcus aureus, 12, 15, 19
Streptococcus pneumoniae, 67
Tyzzer's organism, 19, 43
B_2-microglobulin, 579, 580, 584, 585

Carcinogensis, 423–433
Carcinogens
   dimethylbenzanthracene (DMBA), 191, 193, 285, 423–427
   N-methyl-N-nitro-N-nitrosoguanidine (MNNG), 285, 291, 293
Cell lines
   BHK21, 505, 507, 510–514, 519–521
   CAPAN-1 tumor line, 580
   chinese hamster, 403, 405, 409, 410
   Clouser tumor line, 581–583
   Co115 human tumor cell line, 543–553
   HeLa, 505, 507, 508, 510, 511, 513, 517, 521, 522
   HEp-2, 218–221
   HT-29 tumor line, 580
   human epidermoid cell line, 403, 405, 411
   human lymphoblastoid, 471
   human squamous cell carcinoma, 403, 405, 409, 411
   mouse embryo fibroblasts, 157–165, 405, 407, 409, 411
   mouse L cells, 158, 159, 164
   persistently virus infected, 505
   Sarcoma, 180, 531
   SW-480 tumor line, 580
   3T3, 403, 405, 410, 529, 531
   Y-1 mouse adrenal, 529, 531
Cells
   adoptive transfer
     bone marrow cells, 247, 248, 250
     lymph node cells, 247–249
     lymphocytes, 255–257, 260, 261
     peripheral blood cells, 245, 248
     spleen cells, 93, 94, 97, 98, 111, 113–115, 120, 247–250
     thymus cells, 111, 113, 115, 147–154, 247, 248
   B-cells, 44, 45
   effector, 158, 163, 165, 169, 481
   eosinophils, 4, 6, 133
   fibroblasts, 527–541
   hypoxic cells, 641, 643, 645, 646
   leukocyte, counts, 185, 186, 203

   macrophage, 71, 73, 87, 89, 157–167
   mast cell, 5, 147–154
   natural killer, 386, 387, 413–420, 486, 487, 489, 505–507, 511, 515–519, 522–534
   neutrophil, 67–74
   peritoneal exudate cells, 69, 157–165
   spleen cells, 93, 94, 97, 98, 111, 113–115, 120, 247–250
   stem cells, 197, 198, 200, 204, 205
   T-cells, 1–6, 34, 36, 44, 263, 264, 266, 269, 270
   target cells
     C57BL/6 spleen cells, 232–235
     EBL6, 482, 483, 486–488
     EL-4, 233
     K562, 482, 486–489
     RL ♂ 1, 414, 417
     YAC-1, 414, 416, 417
   transformed cells, EBV, 481–490
Chemotherapeutic agents
   adriamycin, 649–655
   AMSA, 678
   AT 125, 29
   5-azacytidine, 649–655
   azotomycin, 29
   Baker's antifol, 675, 678
   BCNN, 649–655
   cis-platinum, 675, 678, 679, 681
   cyclophosphamide, 29, 579, 641
   cytosine arabinoside, 621, 622, 627
   DBD 675, 678, 681
   dibromodulcitol, 29
   DON, 29
   DTIC, 649–655
   5-fluorouracil, 621, 622, 627, 631
   FT-207 (Futraful), 621–627
   hexamethylmelamine, 29
   hydroxyurea, 657, 662, 663
   isophosphamide, 675, 676, 678–681
   maytansine, 675, 678
   melphalan, 29
   methyl CCNU, 29, 675, 677–681
   mitomycin CCNU, 29, 621, 622, 627, 631, 678, 681
   OK-432, 631
   PALA, 678
   pentamethylmelamine, 29
   PHM, 681
   piperazinedione, 29
   trimethyltrimethylomelamine, 29
   vincristine, 29, 643
Chemotherapy, resistance, 675–682
Chemotherapy-radiation therapy, combined, 641–646

Index

Colony formation, B-lymphocyte, 263–270
Colony stimulating factor, 527, 530, 537, 539, 558, 559, 561–564, 573–578
Crohn's disease, 271–273
Cytotoxicity tests, 157–168, 171, 179, 203, 219, 231, 414, 481, 508, 530, 543

Enzymes
    acid phosphatase, 612, 613, 615–617
    human lactic dehydrogenase, 217–223, 579–584, 589–592, 597, 598, 611, 613, 616, 617
    mouse lactic dehydrogenase, 217–223
Erythopoietin, 557, 559–562

Flow cytometry, 665–671
Fungi
    Aspergillus fumigatus, 77, 78, 82–85, 88, 89
    Candida albicans, 77–82, 88, 89
    Coccidioides immitis, 89
    Cryptococcus neoformans, 77, 78, 89
    Histoplasma capsulatum, 73, 89
    Sporothrix schenckii, 77–79, 85–89

Gompertz function, transformed, 665, 666, 671
Grafts
    allografts
        skin, 179, 189, 190, 192, 194, 275–281
        thymic rudiment, 179
        thymus gland, 113, 114, 147–156, 180, 225, 231–242, 359–364, 424
        thymus gland, cultured, 231–238, 239–242
        tumor, 46–48
    xenografts, non-malignant
        adrenal (human), 299, 303, 305
        brain (human), 299, 303
        breasts (human), 283–294
        esophagus (human), 283–294
        eyeball (human), 299, 303–305
        gastrointestinal tissue (human), 299, 301, 303, 304
        gonad (human), 299, 305
        heart (human), 299, 303
        kidney (human), 298, 303
        liver (human), 297, 299, 303, 304
        lung (human), 297–300, 303
        mammary gland (bovine), 309–314
        pituitary (human), 47
        prostate (human), 283–294
        skin (mouse), 47, 189–192, 194, 263–270, 435–444
        skin (pig), 435–444
Guinea pig
    athymic, 51–57
    Hartley, 52, 56

Haptens (see Antigens and haptens)
Heterografts (see Xenografts)
Histology
    adrenal, 321
    kidney, 321
    of tumor rejection, 481–490, 493–503
    ovary, 321
    submaxillary gland, 321
    testes, 321
    thyroid, 321
Homografts (see Allografts)
Hormones
    adrenalin, 567
    adrenocorticotropic hormone, 557, 567
    dopamine, 567
    estradiol, 309, 313, 317, 319, 323
    growth hormone, 311–313
    human chorionic gonadotropin, 527, 536, 537
    human placental lactogen, 312
    hydrocortisone, 311
    noradrenalin, 567
    progesterone, 311–313
    prolactin (bovine), 311–313
    testosterone, 320, 322
Hyperplasia 355, 413–420, 423–432

Immune elimination
    of fibroblasts from tumor grafts, 527–540
    of tumor cells, 543–553
Immunoglobulin isotypes
    IgA, 4, 33
    IgE, 4, 33, 149, 152, 153, 239–243
    IgG, 4, 33, 103, 105, 106, 108, 109, 149, 152, 153
    IgM, 4, 35, 103, 105–109, 243
    quantitation, 474–476
Immunoregulators
    carrageenan, 79, 84
    Corynebacterium parvum, 165, 218, 220–222
    cyclophosphamide, 169–173, 175, 176
    interferon, 158, 163, 165, 510
    levamisole, 197, 198, 200–205
    Mycobacterium tuberculosis, 158–160, 165
    picibanil, 79
Interferon, 157, 158, 163, 165, 505, 506
Irradiation, 91, 93, 94, 97, 381, 383, 387, 579, 631–639, 641–646, 665–671

Isolation systems
  clean-room system, 13, 17, 18
  laminar flow rack, 17, 53, 284, 348, 360, 380, 436, 632, 666
  plastic isolator, 12–17, 53, 56, 60, 298, 348, 414, 424, 632

Lactalbumin, 309, 311–313

Metastasis, 41–49, 367–375, 379–387, 391–401
  arachnoid space, 395, 397
  distribution pattern, 391–401
  kidney, 391, 393–397
  lung, 391, 393–397
  reproducible, 395
Mice
  irradiated, 91, 93, 94, 97, 481–490
  mutants,
    bg/bg, 154
    LASAT, 391–393, 397, 481
    nu^{str}/nu^{str}, 34–37
    W/W^v, 154
    Xid, 105
  newborn, 391
  splenectomized, 481–490
  strains,
    A, 23, 24
    AKR, 23, 24
    BALB/c, Most chapters
    BALB/c x C57BL/6 x F1, 225–229
    BIO.LP, 275–277
    CBA, 23, 24, 198, 415
    CBA/CaJ, 103–109
    CBA/N, 103–109
    C57BL/6, 1, 23, 24, 225–228, 232–235, 275–279, 359–365, 413, 416
    C57BL/10, 169–177
    C57BL/10J, 167–177
    C57BL/10ScN, 158, 160, 161, 163–165, 169–177
    C57BL/10Sn, 198, 200, 201, 203
    C3H, 276, 277
    C3H/He, 1, 23, 24, 424, 425, 427
    C3H/HeN, 125, 413, 416
    DBA/2, 23, 24
    DDY, 78, 81
    HaLCr, 233
    KK, 23, 24
    NC, 23, 24, 413, 416
    NFR, 481
    NFS/N, 413, 415, 416, 482, 285, 488
    NIH(S), 348, 349, 352, 355
    NZB-NZW, 23, 24, 263–269

    SJL/J, 169–171, 173, 175
    Swiss, 105–109, 169, 284, 463
Mitogens
  concanavalin-A, 45, 150, 179, 186, 194, 208, 213, 214
  lipopolysaccharide, 45, 179, 186
  phytohemagglutinin, 45, 179, 186, 207, 208, 213, 214, 494, 495, 501
  Plasmodium, 91–99
  Pokeweed mitogen, 45, 481
  Staphylococcus, 45
Mitotic activity, tumor, 457–462
Mixed lymphocyte reaction, 232
Mucin, 72, 73

Natural killer cells, 386, 387, 413–420, 486, 487, 489, 505–507, 511, 515–519, 522–534
Neoplasms (see Tumors)
Nitroimidiazole compounds, 646

Parasites
  Angiostrongylus cantonensis, 133–144
  Angiostrongylus costaricensis, 133–144
  Ascaris suum, 3, 4
  Aspiculuris tetraptera, 2, 207, 209, 215
  Babesia microti, 2
  Babesia rodhaini, 3
  Balantidium caviae, 53–55
  Echinococcus multilocularis, 133–144
  Emeria niesculzi, 46
  Fasciola hepatica, 3
  Giardia muris, 2, 5, 11
  Hexamita muris, 2, 11
  Hymenolepis diminuta, 2
  Hymenolepis nana, 2
  Leishmania tropica, 2, 6, 11
  Mesocestoides corti, 2, 4, 6, 11
  Nematospiroides dubius, 2, 4
  Nippostrongylus brasiliensis, 2, 46, 148–150, 154, 208
  Plasmodium berghei, 3, 91
  Plasmodium yoelii, 2
  Pneumocystis carinii, 11, 53–56, 123–131
  Schistosoma mansonii, 3–5, 11
  Strongyloides ratii, 2
  Syphacia obvelata, 2, 11, 207, 209, 215
  Taenia crassiceps, 46
  Taenia taeniaeformis, 2, 4, 6, 11
  Toxoplasma, 11
  Trichinella spiralis, 2, 5, 11
  Trypanosoma cruzi, 2, 5, 11
  Trypanosoma musculi, 2, 11
  Trypanosoma rhodesiense, 3

Index

Proliferation, epidermal, 435–444
Promotors
  croton oil, 423–432, 435–444
  retinoic acid, 435, 436, 439
  tape-stripping, 436, 439
  12-0-tetradecanoylphorbol 13-acetate (TPA), 423–432, 435–444

Quality control
  genetic, 23, 24, 123–131
  microbiological, 11–20, 123–131

Rats
  nude (rnu/rnu), 41–48, 189–195
  nude (rnu^{nz}/rnu^{nz}), 179–187
  strains
    AS, 182
    DA, 182
    Lewis, 190, 191
    W/Fu, 189, 190, 192, 368, 384
Receptors, androgen, 611–618
Reconstitution, immune system, 493–501
Reproductive performance, 317–323
Retrovirus, 169–176, 601–608

Serum proteins
  albumin, 565, 566
  α_1-antitrypsin, 565, 566
  α_1-fetoprotein, 565
  α_2-macroglobulin, 565
  ceruloplasmin, 565
  complement components, 565
  haptoglobin, 565
  hemopexin, 565
  prealbumin, 565
  transferrin, 565, 566

Tolerance, 231–236
Toxins, diphtheria, 582
Tumorigenicity assay, gelatin sponge, 403–411
Tumors, androgen-dependent, 611–618
Tumors, biological effects, 557–570
Tumors, carcinogen-induced, 423–432
Tumors, estimation of growth, 463–470, 579–585, 657–663, 665–671
Tumors, thymus-independent host resistance, 379–387
Tumors, human
  astrocytoma, 601, 603
  bone, 449–452
  breast, 27, 448–452, 457–461, 473, 474, 529, 580, 621–629
  cerebellar medulloblastoma, 590, 592–594
  choriocarcinoma, 529, 530, 539, 559, 598
  colon, 27, 193, 448–452, 481, 494, 496, 497, 499, 543–553, 557–570, 580, 601, 603, 665–671, 675–682
  gastrointestinal (non-colon), 449–452
  genital-urinary tract, 449–452
  germ cell-primitive cell, 449–452
  glioblastoma, 463, 467
  head and neck, 449–452
  hepatoblastoma, 467, 539, 564, 565
  kidney, 449–452, 529, 539, 557
  leiomyosarcoma, 529
  leukemia (K-562 cell line), 391–401
  lung, 29, 193, 449–452, 457–461, 590
  lymphoma, 271, 463, 467, 473, 474, 481–490, 494, 495
  lymphoreticular, 448, 449
  melanoma, 193, 448–452, 463, 467, 473, 474, 494, 499, 539, 590, 601, 603, 649–655, 658, 663, 675–682
  metastases, 447, 449–455
  neuroblastoma, 463, 466, 467, 469, 470, 559
  pancreatic, 193, 534, 539, 580
  primary, 447–452
  rectal, 621–629
  recurrent, 449–452
  retinoblastoma, 641–646
  rhabdomyosarcoma, 463, 465
  soft-tissue sarcoma, 449–452
  stomach, 564, 565, 621
  teratocarcinoma, 601, 603
  thymoma, 463, 467, 539
  yolk sac tumor, 559–561, 564, 565
Tumors, invasiveness, 379–387, 481, 505, 520–523, 543, 544
Tumors, labeling, 463–470, 543–553
Tumors, metastasis, 47, 48, 367–375, 379–387, 391–401, 481–490, 505–524, 543–553
Tumors, mouse
  A9, 379–385
  leukemia, granulocytic, 37
  lymphoma, 37
  reticular cell sarcoma, 37
  thymoma, 37
Tumors, spontaneous
  adenocarcinoma, 349
  colon, 349
  lung, 349

testicular, 349
adenoma, lung, 349
epithelial, 361
gastrointestinal, 361, 363
hepatic, 361–364
lung, 359
lymphoma, 347–356, 363, 364
lymphoreticular, 361, 363
mammary, 363
plasmacytoma, 349, 350, 352, 363
reproductive system, 363
vascular, 361
Tumors, rat
mammary adenocarcinoma
metastasizing (SMT-2A; TMT-081), 368–373, 375, 381, 384–386
non-metastasizing (MT–W9B; MT-100), 368–374, 381, 384
Tumors, transplantation sites

bones, 463–470
eye, 463–470, 641–646
IP vs. IV, 543–553

Virus induction, 601–608
Virus replication, 169–176
Viruses
adenovirus, mouse, 11, 12, 19
cytomegalovirus, 53
encephalomyocarditis virus, 157–165
Friend leukemia virus, 169–176
hepatitis virus, mouse, 11–16, 43
lactic dehydrogenase virus, 218, 219, 221
mink cell focus-inducing virus, 36
murine leukemia virus, 169, 601–608
Sendai virus, 11, 12, 15, 19, 20, 46
Vaccinia virus, 11, 19

Weight, body, 319, 322

Thymusaplastic Nude Mice and Rats in Clinical Oncology

Proceedings of the Symposium at the Klinikum der Johann-Wolfgang-Goethe Universität, Frankfurt am Main, June 11-15, 1979
Edited by Dr. G.B. Bastert, Dr. H.P. Fortmeyer, Prof. Dr. H. Schmidt-Matthiesen.
With contributions by international specialists.
1981. XIV, 556 pp., 270 fig., 77 tables, pb. DM 198,-

The field of oncology gained a new experimental model a few years ago in the form of heterotransplantation of human tumors onto nude mice and rats. The experimental animals tolerate heterotransplanted solid, human tumors without additional immunosuppressive treatment. This permits experimental investigation of a large number of clinically oriented oncological questions which could not be studied to date using, e.g., tissue culture methods. The purpose of the present books is to present the interdisciplinary potential of the new experimental model, based, however, on clinical oncology. With this goal in mind, the results of an international symposium in Frankfurt/Main have been compiled.

Frozen Storage of Laboratory Animals

Proceedings of a Workshop at Harwell/U.K., May 6-9, 1980
Edited by Prof. Dr. G.H. Zeilmaker.
1981. XII, 193 pp., 44 fig., 60 tables, pb. DM 49,-

In this volume the state of the art in embryo freezing and embryo transfer in small laboratory animals is discussed by the foremost experts in the field. Following two authorative papers on the basic aspects of cryopreservation with cryomicroscopic data on intracellular ice formation and analysis of osmotic processes the new two-step-freezing method, oviduct freezing, supcrovulation treatments and embryo transfer are reviewed. Moreover worldwide experiences in establishing mouse embryo banks are discussed as well as the possibility to freeze rat, hamster and rabbit embryos. The recorded discussions focus on practical aspects of the technique such as standardization, registration and future improvements. The book offers an up-to-date insight in the possibility to apply the technique of low-temperature storage of embryos in laboratory animal husbandry.

Gustav Fischer Verlag Stuttgart • New York

Proceedings of the Symposium on the Use of Athymic (Nude) Mice in Cancer Research
Edited by Dr. O.P. Houchens and Dr. A. A. Ovejera.
With contributions by many international specialists.
1978. XXVIII, 289 pp., 34 fig., pb. DM 77,50

Bibliography of the Nude Mouse 1966-1976
Edited by Dr. J. Rygaard and Dr. C.O. Povisen.
1977. 48 pp., pb. DM 18,-

Proceedings of the First International Workshop on Nude Mice
Edited by Dr. J. Rygaard and Dr. C.C. Povlsen.
With contributions by 93 specialists. 1974. 301 pp., pb. DM 79,-

Proceedings of the Second International Workshop on Nude Mice
Edited by T. Nomura, N. Ohsawa, N. Tomaoki, K. Fujiwara.
1977. XIV, 600 pp., many fig. and tab., cloth DM 128,-

Proceedings of the Workshop on Basic Aspects of Freeze Preservation of Mouse Strains
Edited by Prof. Dr. O. Mühlbock.
With contributions by many international specialists.
1976. X,133 pp., 36 fig., 42 tab., pb. DM 52,-

Animal Quality and Models in Biomedical Research
Edited by Prof. Dr. Dr. A. Spiegel, Dr. S. Erichsen and
Dr. H.A. Solleveld.
With contributions by 62 international specialists. 1980.
XIV, 397 pp., 136 fig., 79 tab., DM 136,-

Genetic Variants and Strains of the Laboratory Mouse
Edited by Dr. Margaret C. Green.
With contributions by 11 specialists from USA and Great Britain.
1981. XVI, 476 pp., 10 fig., 28 tab., cloth DM 240,-

Gustav Fischer Verlag Stuttgart · New York